AF126944

Smart Sensors and Devices in Artificial Intelligence

Smart Sensors and Devices in Artificial Intelligence

Editors

Dan Zhang
Xuechao Duan

MDPI • Basel • Beijing • Wuhan • Barcelona • Belgrade • Manchester • Tokyo • Cluj • Tianjin

Editors
Dan Zhang
York University
Canada

Xuechao Duan
Xidian University
China

Editorial Office
MDPI
St. Alban-Anlage 66
4052 Basel, Switzerland

This is a reprint of articles from the Special Issue published online in the open access journal *Sensors* (ISSN 1424-8220) (available at: https://www.mdpi.com/journal/sensors/special_issues/Sensors_AI).

For citation purposes, cite each article independently as indicated on the article page online and as indicated below:

LastName, A.A.; LastName, B.B.; LastName, C.C. Article Title. *Journal Name* **Year**, *Article Number*, Page Range.

ISBN 978-3-03943-669-9 (Hbk)
ISBN 978-3-03943-670-5 (PDF)

© 2021 by the authors. Articles in this book are Open Access and distributed under the Creative Commons Attribution (CC BY) license, which allows users to download, copy and build upon published articles, as long as the author and publisher are properly credited, which ensures maximum dissemination and a wider impact of our publications.

The book as a whole is distributed by MDPI under the terms and conditions of the Creative Commons license CC BY-NC-ND.

Contents

About the Editors vii

Dan Zhang and Bin Wei
Smart Sensors and Devices in Artificial Intelligence
Reprinted from: *Sensors* **2020**, *20*, 5945, doi:10.3390/s20205945 1

Changhui Jiang, Shuai Chen, Yuwei Chen, Boya Zhang, Ziyi Feng, Hui Zhou and Yuming Bo
A MEMS IMU De-Noising Method Using Long Short Term Memory Recurrent Neural Networks (LSTM-RNN)
Reprinted from: *Sensors* **2018**, *18*, 3470, doi:10.3390/s18103470 5

Tianyuan Liu, Jinsong Bao, Junliang Wang and Yiming Zhang
A Hybrid CNN–LSTM Algorithm for Online Defect Recognition of CO_2 Welding
Reprinted from: *Sensors* **2018**, *18*, 4369, doi:10.3390/s18124369 19

Xiangyang Wang, Sheng Guo, Haibo Qu and Majun Song
Design of a Purely Mechanical Sensor-Controller Integrated System for Walking Assistance on an Ankle-Foot Exoskeleton
Reprinted from: *Sensors* **2019**, *19*, 3196, doi:10.3390/s19143196 35

Zhaohui Hao, Guixi Liu, Jiayu Gao and Haoyang Zhang
Robust Visual Tracking Using Structural Patch Response Map Fusion Based on Complementary Correlation Filter and Color Histogram
Reprinted from: *Sensors* **2019**, *19*, 4178, doi:10.3390/s19194178 51

Jisun Park, Mingyun Wen, Yunsick Sung and Kyungeun Cho
Multiple Event-Based Simulation Scenario Generation Approach for Autonomous Vehicle Smart Sensors and Devices
Reprinted from: *Sensors* **2019**, *19*, 4456, doi:10.3390/s19204456 77

Xiaojuan Zhu, Kuan-Ching Li, Jinwei Zhang and Shunxiang Zhang
Distributed Reliable and Efficient Transmission Task Assignment for WSNs
Reprinted from: *Sensors* **2019**, *19*, 5028, doi:10.3390/s19225028 91

Surak Son, YiNa Jeong and Byungkwan Lee
An Audification and Visualization System (AVS) of an Autonomous Vehicle for Blind and Deaf People Based on Deep Learning
Reprinted from: *Sensors* **2019**, *19*, 5035, doi:10.3390/s19225035 115

Medhat Abdel Rahman Mohamed Mostafa, Miljan Vucetic, Nikola Stojkovic, Nikola Lekić and Aleksej Makarov
Fuzzy Functional Dependencies as a Method of Choice for Fusion of AIS and OTHR Data
Reprinted from: *Sensors* **2019**, *19*, 5166, doi:10.3390/s19235166 135

Zhengyu Wang, Daoming Wang, Bing Chen, Lingtao Yu, Jun Qian and Bin Zi
A Clamping Force Estimation Method Based on a Joint Torque Disturbance Observer Using PSO-BPNN for Cable-Driven Surgical Robot End-Effectors
Reprinted from: *Sensors* **2019**, *19*, 5291, doi:10.3390/s19235291 149

Jianqing Wu, Hao Xu, Yongsheng Zhang, Yuan Tian and Xiuguang Song
Real-Time Queue Length Detection with Roadside LiDAR Data
Reprinted from: *Sensors* **2020**, *20*, 2342, doi:10.3390/s20082342 169

Jianqing Wu, Hao Xu, Yuan Tian, Rendong Pi and Rui Yue
Vehicle Detection under Adverse Weather from Roadside LiDAR Data
Reprinted from: *Sensors* **2020**, *20*, 3433, doi:10.3390/s20123433 185

Faraz Malik Awan, Roberto Minerva and Noel Crespi
Improving Road Traffic Forecasting Using Air Pollution and Atmospheric Data: Experiments Based on LSTM Recurrent Neural Networks
Reprinted from: *Sensors* **2020**, *20*, 3749, doi:10.3390/s20133749 203

Saud Altaf, Shafiq Ahmad, Mazen Zaindin and Muhammad Waseem Soomro
Xbee-Based WSN Architecture for Monitoring of Banana Ripening Process Using Knowledge-Level Artificial Intelligent Technique
Reprinted from: *Sensors* **2020**, *20*, 4033, doi:10.3390/s20144033 225

Zilong Wu, Hong Chen and Yingke Lei
Unidimensional ACGAN Applied to Link Establishment Behaviors Recognition of a Short-Wave Radio Station
Reprinted from: *Sensors* **2020**, *20*, 4270, doi:10.3390/s20154270 243

Hongyan Tang, Dan Zhang and Zhongxue Gan
Control System for Vertical Take-Off and Landing Vehicle's Adaptive Landing Based on Multi-Sensor Data Fusion
Reprinted from: *Sensors* **2020**, *20*, 4411, doi:10.3390/s20164411 263

Michal Frniak, Miroslav Markovic, Patrik Kamencay, Jozef Dubovan, Miroslav Benco, Milan Dado
Vehicle Classification Based on FBG Sensor Arrays Using Neural Networks
Reprinted from: *Sensors* **2020**, *20*, 4472, doi:10.3390/s20164472 283

Jen-Wei Huang, Meng-Xun Zhong and Bijay Prasad Jaysawal
TADILOF: Time Aware Density-Based Incremental Local Outlier Detection in Data Streams
Reprinted from: *Sensors* **2020**, *20*, 5829, doi:10.3390/s20205829 301

About the Editors

Dan Zhang, Professor and Tier 1 York Research Chair in Advanced Robotics and Mechatronics at York University. Dr. Zhang's research interests include robotics and mechatronics; high-performance parallel robotic machine development; sustainable/green manufacturing systems; and rehabilitation robots and rescue robots. Dr. Zhang's contributions to and leadership within the field of robotic and automation have been recognized with several prestigious awards, within his own university (Kaneff Professorship, Tier 1 York Research Chair in Advanced Robotics and Mechatronics, Research Excellence Award both from university level and faculty level), the Province of Ontario (Early Researcher Award), professional societies (a Fellow of the Canadian Academy of Engineering (CAE), a Fellow of the Engineering Institute of Canada (EIC), a Fellow of American Society of Mechanical Engineers (ASME), and a Fellow of Canadian Society for Mechanical Engineering (CSME), a Senior Member of IEEE, and a Senior Member of SME.), and federal government (Canada Research Chair in January 2009 and renewed in January 2014).

Xuechao Duan is a professor and deputy director of the Research Institute of Mechatronics at Xidian University, China. Dr. Duan's research interests include robotics and mechatronics, innovative design of electronic equipments, and intelligent measuring and control in smart manufacturing. At present, he is a member of the Key Laboratory of Electronic Structure Design, Ministry of Education, China. He is also a senior member of the Chinese Mechanical Engineering Society, a member of the Institution of Engineering Technology and a member of the Chinese Institute of Electronics.

Editorial

Smart Sensors and Devices in Artificial Intelligence

Dan Zhang [1,*] and Bin Wei [2]

1. Department of Mechanical Engineering, York University, Toronto, ON M3J 1P3, Canada
2. Department of Computer Science, Algoma University, Sault Ste Marie, ON P6A 2G4, Canada; bin.wei@algomau.ca
* Correspondence: dzhang99@yorku.ca; Tel.: +1-647-209-0959

Received: 19 October 2020; Accepted: 20 October 2020; Published: 21 October 2020

As stated in the Special Issue call, "sensors are eyes or/and ears of an intelligent system, such as Unmanned Aerial Vehicle (UAV), Automated Guided Vehicle (AGV) and robots. With the development of material, signal processing and multidisciplinary interactions, more and more smart sensors are proposed and fabricated under increasing demands for homes, industry and military fields. Networks of sensors will be able to enhance the ability to obtain huge amounts of information (big data) and improve precision, which also mirrors the developmental tendency of modern sensors. Moreover, artificial intelligence is a novel impetus for sensors and networks, which gets sensors to learn and think and feed more efficient results back."

The current development of sensors is purely based on mathematical models, in which one needs to know the parameters exactly. However, sometimes there are uncertainties and variation which are impossible to anticipate, and under this situation, the sensors will not work properly. By combining sensors with artificial intelligence, the uncertainties and variation can be addressed and handled. The essential thing about artificial intelligence is to predict the "future". For example, among the applications in artificial intelligence is the learning control currently under developing. By using a learning control, uncertainties and variations can be effectively handled [1], whereas these cannot be handled traditional control systems. From another perspective, driving a vehicle at night can be used as an example. Supposing the vehicle light is not working, one can imagine that it will be difficult to drive fast as the driver's eyes may not detect the road condition ahead. However, supposing that one drives in the daytime, it is very easy to drive fast because the driver can predict what it is going on in front of them. Our human eyes are sensors, and without the sensors, the associated systems will not work properly. To go one step further, supposing there is a GPS installed in the car, and the GPS here is considered to be artificial intelligence, by resorting to the GPS, it can help drivers drive fast even without the vehicle light at night. It is known that by integrating artificial intelligence into sensors and devices, uncertainties can be adaptively addressed so as to, for example, reduce tracking errors or improve accuracy in robotics applications, or predict the future as discussed in the previous driving case.

This Special Issue focuses on the smart sensors and networks, especially sensing technologies utilizing artificial intelligence. This editorial summarizes the whole Special Issue. We received 43 papers in total, and 17 of them were published.

Summary of the Special Issue

In [2], a new algorithmic rule for the purpose of streaming data, referred to as "time-aware density-based incremental local outlier detection", was proposed to conquer variations in data that change as time goes on. The results show that the proposed "time-aware density-based incremental local outlier detection" performs better than that of the existing candidates in the sense of the AUC in most of the cases on different kinds of datasets. In [3], the study was focused on the vehicle classification with (fiber Bragg grating) FBG sensor arrays by employing AI from partial records. The developed neural network was trained by resorting to a dataset which lacked vehicle velocity data, which is generated by the visual identification of a vehicle going over the testing platform. The result indicates that the developed neural network can successfully separate trucks from other vehicles. The study shows that by using the artificial intelligence, the system can handle uncertainties even in a situation of unknown data by predicting the past events. Similarly, in [4], the study introduced a type of adaptive control algorithm for the landing gear mechanism under an unknown condition. Based on the information from the optical flow sensor and depth camera, the control system accomplished multi-sensor data fusion for the purpose of expanding the functionalities of landing and taking off for UAVs in an unknown condition and environment.

In [5], a new unidimensional auxiliary classifier generative adversarial network was developed in order to get more signals for short-wave radio stations and the unidimensional DenseNet is employed to perceive link establishment behaviors for electronic countermeasures even with no communication protocol standard. In [6], for the purpose of fruit ripeness monitoring in real time and maximizing the fruit quality during storage via the forecasting of the current fruits condition and therefore minimizing financial loss, the study illustrated a neural network architectural design, i.e., "Xbee-based wireless sensor nodes network", and subsequently the resulting data were used for training in the "artificial neural network" for validating the data. The results indicated that the proposed wireless sensor nodes network architecture was able to recognize the fruit condition.

In [7], the authors developed a traffic forecasting methodology which used air pollution and atmospheric parameters and the timestamped traffic intensity data for the purpose of predicting the flow of the traffic. The "long short-term memory recurrent neural network" was employed here to help predict the flow of the traffic, with the help of the data from the air pollution and atmospheric condition. Similarly in [8,9], the performance of several data processing algorithms that is geared to roadside light detection and ranging under unknown weather conditions was evaluated, and a background filtering and object clustering method was developed for the purpose of processing the roadside light detection and ranging data under the unknown weather conditions. It was shown that the current processing algorithm for the roadside light detection and ranging was based on assuming known weather conditions. Unknown weather conditions are the major challenges for data processing.

In [10], a one-dimensional clamping force sensing approach which does not need internal force sensors was proposed for a cable-driven surgical robot. The clamping force estimation approach was developed on the basis of a "joint torque disturbance observer", which basically examines the differences among the real-time estimated cable tension and the actual cable tension by resorting to a "PSO Back Propagation Neural Network". The advantage of the proposed approach was that it only uses the known data of the motor displacement and without the information from the internal force sensors. In [11], a fuzzy functional dependency was used under the situation of data fusion from an automatic identification system and over-the-horizon radars sources. The fuzzy logic approach proposed in this study was proven to be a favorable tool in handling uncertainties from different sensors.

In [12], an audification and visualization system used for people who are deaf and blind in autonomous vehicles was developed by using deep learning. The audification and visualization system has three different sectors. The data collection and management sector keeps the vehicle data, the audification conversion sector contains a speech-to-text sub-sector which can accept user speech and subsequently transform it into text data, and the data visualization sector can conjure up the collected data and set the envisaged data according to the vehicle display size. Similarly, in [13], a scenario-generation approach contingent on deep learning was developed for the purpose of automatically generating scenarios to train autonomous vehicle smart sensors. To create different situations, the developed approach extracts several events from a video that were captured on a real road based on deep learning and creates different scenarios in a virtual simulator. The method developed here contains different scenarios by extracting them from one driving event and allows interactions between objects.

In [14], the authors developed two distributed task assignments for wireless sensor networks on the basis of the "transmission-oriented reliable and energy-efficient task allocation" for the purpose of solving the distributed task allocation issue in wireless sensor networks. In the first distributed task assignments for wireless sensor networks, the sink allocates reliability to all cluster heads based on the requirements of the reliability, in this way the cluster head carries out the local task allocation. Similarly, the global view was achieved through collecting local views from multiple sink nodes. In [15], a tracker derived from the "structural patch response fusion under correlation filter and color histogram" was developed. The developed approach contains different sub-trackers that can adaptively address illumination variation. To recognize and fully use the patches, an adaptive hedge algorithm was developed for the purpose of hedging the patches responses into a much more reliable one in each component tracker.

In [16], a mechanical sensor-controller-integrated system was developed for the purpose of reaching the objective of identifying gait. The system contained a sensing section and a mechanical executing section. The sensing section contained a sensor which can possess two input channels. Because the system was linked with the spring, the sensor was evaluated along with the controller.

In [17], the authors developed a convolutional neural network combined with long short-term memory networks algorithm, which had the merits of both. The developed algorithm initiated a shallow convolutional neural network to obtain the main characteristics of the molten pool image. After that, the main characteristics were converted into the feature matrix. The good aspect about the algorithm lies in the fact that it is able to learn the best hybrid characteristics via the "error back propagation algorithm" in the shallow convolutional neural networks for a single molten pool image in order to have the engineering requirements for the purpose of observing the welding process in real time. Very similarly in [18], an AI-based approach was developed for the purpose of reducing the noise in the MEMS inertial measurement unit output signals. Particularly, a derivative of the "recurrent neural network long short-term memory" was used in order to filter the MEMS gyroscope outputs, the signals of which are considered as time series.

All of the above studies indicate the significance of combining AI with sensors to address the uncertainties and unknown data. This editorial is to summarize the whole Special Issue and act as a formal closure.

Funding: This research was funded by the Natural Sciences and Engineering Research Council of Canada grant number: RGPIN-2016-05030 and York Research Chairs program.

Acknowledgments: We would like to thank all authors who have contributed their work to the Special Issue "Smart Sensors and Devices in Artificial Intelligence". Thanks are also given to all the hard working reviewers for their detailed comments and suggestions. The papers in this Special Issue illustrate the breadth and depth of sensor technologies applied for solving different problems.

Conflicts of Interest: Page: 3The authors declare no conflict of interest.

References

1. Zhang, D.; Wei, B. On the Development of Learning Control for Robotic Manipulators. *Robotics* **2017**, *6*, 23. [CrossRef]
2. Huang, J.-W.; Zhong, M.-X.; Jaysawal, B.P. TADILOF: Time Aware Density-Based Incremental Local Outlier Detection in Data Streams. *Sensors* **2020**, *20*, 5829. [CrossRef] [PubMed]
3. Frniak, M.; Markovic, M.; Kamencay, P.; Dubovan, J.; Benco, M.; Dado, M. Vehicle Classification Based on FBG Sensor Arrays Using Neural Networks. *Sensors* **2020**, *20*, 4472. [CrossRef] [PubMed]
4. Tang, H.; Zhang, D.; Gan, Z. Control System for Vertical Take-Off and Landing Vehicle's Adaptive Landing Based on Multi-Sensor Data Fusion. *Sensors* **2020**, *20*, 4411. [CrossRef] [PubMed]
5. Wu, Z.; Chen, H.; Lei, Y. Unidimensional ACGAN Applied to Link Establishment Behaviors Recognition of a Short-Wave Radio Station. *Sensors* **2020**, *20*, 4270. [CrossRef] [PubMed]
6. Altaf, S.; Ahmad, S.; Zaindin, M.; Soomro, M.W. Xbee-Based WSN Architecture for Monitoring of Banana Ripening Process Using Knowledge-Level Artificial Intelligent Technique. *Sensors* **2020**, *20*, 4033. [CrossRef] [PubMed]

7. Awan, F.M.; Minerva, R.; Crespi, N. Improving Road Traffic Forecasting Using Air Pollution and Atmospheric Data: Experiments Based on LSTM Recurrent Neural Networks. *Sensors* **2020**, *20*, 3749. [CrossRef] [PubMed]
8. Wu, J.; Xu, H.; Tian, Y.; Pi, R.; Yue, R. Vehicle Detection under Adverse Weather from Roadside LiDAR Data. *Sensors* **2020**, *20*, 3433. [CrossRef] [PubMed]
9. Wu, J.; Xu, H.; Zhang, Y.; Tian, Y.; Song, X. Real-Time Queue Length Detection with Roadside LiDAR Data. *Sensors* **2020**, *20*, 2342. [CrossRef] [PubMed]
10. Wang, Z.; Wang, D.; Chen, B.; Yu, L.; Qian, J.; Zi, B. A Clamping Force Estimation Method Based on a Joint Torque Disturbance Observer Using PSO-BPNN for Cable-Driven Surgical Robot End-Effectors. *Sensors* **2019**, *19*, 5291. [CrossRef] [PubMed]
11. Mohamed Mostafa, M.A.R.; Vucetic, M.; Stojkovic, N.; Lekić, N.; Makarov, A. Fuzzy Functional Dependencies as a Method of Choice for Fusion of AIS and OTHR Data. *Sensors* **2019**, *19*, 5166. [CrossRef] [PubMed]
12. Son, S.; Jeong, Y.; Lee, B. An Audification and Visualization System (AVS) of an Autonomous Vehicle for Blind and Deaf People Based on Deep Learning. *Sensors* **2019**, *19*, 5035. [CrossRef]
13. Park, J.; Wen, M.; Sung, Y.; Cho, K. Multiple Event-Based Simulation Scenario Generation Approach for Autonomous Vehicle Smart Sensors and Devices. *Sensors* **2019**, *19*, 4456. [CrossRef]
14. Zhu, X.; Li, K.-C.; Zhang, J.; Zhang, S. Distributed Reliable and Efficient Transmission Task Assignment for WSNs. *Sensors* **2019**, *19*, 5028. [CrossRef]
15. Hao, Z.; Liu, G.; Gao, J.; Zhang, H. Robust Visual Tracking Using Structural Patch Response Map Fusion Based on Complementary Correlation Filter and Color Histogram. *Sensors* **2019**, *19*, 4178. [CrossRef] [PubMed]
16. Wang, X.; Guo, S.; Qu, H.; Song, M. Design of a Purely Mechanical Sensor-Controller Integrated System for Walking Assistance on an Ankle-Foot Exoskeleton. *Sensors* **2019**, *19*, 3196. [CrossRef]
17. Liu, T.; Bao, J.; Wang, J.; Zhang, Y. A Hybrid CNN–LSTM Algorithm for Online Defect Recognition of CO_2 Welding. *Sensors* **2018**, *18*, 4369. [CrossRef] [PubMed]
18. Jiang, C.; Chen, S.; Chen, Y.; Zhang, B.; Feng, Z.; Zhou, H.; Bo, Y. A MEMS IMU De-Noising Method Using Long Short Term Memory Recurrent Neural Networks (LSTM-RNN). *Sensors* **2018**, *18*, 3470. [CrossRef] [PubMed]

Publisher's Note: MDPI stays neutral with regard to jurisdictional claims in published maps and institutional affiliations.

© 2020 by the authors. Licensee MDPI, Basel, Switzerland. This article is an open access article distributed under the terms and conditions of the Creative Commons Attribution (CC BY) license (http://creativecommons.org/licenses/by/4.0/).

Article

A MEMS IMU De-Noising Method Using Long Short Term Memory Recurrent Neural Networks (LSTM-RNN)

Changhui Jiang [1,2], Shuai Chen [1,*], Yuwei Chen [2], Boya Zhang [1], Ziyi Feng [2], Hui Zhou [3] and Yuming Bo [1]

1. School of Automation, Nanjing University of Science and Technology, Nanjing 210094, China; changhui.jiang1992@gmail.com (C.J.); soochow_njust@sina.com (B.Z.); byming@mail.njust.edu.cn (Y.B.)
2. Centre of Excellence in Laser Scanning Research, Finnish Geospatial Research Institute (FGI), Geodeetinrinne 2, FI-02431 Kirkkonummi, Finland; yuwei.chen@nls.fi (Y.C.); ziyi.feng@nls.fi (Z.F.)
3. Department of Photogrammetry and Remote Sensing, Wuhan University, 129 Luoyu Road, Wuhan 430079, China; zhouhui@whu.edu.cn
* Correspondence: c1492@163.com; Tel.: +86-138-1391-5826

Received: 29 September 2018; Accepted: 13 October 2018; Published: 15 October 2018

Abstract: Microelectromechanical Systems (MEMS) Inertial Measurement Unit (IMU) containing a three-orthogonal gyroscope and three-orthogonal accelerometer has been widely utilized in position and navigation, due to gradually improved accuracy and its small size and low cost. However, the errors of a MEMS IMU based standalone Inertial Navigation System (INS) will diverge over time dramatically, since there are various and nonlinear errors contained in the MEMS IMU measurements. Therefore, MEMS INS is usually integrated with a Global Positioning System (GPS) for providing reliable navigation solutions. The GPS receiver is able to generate stable and precise position and time information in open sky environment. However, under signal challenging conditions, for instance dense forests, city canyons, or mountain valleys, if the GPS signal is weak and even is blocked, the GPS receiver will fail to output reliable positioning information, and the integration system will fade to an INS standalone system. A number of effects have been devoted to improving the accuracy of INS, and de-nosing or modelling the random errors contained in the MEMS IMU have been demonstrated to be an effective way of improving MEMS INS performance. In this paper, an Artificial Intelligence (AI) method was proposed to de-noise the MEMS IMU output signals, specifically, a popular variant of Recurrent Neural Network (RNN) Long Short Term Memory (LSTM) RNN was employed to filter the MEMS gyroscope outputs, in which the signals were treated as time series. A MEMS IMU (MSI3200, manufactured by MT Microsystems Company, Shijiazhuang, China) was employed to test the proposed method, a 2 min raw gyroscope data with 400 Hz sampling rate was collected and employed in this testing. The results show that the standard deviation (STD) of the gyroscope data decreased by 60.3%, 37%, and 44.6% respectively compared with raw signals, and on the other way, the three-axis attitude errors decreased by 15.8%, 18.3% and 51.3% individually. Further, compared with an Auto Regressive and Moving Average (ARMA) model with fixed parameters, the STD of the three-axis gyroscope outputs decreased by 42.4%, 21.4% and 21.4%, and the attitude errors decreased by 47.6%, 42.3% and 52.0%. The results indicated that the de-noising scheme was effective for improving MEMS INS accuracy, and the proposed LSTM-RNN method was more preferable in this application.

Keywords: microelectromechanical systems; inertial measurement unit; long short term memory recurrent neural networks; artificial intelligence

1. Introduction

Global Navigation Satellite System (GNSS) and Inertial Navigation System (INS) have been set up in various vehicles and carriers for navigation and tracking [1–5]. A GNSS receiver is usually a chip, which is small, low-cost and precise. With continuously receiving the signal from the navigation satellites in orbit, the GNSS receiver is able to provide reliable and constant positioning, navigation and timing (PNT) information [6–10]. However, limited by the principle that at least four satellites are essential for computing positioning and velocity, the GNSS receiver will fail to work normally under challenging signal conditions [6–10]. The navigation signal transmits from the satellites a long way, and becomes too weak when reaching the ground, therefore, it can be easily blocked temporarily by the environment. For bridging the signal outages, Inertial Navigation System (INS) is employed to output positioning information during the signal outage. Traditional fiber or laser Inertial Measurement Units (IMU) are precise, but too big and expensive for vehicles or handheld devices [11–13].

Recently, Microelectromechanical Systems (MEMS) IMU has gained a boom in applications of position and navigation, especially, vehicles, handheld devices, and precise-guidance bombs, due to its low cost and small size brought by the advanced MEMS manufacturing technology [14–16]. With proper circuit and structure design, the accuracy of MEMS IMU has gradually improved [14–16]. Although, the MEMS IMU obtained a reduction in volume, and cost compared with conventional fiber or laser IMU, the MEMS IMU experiences more non-linear or random errors, which leads to the MEMS INS navigation solutions diverging dramatically over time [14–18]. Commonly, the GNSS receiver is employed as an outer aiding to calibrating the INS, thus, the GNSS and INS integrated navigation system is able provide reliable and continuous navigation solutions, even during short-term signal outage. During the outage, the problem is that the errors of INS increase quickly without outer sensors or reference aiding. Under this condition, the modeling or de-noising of the MEMS IMU outputs will be a key step to improve the accuracy of MEMS INS. Researchers are devoted to identifying and modeling the errors contained in the MEMS IMU raw signals, which can be divided into two parts: System errors and random errors. The system part refers to the bias, and scale factor errors. These errors can be calibrated or quantified by certain experiments in a laboratory. The calibration process has been investigated and reported by a number of researchers [19–22]. However, the random part can lead to the drifts and instabilities in bias or scale factor over time, which is the key component leading to the INS errors divergence [17–26].

Therefore, before deploying MEMS IMU for navigation, an accurate model of the random and systemic is requisite to ensure the accuracy. For identifying and modeling the random errors, researchers have devoted to proposing some de-nosing techniques in this random signals processing, and overall, the approaches can be divided to statistical modeling methods represented by Wavelet De-noising (WD), Allan Variance (AV), Auto Regressive and Moving Average (ARMA), and Artificial Intelligence (AI) methods represented by Support Vector Machine (SVM), Neural Networks (NN) [17–26]. Generally, the random errors contain high frequency (long-term) and low frequency (short-term) parts. Among statistical methods, the WD method performs significantly in removing the high frequency part [17–26], but it has restricted ability in removing low frequency errors. The AV method is another statistical method, and has been widely used in MEMS IMU errors analysis in time domain [17,18]. In the AV method, the stability of MEMS IIMU measurements are presented as a function of average time, and the intrinsic noise is described by five basic parts termed as: Quantization noise, angle random walk, bias instability, rate random walk, and rate ramp [17,18]. Usually, the AV method is employed to exploit the noise characteristics and obtain first-order Gauss Markov (GM) or ARMA parameters [17,18]. Traditional or conventional approaches are unsatisfactory for this application. Moreover, the unsatisfactory estimation or compensation of the random errors will lead to the failure to provide reliable navigation information estimation in short time. Another approach is AI methods including SVM and NN, which have been utilized in MEMS IMU modeling, and found to be better than other conventional methods [27–36]. These methods operate the signal de-noising or modeling as sequence prediction problem, and the MEMS IMU measurements are treated as time series.

Generally, in the data science community, sequence prediction problems have been around for a long period of time in a wide range of applications, including stock price prediction and sales pattern finding, language translation and speech recognization [37–41]. Recently, a new breakthrough has happened in the data science community, and a Long Short Term Memory Recurrent Neutral Networks (LSTM-RNN) has been proposed and has been demonstrated more effective for almost all of these sequence prediction problems [37–41]. Compared with conventional RNN, LSTM-RNN introduces the "gate structure" to address the long-term memory, which allows it to have the pattern of selectively remembering for a long time. This special design or structure makes it more suitable for predicting or processing time based series data. In this paper, the LSTM-RNN is incorporated in MEMS IMU gyroscope raw signal de-noising. LSTM-RNN has performed excellently in time series signal processing, for instance stock price prediction, speech single processing, and others [37–41]. A MEMS Inertial Measurement Unit (IMU) manufactured by MT Microsystems Company known as MSI2000 IMU is employed in the experiments for testing [42]. Firstly, a common ARMA is employed to process the raw signal, then the order and parameters are determined through the auto-correlation and partial correlation operation; secondly, a single LSTM and multi-layer LSTM are compared in the MEMS gyroscope raw signal de-nosing in aspects of average training loss, training time and de-noising performance; finally, the three-axis attitude errors of raw signals, ARMA, and LSTM-RNN are compared and analyzed.

The remainder of this paper is organized as: (1) Section 2 introduces the methods including Auto Regressive Moving Average Method (ARMA), and the proposed LSTM-RNN; (2) Section 3 presents the experiments, results and comparison (3) the following are the discussion, conclusion and future work.

2. Method

In this section, the conventional ARMA representing the statistical methods and the proposed LSTM-RNN representing AI methods are presented. The principles, basic equations and information flow are briefly introduced.

2.1. ARMA Model

As illustrated in previous papers [25,26], the following two steps are essential for setting up an ARMA model: (1) After the obtaining the raw gyroscope signals, auto-correlation and partial correlation are operated to characterize the noise and select the suitable time series model; (2) estimating the parameters of the ARMA model.

The auto-correlation of a signal is a product operation of the signal and a time-shifted version of the signal itself. Assuming $r(t)$ is a random signal sequence, and the auto-correlation can be modelled as [25,26]:

$$R(\tau) = E(r(t)r(t+\tau)) \tag{1}$$

where, $E(\cdot)$ is the expectation operator, τ is the time delay or shift. The partial correlation process is defined as [25,26]:

$$P(\tau) = \frac{\sum (r(t) - E(r(t)))(r(t+\tau) - E(r(t+\tau)))}{\sqrt{\sum (r(t) - E(r(t)))^2 \sum (r(t+\tau) - E(r(t+\tau)))^2}} \tag{2}$$

where, $E(\cdot)$ is the expectation operator, τ is the time delay or shift.

The ARMA model is defined as:

$$z(k) = \sum_{i=1}^{p} a_i z(k-i) + \sum_{j=1}^{q} b_j \varepsilon(k-j) + \varepsilon(k) \tag{3}$$

where, $\varepsilon(k)$ is a zero mean and unknown white noise, p and q are the order of the ARMA model, $z(k-i)$ is the input time series data, and the a_i and b_j are the related parameters, some approaches are published for obtaining the values of these parameters, for instance Kalman filter and the least square estimation method [25,26]. Generally, the auto-correlation and partial correlation function is used to decide the order of the AMRA model.

2.2. LSTM-RNN Method

Long Short Term Memory (LSTM) is a popular variant of the common Recurrent Neural Network (RNN). An RNN composed of LSTM units is often called an LSTM network. Different from RNN, new structure termed as "gate" is added to LSTM. Commonly, a LSTM unit is composed of a cell, an "input gate", "output gate" and a "forget gate." The basic structure of a single layer LSTM unit is shown as Figure 1. The cell remembers values over arbitrary time intervals and the three different gates regulate and control the flow of information into and out of the cell [38–40]. Following is the detailed description of the different gates and relative equations.

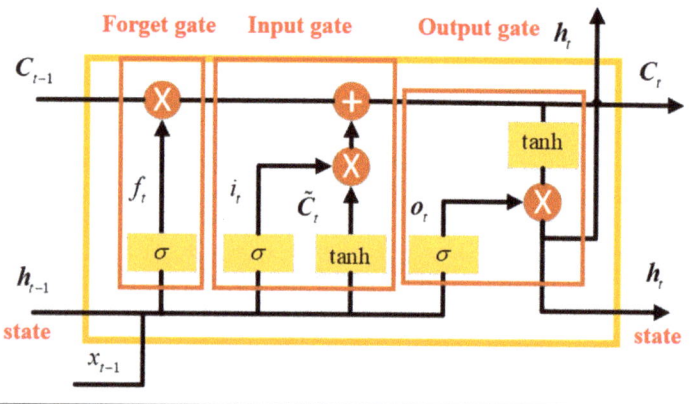

Figure 1. Basic structure of a Long Short Term Memory (LSTM) Unit.

As illustrated in Figure 1, the first part of the LSTM is the "forget gate", which is employed to decide what information is going to get thrown away from the cell state, the decision is made by a sigmoid layer called "forget gate layer". h_{t-1} and x_t are input to the function, and outputs a value ranging from 0 to 1 for each number in the cell state C_{t-1}. The values represent the forgetting degree of each number in the cell state, and "1" represents "completely keep this" while "0" represents "completely get rid of this". The operation equation f_t is as:

$$f_t = \sigma\left(W_f \cdot [h_{t-1}, x_t] + b_f\right) \tag{4}$$

where, $\sigma(\cdot)$ is a sigmoid function, W_f is the updating weights, b_f is the bias, h_{t-1} is the hidden state, and x_t is the input vector.

After deciding the memory of the previous hidden state, and the second part is the "input gate", which is utilized to decide what new information is going to be stored in the current cell state. This gate is composed of two parts: (1) A sigmoid layer to decide what values are going to be updated, the output values i_t range from 0 to 1, and they represent the updating degree of each number in input; (2) another part is a *tanh* layer which creates a vector of new candidate values $\tilde{C}_t$, which will be added to the cell state after multiplied with the decision vector i_t. The relative equations are as following:

$$i_t = \sigma(W_i \cdot [h_{t-1}, x_t] + b_i) \tag{5}$$

$$\tilde{C}_t = \tanh(W_C \cdot [h_{t-1}, x_t] + b_C) \tag{6}$$

where, $\sigma(\cdot)$ is a sigmoid function, W_i is the updating weights, b_i is the bias in the input gate, h_{t-1} is the hidden state at time $t-1$, W_C is the updating weights, b_C is the bias, and x_t is the input vector.

The last part is the "output gate", which is employed to decide what is going to output, similarly, a sigmoid layer outputs values o_t, which is employed to decide what parts of the cell state will be output, then the cell state is put through a *tanh* function. After this operation, the cell state values are pushed to be between -1 and 1. Finally, the results are multiplied by the output of the sigmoid gate, and the output parts are decided. The related equations are as:

$$o_t = \sigma(W_o \cdot [h_{t-1}, x_t] + b_o) \tag{7}$$

$$h_t = o_t * \tanh(C_t) \tag{8}$$

where, W_o is the updating weights, and b_o is the bias in the output gate, C_t is the cell state at time t.

The above Equations (4)–(8) describes the basic LSTM unit for RNN, which is just a single LSTM unit. Figure 2 presents a sequence of LSTM-RNN units in time domain. In Figure 2a, is a simple description of the LSTM-RNN working flow. The output of LSTM-RNN is decided by not only current state, but a long-term memory. Figure 2b gives the details. The cell state and hidden layer is covered to the next LSTM Unit, and the inner gate will decide the memory degree of past information. In addition, before it is employed for prediction, a training procure is necessary for determining the unknown parameters in the above Equations [37–40].

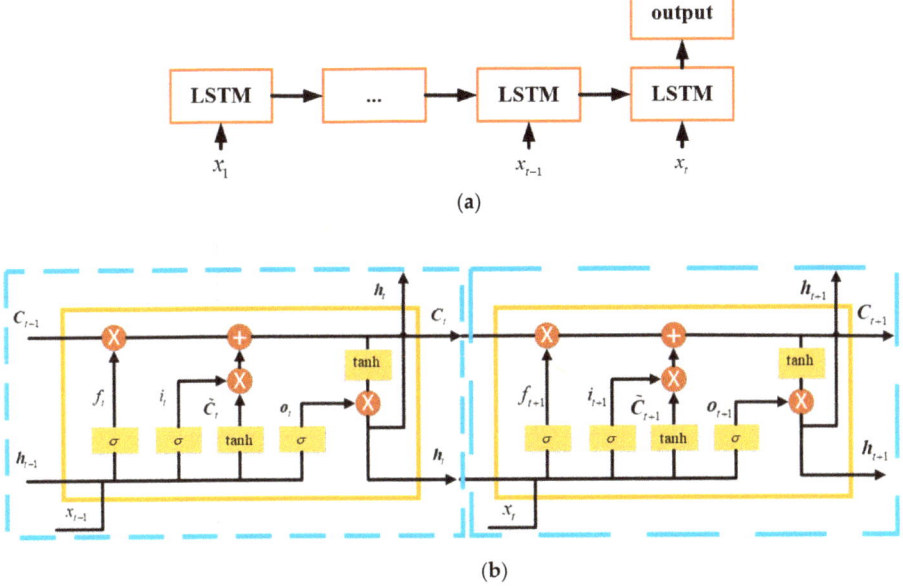

Figure 2. Working flow of the LSTM-RNN. (**a**) brief working flow of LSTM-RNN (Recurrent Neural Network); (**b**) a sequence of LSTM Unit.

3. Experiments and Results

This section will present experiments and the relative analyses for evaluating the performance of the proposed LSTM-RNN method. The laboratory experiments are conducted using the data which was collected from a MEMS IMU (MSI3200) manufactured by MT Microsystems Company, Shijiazhuang, China [29]. The real picture and the specifications of the IMU are as Figure 3 and Table 1 respectively. The gyroscope bias stability is $\leq 10°/h$, and the random walk is $\leq 10°/\sqrt{h}$. The accelerometer bias was 0.5 mg, and the bias stability was 0.5 mg. The IMU was placed statically on the table, and the sampling frequency was 400 Hz. Thus, the amount of data was 48,000. The gyroscope output data unit was in degree/s. The raw noisy data of the X-axis gyroscope output is shown as Figure 3 (red line representing the raw data), and the bias was reduced before modeling the errors. After removing the bias, the result is presented in Figure 4 (blue line representing the data excluding bias). Note that the program in this experiment was developed in Python with the Tensorflow package, which is operated in an Alienware R2 PC installed an i7 Intel CPU and 16 GB random memory.

Figure 3. MSI3200 Inertial Measurement Unit.

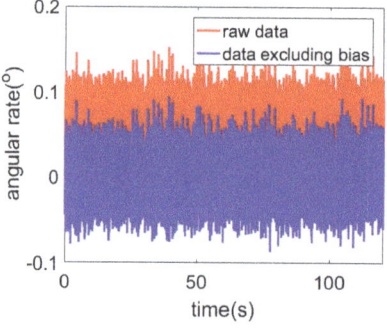

Figure 4. X-axis gyroscope signals.

Table 1. Specifications of MSI3200 IMU (Inertial Measurement Unit).

		Range	±300°/s
	Gyroscope	Bias stability (1σ)	≤10°/h
		Bias stability (Allan)	≤2°/h
		Angle random walk	≤10°/$\sqrt{h}$
MEMS IMU	Accelerometer	range	±15 g
		Bias stability (1σ)	0.5 mg
		Bias stability (Allan)	0.5 mg
	Power consumption		1.5 W
	Weight		250 g
	Size		70 mm × 54 mm × 39 mm
	Sampling rate		400 Hz

The remainder of his section is divided into three parts: (1) The first part describes the results using the ARMA method to model the errors with presenting the auto-correction and partial correction results. The ARMA models are given according to the auto-correction and partial correction results. The standard deviation (STD) values of the signals are compared with the corresponding raw gyroscope signals; (2) the second part is the results using the LSTM-RNN to model the errors, and presenting the training time and prediction accuracy for different input vector length. Moreover, a multi-layer LSTM-RNN is designed and compared with a single-layer LSTM-RNN in terms training time, computation load and performance; (3) the last part presents the comparisons conducted between the ARMA and LSTM-RNN, including statistical results and position results.

3.1. Error Modeling Using ARMA

For time series analysis, Auto-Correlation Function (ACF) and Partial Auto-Correlation Function (PACF) characteristics are usually employed to select the proper model. As aforementioned in Section 2, the ACF and PACF are presented as Equations (1) and (2), and then Figures 4–6 show the ACF and PACF results of the X-axis, Y-axis, and Z-axis gyroscope raw signals respectively, which are processed according to the Equations (1) and (2). From Figures 5–7, it is evident and obvious that ACF and PACF of the three-axis gyroscope are tail off. Thus, the ARMA model is suitable for this application, and the order of the ARMA model is determined using the results. More details about parameters determination can be found in the references [25,26]. Therefore, ARMA models for these three-axis gyroscope signals are presented as:

$$z(k) = 0.3475z(k-1) + 0.163z(k-2) - 0.0508\varepsilon(k-1) + \varepsilon(k) \tag{9}$$

$$z(k) = 0.5065z(k-1) + 0.2583z(k-2) - 0.2371\varepsilon(k-1) + \varepsilon(k) \tag{10}$$

$$z(k) = 0.3309z(k-1) + 0.2592z(k-2) + 0.132\varepsilon(k-1) + \varepsilon(k) \tag{11}$$

where, the $z(k)$ is the data at time $\varepsilon(k)$ is the white noise at time k. The results from the ARMA is listed in Table 2. Compared with raw signal, the standard deviation (STD) of the three-axis gyroscope outputs decrease by 31.1%, 20.0% and 25.0%. The results show the ARMA performs effectively for de-noising MEMS gyroscope raw signals. Specifically, the parameters or orders are fixed in this experiment.

Table 2. Standard deviation of Auto Regressive and Moving Average (ARMA) modeling results for the three-axis gyroscope.

	X	Y	Z
Raw data	0.0247	0.035	0.056
ARMA	0.017	0.028	0.042

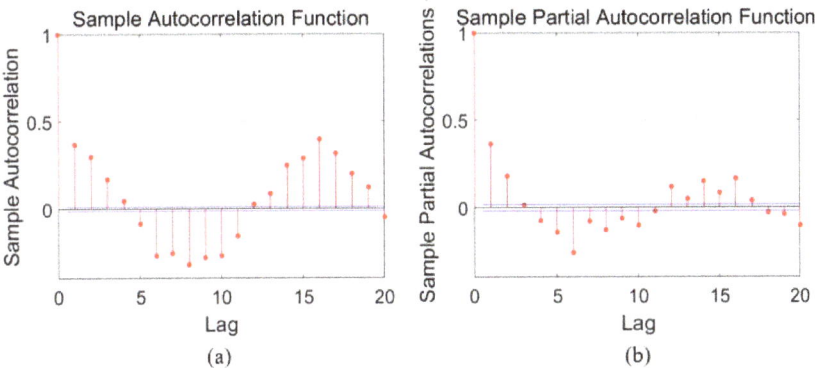

Figure 5. Auto-correlation and partial correlation analysis results of X-axis gyroscope signals. (a) autocorrelation analysis diagram; (b) Partial correlation analysis diagram.

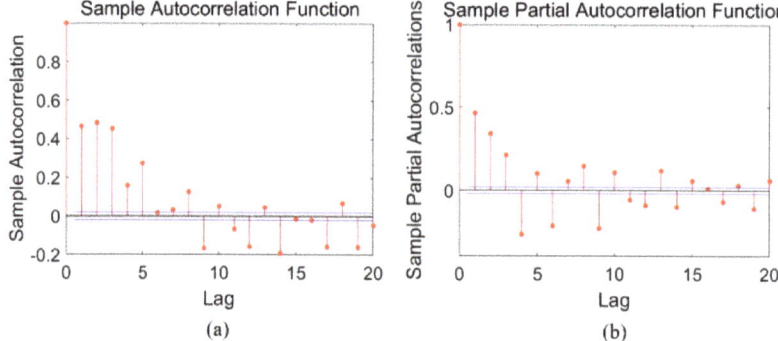

Figure 6. Auto-correlation and partial correlation analysis results of Y-axis gyroscope signals. (**a**) autocorrelation analysis diagram; (**b**) Partial correlation analysis diagram.

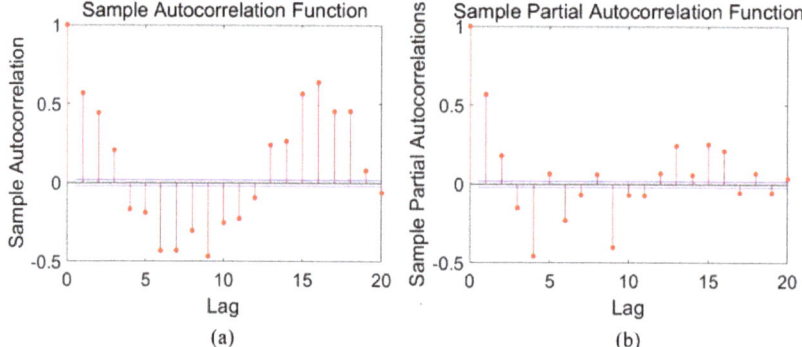

Figure 7. Auto-correlation and partial correlation analysis results of Z-axis gyroscope signals. (**a**) autocorrelation analysis diagram; (**b**) Partial correlation analysis diagram.

3.2. Error Modeling Using LSTM-RNN

In this proposed LSTM-RNN method, the employed MEMS gyroscope dataset is labeled as $[x_1, x_2, \ldots, x_N]$, the subscript N is termed as the amount of the IMU data samples. The dataset is divided into a training and testing part. The training part is used to build the model and the testing part is used to verify the model. The input data vector for training is defined as:

$$Input_i = [x_i, x_{i+1}, \ldots, x_{i+step}], i \in [1, N - step] \tag{12}$$

The output data vector is defined as:

$$Onput_i = [x_{i+step+1}], i \in [1, N - step] \tag{13}$$

In above equations, the variable *step* is the length of the input data vector for training procedure. Suitable values of the input vector length *step* is identified to realize a tradeoff between training time and the prediction performance. Table 3 shows the three axis gyroscope data training results and standard deviation (STD) of the prediction. In this test, the 5, 10, 15, 20 and 30 are the selected values of the vector length. Table 4 shows the comparison results. The training dataset length is 1000, and the testing dataset length is 48,000 (2 min with 400 Hz sampling rate). The specifications of the LSTM-RNN are presented in Table 5. The consumption time increases with the input vector length, and the STD values decrease first, and then increase. Hence, 20 is selected as the length of input vector, which is the best tradeoff between the STD and the computation time.

Table 3. Standard deviation of the raw and predicted datasets by single LSTM-RNN.

	X	Y	Z
Raw data	0.0247	0.035	0.056
Single LSTM-RNN	0.0098	0.022	0.031

Table 4. Performance of gyroscope X-axis with varying values of input vector.

Length	STD	Time (s)
5	0.0096	1.25
10	0.0095	1.33
15	0.0063	2.13
20	0.0052	2.90
30	0.0094	2.98

Table 5. Specifications of LSTM-RNN.

Batch size	128
Training epoch	50
Learning rate	0.01
Hidden unit amount	1

As shown in Table 5, the noises are considerably decreased using the LSTM-RNN, and the STD values decrease by 60.3%, 37%, and 44.6%. As aforementioned, the training epoch was set to 50, and a multi-layer LSTM-RNN was designed and compared with single LSTM-RNN. Table 6 shows the results. The multi-layer LSTM-RNN (two hidden units) has a lower average training loss at epoch of 50, which decreases by 55.4%, 34.2% and 32.1%, However, it seems that there is no obvious advance in filtering performance. The STD values of the filters data have no reduction, and are even a little higher. The operation time consumption is almost twice that of single LSTM-RNN.

Figure 8 shows the training loss comparison of single-layer LSTM RNN and multi-layer LSTM-RNN, and they have identical accuracy at the 20th epochs. Thus, multi-layer LSTM-RNN was trained with 20 epochs, and Table 7 shows the results for the multi-layer LSTM-RNN. The results were compared with single LSTM-RNN with 50 training epochs. The average training losses have a slight increase, and the STD values are almost identical to that of multi-layer LSTM-RNN with 50 training epochs. However, compared with the single LSTM-RNN, the time consumption of multi-layer LSTM-RNN is less than that of single LSTM-RNN, which is initialed by the training epochs reducing in multi-layer LSTM-RNN. In aspects of the STD values, the de-noised three-axis gyroscope outputs have an improvement of 22.4%, 9.1%, and 22.6% respectively compared with single-layer LSTM-RNN trained after 50 epochs. This decline in accuracy means the multi-layer LSTM-RNN with fewer training epochs have weaker generation ability, since the multi-layer LSTM-RNN has more parameters which need more training epochs. Thus, while the training epochs are set to 20, the multi-layer has slightly worse STD values compared with the single-layer LSTM-RNN.

Table 6. Comparison of Single LSTM-RNN and multi-layer LSTM-RNN.

	X			Y			Z		
	Training Loss	STD	Time	Training Loss	STD	Time	Training Loss	STD	Time
Single LSTM-RNN	0.00053	0.0098	4.52	0.0010	0.022	4.94	0.022	0.031	4.56
Multi-layer LSTM-RNN	0.000467	0.011	9.31	0.0009	0.023	9.21	0.014	0.038	8.65
Raw data	/	0.0246	/	/	0.0352	/	/	0.056	/

Table 7. Comparison of single LSTM-RNN with 50 training epochs and multi-layer LSTM-RNN with 20 training epochs.

	X			Y			Z		
	Training Loss	STD	Time	Training Loss	STD	Time	Training Loss	STD	Time
Single LSTM-RNN	0.00053	0.0098	4.52	0.0010	0.022	4.94	0.022	0.031	4.56
Multi-layer LSTM-RNN	0.00045	0.012	3.68	0.0009	0.024	3.82	0.017	0.038	3.76
Raw data	/	0.0246	/	/	0.0352	/	/	0.056	/

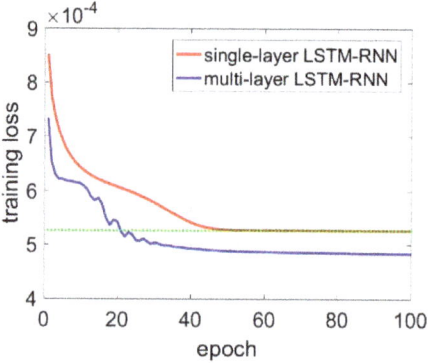

Figure 8. Training loss comparison between single-layer LSTM and multi-layer LSTM.

3.3. Comparisons of ARMA and LSTM-RNN

This part presents the comparisons between ARMA and LSTM-RNN. Table 8 shows the STD results from the ARMA and LSTM-RNN de-noising methods. Compared with raw signals, STD values of the three-axis gyroscope data from the ARMA method perform a 31.2%, 20.0% and 25.0% improvement, and the STD values of the single-layer LSTM-RNN de-noised signals decrease by 42.4%, 21.4% and 21.4% respectively. With the same MEMS IMU dataset, the LSTM-RNN has an obvious improvement of 42.3%, 21.4% and 26.2% respectively for the three-axis gyroscope dataset.

Further, Figure 9 shows the attitude errors of the ARMA and LSTM-RNN de-noised MEMS IMU data. In the Figure 9, the blue line represents the position errors of the raw signals from MEMS IMU, the green line represents the position errors of AMMA de-noised MEMS IMU, and the red line represents the position errors of the designed single LSTM-RNN de-noised MEMS IMU. Table 9 shows the maximum errors of the three-axis gyroscope, compared with raw signal. The pitch, roll angle and yaw angle errors decreased by 15.8%, 18.3% and 51.3% respectively. Specifically, the pitch error decreases from $-5.07°$ to $-4.27°$, the roll error decreases from $-1.95°$ to $-1.60°$, and the yaw angle error decreases from $-3.85°$ to $-1.87°$. Moreover, the errors from signals de-noised by LSTM-RNN decreased by 47.6%, 42.3% and 52.0%, further compared with ARMA results. To be specific, the pitch, roll and yaw angles have an extra improvement of $2.04°$, $0.66°$ and $0.96°$ compared with that of ARMA. The ARMA employed in this experiment was operated with fixed parameters with selected part of the dataset. Thus, in the testing dataset, more feasible parameters are available and suitable for better performance. However, the single-layer LSTM and ARMA were tested with the identical dataset, this might demonstrate that LSTM-RNN have better generation ability in this application. This is what we think might account for the accuracy improvement of the single-layer LSTM-RNN compared with the common ARMA method. In addition, the yaw errors from the LSTM-RNN de-noised signals have an upward trend which is different from the errors from the ARMA and raw signals. We think the principle of the LSTM-RNN may account for this, and the specifications need more and further investigation. Overall, the results demonstrate the effectiveness of LSTM-RNN in MEMS gyroscope signals de-nosing.

Table 8. Standard deviation of ARMA modeling results for the three-axis gyroscope.

	X	Y	Z
Raw data	0.0247	0.035	0.056
ARMA	0.017	0.028	0.042
Single LSTM-RNN	0.0098	0.022	0.031

Table 9. Maximum attitude errors from raw, ARMA and LSTM-RNN

	Pitch	Roll	Yaw
Raw data	−5.070	−1.952	−3.853
ARMA	−4.268	−1.601	−1.873
Single LSTM-RNN	−2.231	−0.942	0.826

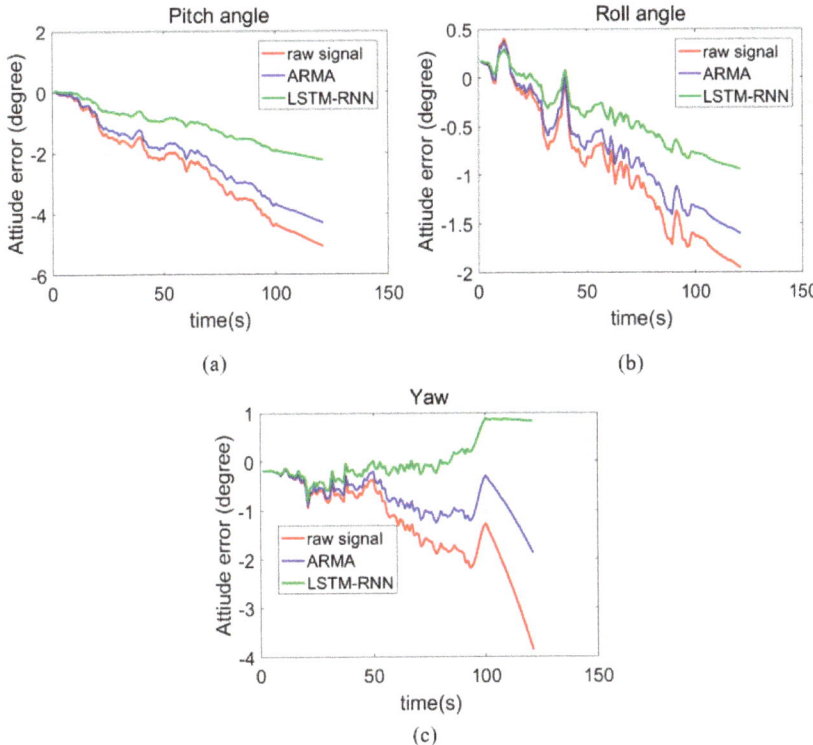

Figure 9. LSTM-RNN MEMS IMU attitude errors. (a) Pitch angle; (b) Roll angle; (c) Yaw angle

4. Discussion

1. In this paper, limited by the computing capacity of the employed computer, the LSTM-RNN had a limited amount of layers, which might have a negative influence on the generation ability of LSTN-RNN and the prediction performance in the long term.
2. In this paper, just one of the RNN variants LSTM-RNN were employed and evaluated in this application, and it has significant meaning to explore different LSTM-RNN structures more suitable for MEMS IMU errors modeling and de-noising.

3. This method was tested only using static data, and dynamic trajectory data should be included for fully evaluating the proposed method. The noise characteristics in dynamic environment may be different from that in dynamics.

5. Conclusions

This paper discussed a LSTM-RNN based MEMS IMU errors modelling method. A MEMS IMU (MSI 3200) was employed for testing the proposed method. Through the comparisons, three major conclusions were drawn as: (1) LSTM-RNN outperformed the ARMA in this application. Compared with the ARMA model, the standard deviation of the single LSTM-RNN de-noised signals decreased by 42.4%, 21.4% and 21.4% respectively, and the attitude errors decreased by 47.6%, 42.3% and 52.0%; (2) multi-layer LSTM-RNN was able to realize the settled average training loss with less training epochs. However, the multi-layer LSTM-RNN did not outperform the single LSTM-RNN in standard deviation values of the prediction. When the training epoch was set to 20, the multi-layer had a slightly better prediction accuracy with less computation time than the single LSTM-RNN.

In addition, we think some more details are worthy of being investigated further in the future: (1) It is meaningful to investigate the deep LSTM-RNN network, which should be trained with a large amount data. Well trained deep LSTM-RNN has been demonstrated to be more feasible in some applications. A deep LSTM-RNN will be implemented and presented in future; (2) many variants of RNN are published and have been demonstrated effectively in solving time series prediction problems. It is meaningful to further investigate and compare their performance, and find more preferable neural networks suitable for this particular application. Comparison of several popular variants of RNN will be presented in future.

Author Contributions: C.J. proposed the idea, developed the LSTM-RNN Python software, and written the first version of this paper. S.C. guided the paper writing, reviewed the paper and offered the funding. Y.C. revised the paper, discussed the idea and guided the paper writing. B.Z. collected and processed the data. Z.F., H.Z. and Y.B. reviewed the paper before submission.

Funding: This research was funded by the Fundamental Research Funds for the Central Universities (Grant No. 30917011105); the National Defense Basic Scientific Research program of China (Grant No. JCKY2016606B004); Jiangsu Planned Projects for Postdoctoral Research Funds, grant number (Grant No. 1501050B); the China Postdoctoral Science Foundation, grant number (Grant No. 2015M580434) and the special grade of the financial support from the China Postdoctoral Science Foundation with grant number (Grant No. 2016T90461).

Acknowledgments: The author gratefully acknowledges the financial support from China Scholarship Council (CSC, Grant No. 201806840087) and excellent doctor training fund of Nanjing University of Science and Technology (NJUST). If you want the source code, please e-mail me, and I will share the all the materials.

Conflicts of Interest: The authors declare no conflict of interest.

References

1. Brown, A.K. GPS/ins uses low-cost mems IMU. *IEEE Aerosp. Electron. Syst. Mag.* **2005**, *20*, 3–10. [CrossRef]
2. Noureldin, A.; Karamat, T.B.; Eberts, M.D.; El-Shafie, A. Performance enhancement of MEMS-based INS/GPS integration for low-cost navigation applications. *IEEE Trans. Veh. Technol.* **2009**, *58*, 1077–1096. [CrossRef]
3. Chen, Y.; Tang, J.; Jiang, C.; Zhu, L.; Lehtomäki, M.; Kaartinen, H.; Kaijaluoto, R.; Wang, Y.; Hyyppä, J.; Hyyppä, H.; et al. The accuracy comparison of three simultaneous localization and mapping (SLAM)-Based indoor mapping technologies. *Sensors* **2018**, *18*, 3228. [CrossRef] [PubMed]
4. Zhang, X.; Zhu, F.; Tao, X.; Duan, R. New optimal smoothing scheme for improving relative and absolute accuracy of tightly coupled GNSS/SINS integration. *GPS Solut.* **2017**, *21*, 861–872. [CrossRef]
5. Tang, J.; Chen, Y.; Chen, L.; Liu, J.; Hyyppä, J.; Kukko, A.; Kaartinen, H.; Hyyppä, H.; Chen, R. Fast fingerprint database maintenance for indoor positioning based on UGV SLAM. *Sensors* **2015**, *15*, 5311–5330. [CrossRef] [PubMed]
6. Jiang, C.; Chen, S.; Bo, Y.; Sun, Z.; Lu, Q. Implementation and performance evaluation of a fast relocation method in a GPS/SINS/CSAC integrated navigation system hardware prototype. *IEICE Electron. Express* **2017**, *14*, 20170121. [CrossRef]

7. Ma, L.; You, Z.; Liu, T.; Shi, S. Coupled integration of CSAC, MIMU, and GNSS for improved PNT performance. *Sensors* **2016**, *16*, 682. [CrossRef] [PubMed]
8. Petritoli, E.; Leccese, F. Improvement of altitude precision in indoor and urban canyon navigation for small flying vehicles. In Proceedings of the 2015 IEEE Metrology for Aerospace (MetroAeroSpace), Benevento, Italy, 4–5 June 2015.
9. Jiang, C.; Chen, S.; Bo, Y.; Sun, Z.; Lu, Q. Performance Analysis of GNSS Vector Tracking Loop Based GNSS/CSAC Integrated Navigation System. *J. Aeronaut. Astronaut. Aviat.* **2017**, *49*, 289–297.
10. Fernández, E.; Calero, D.; Parés, M.E. CSAC Characterization and Its Impact on GNSS Clock Augmentation Performance. *Sensors* **2017**, *17*, 370. [CrossRef] [PubMed]
11. Lee, B. Review of the present status of optical fiber sensors. *Opt. Fiber Technol.* **2003**, *9*, 57–79. [CrossRef]
12. Narasimhappa, M.; Sabat, S.L.; Nayak, J. Fiber-optic gyroscope signal denoising using an adaptive robust Kalman filter. *IEEE Sens. J.* **2016**, *16*, 3711–3718. [CrossRef]
13. Narasimhappa, M.; Nayak, J.; Terra, M.H.; Sabat, S.L. ARMA model based adaptive unscented fading Kalman filter for reducing drift of fiber optic gyroscope. *Sens. Actuators A Phys.* **2016**, *251*, 42–51. [CrossRef]
14. Jiang, C.; Chen, S.; Chen, Y.; Bo, Y. Research on Chip Scale Atomic Clock Driven GNSS/SINS Deeply Coupled Navigation System for Augmented Performance. *IET Radar Sonar Navig.* **2018**. [CrossRef]
15. Eling, C.; Klingbeil, L.; Kuhlmann, H. Real-time single-frequency GPS/MEMS-IMU attitude determination of lightweight UAVs. *Sensors* **2015**, *15*, 26212–26235. [CrossRef] [PubMed]
16. Jiang, C.; Chen, S.; Bo, Y.; Qu, Y.; Han, N. Research of fast relocation technology assisted by IMU in the GPS/SINS ultra-tightly coupled navigation system. *J. Aeronaut. Astronaut. Aviat.* **2016**, *48*, 253–259.
17. El-Sheimy, N.; Hou, H.; Niu, X. Analysis and modeling of inertial sensors using Allan variance. *IEEE Trans. Instrum. Meas.* **2008**, *57*, 140–149. [CrossRef]
18. Quinchia, A.G.; Falco, G.; Falletti, E.; Dovis, F.; Ferrer, C. A comparison between different error modeling of MEMS applied to GPS/INS integrated systems. *Sensors* **2013**, *13*, 9549–9588. [CrossRef] [PubMed]
19. Vaccaro, R.J.; Zaki, A.S. Statistical modeling of rate gyros. *IEEE Trans. Instrum. Meas.* **2012**, *61*, 673–684. [CrossRef]
20. Syed, Z.; Aggarwal, P.; Goodall, C.; Niu, X.; El-Sheimy, N. A new multi-position calibration method for MEMS inertial navigation systems. *Meas. Sci. Technol.* **2007**, *18*, 1897. [CrossRef]
21. Aggarwal, P.; Syed, Z.; Niu, X.; El-Sheimy, N. A standard testing and calibration procedure for low cost MEMS inertial sensors and units. *J. Navig.* **2008**, *61*, 323–336. [CrossRef]
22. Bekkeng, J.K. Calibration of a novel MEMS inertial reference unit. *IEEE Trans. Instrum. Meas.* **2009**, *58*, 1967–1974. [CrossRef]
23. Kang, C.H.; Kim, S.Y.; Park, C.G. Improvement of a low cost MEMS inertial-GPS integrated system using wavelet denoising techniques. *Int. J. Aeronaut. Space Sci.* **2011**, *12*, 371–378. [CrossRef]
24. Chen, D.; Han, J. Application of wavelet neural network in signal processing of MEMS accelerometers. *Microsyst. Technol.* **2011**, *17*, 1–5. [CrossRef]
25. Huang, L. Auto regressive moving average (ARMA) modeling method for Gyro random noise using a robust Kalman filter. *Sensors* **2015**, *15*, 25277–25286. [CrossRef] [PubMed]
26. Huang, L.; Li, Z.; Xie, F.; Feng, K. Novel time series modeling methods for gyro random noise used in Internet of Things. *IEEE Access* **2018**, *6*, 47911–47921. [CrossRef]
27. Wang, L.; Wei, G.; Zhu, Y.; Liu, J.; Tian, Z. Real-time modeling and online filtering of the stochastic error in a fiber optic current transducer. *Meas. Sci. Technol.* **2016**, *27*, 105103. [CrossRef]
28. Bhatt, D.; Aggarwala, P.; Devabhaktunia, V.; Bhattacharyab, P. A novel hybrid fusion algorithm to bridge the period of GPS outages using low-cost INS. *Expert Syst. Appl.* **2014**, *41*, 2166–2173. [CrossRef]
29. Kopáčik, A.; Kajánek, P.; Lipták, I. Systematic error elimination using additive measurements and combination of two low cost IMSs. *IEEE Sens. J.* **2016**, *16*, 6239–6248. [CrossRef]
30. Yang, H.; Li, W.; Luo, T.; Liang, H.; Zhang, H.; Gu, Y.; Luo, C. Research on the Strategy of Motion Constraint-Aided ZUPT for the SINS Positioning System of a Shearer. *Micromachines* **2017**, *8*, 340. [CrossRef]
31. Ning, Y.; Wang, J.; Han, H.; Tan, X.; Liu, T. An Optimal Radial Basis Function Neural Network Enhanced Adaptive Robust Kalman Filter for GNSS/INS Integrated Systems in Complex Urban Areas. *Sensors* **2018**, *18*, 3091. [CrossRef] [PubMed]
32. Jerath, K.; Brennan, S.; Lagoa, C. Bridging the gap between sensor noise modeling and sensor characterization. *Measurement* **2018**, *116*, 350–366. [CrossRef]

33. Bhatt, D.; Aggarwal, P.; Bhattacharya, P.; Devabhaktuni, V. An enhanced mems error modeling approach based on nu-support vector regression. *Sensors* **2012**, *12*, 9448–9466. [CrossRef] [PubMed]
34. Chu, Y.; Fei, J. Adaptive global sliding mode control for MEMS gyroscope using RBF neural network. *Math. Probl. Eng.* **2015**. [CrossRef]
35. El-Rabbany, A.; El-Diasty, M. An efficient neural network model for de-noising of MEMS-based inertial data. *J. Navig.* **2004**, *57*, 407–415. [CrossRef]
36. Xing, H.; Hou, B.; Lin, Z.; Guo, M. Modeling and Compensation of Random Drift of MEMS Gyroscopes Based on Least Squares Support Vector Machine Optimized by Chaotic Particle Swarm Optimization. *Sensors* **2017**, *17*, 2335. [CrossRef] [PubMed]
37. Hochreiter, S.; Schmidhuber, J. Long short-term memory. *Neural Comput.* **1997**, *9*, 1735–1780. [CrossRef] [PubMed]
38. Gallicchio, C. Short-term memory of deep rnn. *arXiv* **2018**, arXiv:1802.00748.
39. Li, X.; Wu, X. Constructing long short-term memory based deep recurrent neural networks for large vocabulary speech recognition. In Proceedings of the 2015 IEEE International Conference on Acoustics, Speech and Signal Processing (ICASSP), South Brisbane, Australia, 19–24 April 2015.
40. Understanding LSTM Networks. Available online: https://colah.github.io/posts/2015-08-Understanding-LSTMs (accessed on 27 August 2015).
41. Hosseinyalamdary, S. Deep Kalman Filter: Simultaneous Multi-Sensor Integration and Modelling; A GNSS/IMU Case Study. *Sensors* **2018**, *18*, 1316. [CrossRef] [PubMed]
42. MT Microsystems. Available online: http://www.mtmems.com/product_view.asp?id=28 (accessed on 30 September 2018).

© 2018 by the authors. Licensee MDPI, Basel, Switzerland. This article is an open access article distributed under the terms and conditions of the Creative Commons Attribution (CC BY) license (http://creativecommons.org/licenses/by/4.0/).

Article

A Hybrid CNN–LSTM Algorithm for Online Defect Recognition of CO_2 Welding

Tianyuan Liu [1], Jinsong Bao [1,*], Junliang Wang [1] and Yiming Zhang [2]

1 College of Mechanical Engineering, Dong Hua University, Shanghai 201620, China; tyliu@mail.dhu.edu.com (T.L.); junliangwang@dhu.edu.cn (J.W.)
2 College of Literature, Science and the Arts, The University of Michigan, Ann Arbor, MI 48109, USA; yimingz@umich.edu
* Correspondence: bao@dhu.edu.cn; Tel.: +86-216-779-2562

Received: 11 October 2018; Accepted: 6 December 2018; Published: 10 December 2018

Abstract: At present, realizing high-quality automatic welding through online monitoring is a research focus in engineering applications. In this paper, a CNN–LSTM algorithm is proposed, which combines the advantages of convolutional neural networks (CNNs) and long short-term memory networks (LSTMs). The CNN–LSTM algorithm establishes a shallow CNN to extract the primary features of the molten pool image. Then the feature tensor extracted by the CNN is transformed into the feature matrix. Finally, the rows of the feature matrix are fed into the LSTM network for feature fusion. This process realizes the implicit mapping from molten pool images to welding defects. The test results on the self-made molten pool image dataset show that CNN contributes to the overall feasibility of the CNN–LSTM algorithm and LSTM network is the most superior in the feature hybrid stage. The algorithm converges at 300 epochs and the accuracy of defects detection in CO_2 welding molten pool is 94%. The processing time of a single image is 0.067 ms, which fully meets the real-time monitoring requirement based on molten pool image. The experimental results on the MNIST and FashionMNIST datasets show that the algorithm is universal and can be used for similar image recognition and classification tasks.

Keywords: deep learning; CNN; LSTM; CO_2 welding; molten pool; online monitoring

1. Introduction

Welding is a dynamic, interactive, and non-linear process. The monitoring of welding defects is a difficult problem due to these characteristics of welding. The main difficulties in this task include deciding when a defect occurred and which type of defect occurred. In the actual welding process, skilled welders can dynamically adjust the welding process from observing the state of molten pool to prevent welding defects. That gave rise to our idea that we could adjust the welding process by observing the molten pool. An accurate mapping model between the molten pool image and the weld quality is a vital part of this method [1,2]. The molten pool images are independently used as inputs to this model. A typical molten pool image contains many objects, such as welding wire, arc, molten pool, weld seam, metal accumulation, splash, smoke, etc. Although the molten pool is the main part of the whole image, it is necessary to consider all objects and the relationship between various objects for the purpose of accurately reflecting the welding information through the molten pool image. Therefore, extracting the features from all objects in the molten pool image and hybridizing the different features are the key to establishing an accurate mapping model.

The research of molten pool image can be divided into two categories. One of them is based on multiple molten pool images. The main idea is to find the mutation rule of molten pool characteristic signals when welding defects occur by multi-level (time domain, frequency domain) statistical analysis of the characteristics of multiple molten pool images [3–8]. This idea can synthetically consider the molten pool images' information of the whole welding process, and the features used for analysis are generally primary features, which are relatively easy to design and obtain. However, this idea can only be used to analyze the overall welding quality and locate the welding defects after welding is complete and does not satisfy requirements of real-time monitoring of welding. Another idea is based on single molten pool images, which is more suitable for an online monitoring process [9–13]. In studying molten pool images of the welding process, the most original method is to manually design and identify the statistics of the characteristics of molten pool (length, width, area, spatter number, etc.) and then identify the molten pool state. Although this method is highly interpretable, it requires a lot of prior knowledge and is very time-consuming. Furthermore, such a model poorly adapts to other image classification problems. With the development of deep learning, convolutional neural network (CNN) replaced the process of human design and the extraction of primary features, achieving great results [14–16]. However, for the purpose of further improving the accuracy of defect recognition, more convolutional layers need to be stacked in the feature fusion stage, which will bring huge computational cost, making real-time monitoring infeasible. In view of the existing problems in the feature hybrid stage, principal component analysis (PCA) and fully connected layers are widely used to combine the results of feature extraction [17–21]. Although PCA has great interpretability, such a deterministic process may leave out features with small contributions, and these features may entail important information about welding quality. Therefore, the process of feature information fusion lacks the ability of intelligent fusion. Adding a fully connected layer and adjusting the weight of each shallow feature using back propagation can play a certain role in intelligent hybrid of features; however, this hybrid method is often too simple and insufficient to extract high-level abstract information.

Traditional neural networks (including CNN) assume that all inputs and outputs are independent of each other, while the basic assumption of recurrent neural network (RNN) is that there is an interaction between the input sequences, and this feature of RNN provides a new approach to feature hybrid [22,23]. References [24–30] propose a method for intelligently hybridizing the features of each individual in the input sequence using the long short-term memory network (LSTM, a variant of RNN), which can extract the long-term dependencies of the data features in the sequence to improve the recognition accuracy. However, the original input of the whole online molten pool status recognition task is a single molten pool image at a certain moment rather than a sequence of images. Therefore, in view of the above problems and the complexity of the molten pool images, this paper proposes an innovative strategy. In the feature extraction stage, multiple convolutional kernels are used to scan the whole molten pool image to obtain the redundant features of all objects in the molten pool image. Due to the distance that the convolution kernel slides each time is less than the size of the convolution kernel itself, and there are overlapping parts in each scan area of the convolution kernel, so the feature blocks extracted by the convolution kernel also depend strongly on each other. When describing a thing, we often hope to construct a set of bases, which can form a complete description of a thing. The same is true in the same level of a convolution network, that is, the relationship between feature maps extracted from the same level of convolution kernels lies in the formation of a description of images on different bases at the same level. So in the stage of feature fusion, several feature images extracted by CNN are unified and reconstructed into a two-dimensional feature matrix that contains the correlation information from the interior of a single feature image and from multiple feature images. In order to improve the accuracy of molten pool image recognition, each row of the feature matrix is considered as a basic unit to be hybridized, and the number of rows is considered as the length of a sequence. In this way, the single image of molten pool is converted into "sequential" data in this sense. Then, the long-term dependencies property of the LSTM network is used to filter and fusion the rows of the feature matrix to obtain high-level abstract information. In this case, the model is transformed into a multi-input single-output model like text sentiment analysis. Each input can be understood as a contribution of the feature vector at this time step to the overall molten pool image identification task in the context. The CNN–LSTM algorithm proposed in this paper establishes the end-to-end mapping relationship between molten pool image and welding defects. The advantage of this algorithm is that it can intelligently learn the best hybrid features through the error back propagation algorithm in the shallow CNN network for a single molten pool image to meet the engineering requirements for real-time monitoring of the welding process. In this paper, the molten pool image is obtained by a CO_2 welding test. The feasibility, superiority to other models, and contribution sources of the proposed algorithm are tested and studied. The experiment is carried out on the MNIST and FashionMNIST datasets to illustrate the versatility of the CNN–LSTM algorithm. The feature hybrid method in this paper also has certain reference significance for similar image recognition tasks.

2. Deep Learning Model Based on CNN–LSTM

A CNN is a neural network that uses convolution operation instead of traditional matrix multiplication in at least one layer of the network. It is especially used to deal with data with similar grid structures, a data structure common in computer vision and image processing [14]. The 2D image data can be directly used as the bottom-level input of a CNN, and then the essential features of the image are extracted layer-by-layer through convolution and pooling operations. These features have the invariance of translation, rotation, and scaling. However, the output layer of the traditional CNN is fully connected with the hidden layer. This feature fusion method which takes all outputs of the convolutional layer is far too simple for the purpose of our model. Problems with this method include bad kernels, multiple kernels extracting the same information, and unnecessary information extracted by kernels. It is possible to extract deeper image features and improve recognition accuracy by increasing the number of convolutional kernels, convolutional layers, and pooling layers. But it will undoubtedly lead to a huge network, thereby increasing the cost of computation, and also facing the risk of overfitting [14,15]. As a time recurrent neural network, LSTM is suitable for processing the sequence problem with time dependence. The input feature tensor is selectively forgotten, input and output through three threshold structures. It can filter and fuse the empty input, similar information, and unnecessary information extracted by the convolutional kernels, so that the effective feature information can be stored in the state cell for a long time. Therefore, an algorithm combining CNN and LSTM was proposed in literature [24–30], which has achieved good results in gesture recognition, voice recognition, rainfall prediction, machine health condition prediction, text analysis, and other fields. However, the above literature is targeted at prediction tasks, and the input of LSTM is also a batch of images in time series. But the molten pool online monitoring process is faced with the identification task. The original input of this task is a single molten pool image taken by the camera at a certain moment. The ideas of sequence dependency are clearly inapplicable to this problem. Therefore, in view of the above problems, this paper proposes an algorithm named CNN–LSTM for the online monitoring task of the molten pool, which hybridizes the advantages of CNN and LSTM. The overall architecture of CNN–LSTM is shown in Figure 1.

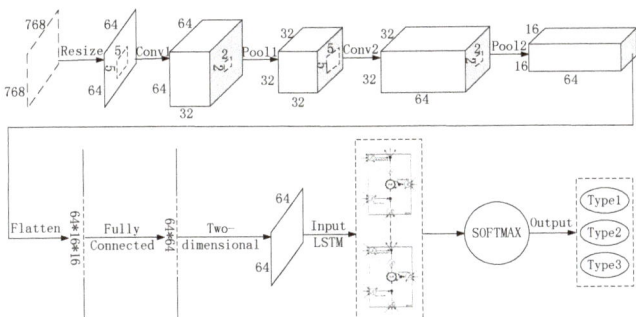

Figure 1. Convolutional neural network and long short-term memory network (CNN–LSTM) algorithm overall architecture.

The CNN–LSTM algorithm is designed for the recognition task of a single image. Since CNN's feature extraction is adaptive and self-learning, our model can overcome the reliance of feature extraction and data reconstruction relying on human experience and subjective consciousness in traditional recognition algorithms. It uses multiple convolutional kernels to scan the entire molten pool image to obtain redundant features of all objects as candidates. In the feature hybrid stage, the three-dimensional feature tensor output from the last layer of CNN is firstly stretched into a one-dimensional feature vector. As mentioned earlier, this vector has all feature information extracted by convolutional kernels, which includes some blank information, similar information, unnecessary information, and so on. Then the feature vector is mapped to two-dimensional space as the input of LSTM. Each row of the feature matrix is considered as a basic unit to be hybridized. Each time step reads a row of feature information and divides a feature matrix into several time steps to read. In this way, the single image of molten pool is converted into "sequential" data in this sense. The LSTM network is used to extract the dependencies between each row of feature matrix, so as to filter and hybridize the features extracted from the CNN network. Figure 2 shows the innovation of the CNN–LSTM network. In the time interval of the CNN–LSTM network identification molten pool image, the input of LSTM network at time t includes the output h_{t-1} and unit state c_{t-1} at time $t-1$, and the network's input x_t of current time. The feature tensor and the cell state can be filtered and hybridized by three carefully designed threshold structures, so that the effective features extracted from the CNN can be stored in cell state for a long time and the invalid features are forgotten.

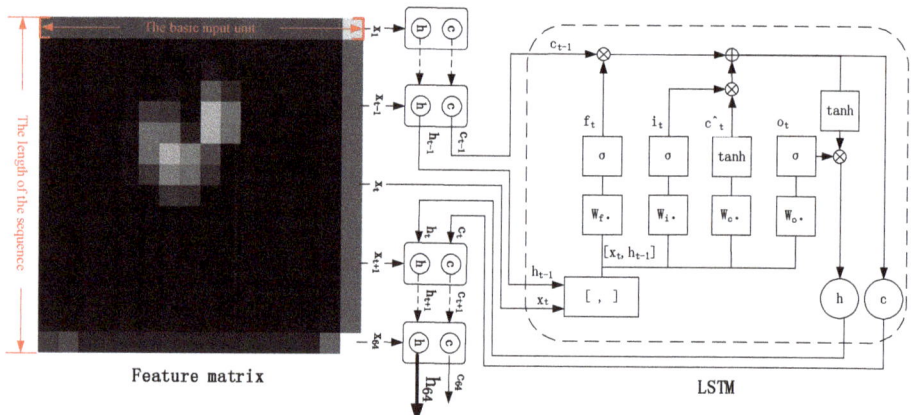

Figure 2. Feature hybrid mechanism of CNN-LSTM.

3. Model Implementation and Parameter Details

3.1. Model Implementation Process

It can be seen from Figure 1 that this algorithm is mainly divided into feature extraction stage based on CNN and feature fusion stage based on LSTM. In the feature extraction stage, the forward propagation process of the image signal is as follows: it is assumed that the l layer is a convolutional layer, and the $l-1$ layer is a pooling layer or an input layer. Then the calculation formula of the l layer is:

$$x_j^l = f(\sum_{i \in M_j} x_i^{l-1} \times k_{ij}^l + b_j^l) \qquad (1)$$

The x_j^l on the left of the above equation represents the jth feature image of the l layer. The right side shows the convolution operation and summation for all associated feature maps x_i^{l-1} of the $l-1$ layer and the jth convolutional kernel of the lth layer, and then adds an offset parameter, and finally passes the activation function $f(*)$. Among them, l is the number of layers, f is the activation function, M_j is an input feature map of the upper layer, b is offset, and k is convolutional kernel.

Assuming that the l layer is pooling layer (down sampling layer), the $l-1$ layer is the convolutional layer. The formula for the l layer is as follows:

$$x_j^l = f(\beta_j^l down(x_j^{l-1}) + b_j^l) \qquad (2)$$

In the above formula, l is the number of pooling layer, f is the activation function, $down(*)$ is the down sampling function; β is the down sampling coefficient, and b is the offset.

In the feature hybrid stage, the network uses three threshold structures to control the state of the cell that preserves long-term memory. The meaning of long short-term memory is: c_t corresponds to long-term memory, and $\tilde{c}_t$ corresponds to short-term memory. The $\sigma(*)$ in Expressions (3), (4), and (7) is a Sigmoid function. If the output of Sigmoid function is 1, then the information is fully remembered. If the output is 0, then it is completely forgotten. If the output is the value between 0 and 1, it is the proportion of information to be remembered. The gate is actually equivalent to a fully connected layer and its input is a vector and output is a real vector between 0 and 1. It uses the output vector of the "gate" multiplied by the vector we want to control. The forgetting gate f_t determines how much historical information can be remained in a long-term state c_t; $\tilde{c}_t$ is used to describe the short-term state of current input. The input gate i_t determines how much of the current network input information can be added to the long-term state c_t; the output gate o_t controls how much of the aggregated information is available as the current output. The expressions are as follows:

$$f_t = \sigma\left(W_f \bullet [h_{t-1}, x_t] + b_f\right), \qquad (3)$$

$$i_t = \sigma(W_i \bullet [h_{t-1}, x_t] + b_i), \qquad (4)$$

$$\tilde{c}_t = tanh(W_c \bullet [h_{t-1}, x_t] + b_c), \qquad (5)$$

$$c_t = f_t \circ c_{t-1} + i_t \circ \tilde{c}_t, \qquad (6)$$

$$o_t = \sigma(W_o \bullet [h_{t-1}, x_t] + b_o), \qquad (7)$$

$$h_t = o_t \circ tanh(c_t), \qquad (8)$$

The above are the formulas of the forward propagation process of the image signal. "$\bullet$" means matrix multiplication, and "$\circ$" means multiplication by elements of the same position. The output of the last time step of the LSTM network includes current unit state c_{64} and current output h_{64}. We take h_{64} as the overall output of the LSTM part, which is the input of SOFTMAX. After the signal passed through the SOFTMAX, the judgment of the category is given in the form of probability. In the algorithm training stage, the network adopts the error back propagation method to iteratively update the weights and offsets until the number of epochs is reached.

3.2. Model Parameter Details

Tensorflow is a deep learning framework developed by Google. It provides a visual tool Tensorboard that can display the learning process of algorithms. In order to realize the CNN–LSTM algorithm proposed in this paper, the relevant hyper-parameters under this deep learning framework are set as follows: in view of the fact that the gray image of the molten pool taken by the charge coupled device (CCD) camera is too large (768 × 768), it brings great difficulty to the network operation. Therefore, the gray image size is first converted to 64 × 64. In the first convolutional layer (Conv1), there are 32 convolutional kernels with a size of 5 × 5. The convolution stride is 1, and the padding method is same to ensure that the image size is unchanged after convolution. At this time, the image data is converted to 64 × 64 × 32. The first pooling layer (Pool1) uses the maximum pooling. The pooling window with a size of 2 × 2, the pooling stride is 1, and the padding method is same.

At this time, the image data is converted to 32 × 32 × 32. In the second convolution layer (Conv2), there are 64 convolution kernels with a size of 5 × 5. The convolution stride is 1, and the padding method is same. At this time, the image data is converted to 32 × 32 × 64. The second pooling layer (Pool2) uses the maximum pooling. The pooling window with a size of 2 × 2, the pooling stride is 1, and the padding method is same. At this time, the image data is converted to 16 × 16 × 64. The fully connected layer adjusts the feature matrix to 64 × 64, and each time step takes one row as the input to the LSTM network. There are 64 time steps in the total. There are 100 hidden units in the LSTM network. Finally, the classification results of defects are obtained through SOFTMAX. In addition, the learning rate of this network is set to 10^{-4}, and Adam is chosen as the optimizer.

Considering the small sample size, in order to prevent overfitting and reduce the amount of calculation, the first layer convolution result, the second layer convolution result and the fully connected layer result all use ReLU (Rectified Linear Units) activation function: ReLU(x) = max(0,x), which is shown in Figure 3. The ReLU activation function is more expressive than the linear function. The convergence rate of ReLU is faster than that of nonlinear activation functions such as Sigmoid and Tanh. Moreover, since the derivative of ReLU activation function is equal to 1, it can help with vanishing gradient problem [31,32]. In order to further reduce the possibility of overfitting caused by the small sample size, a random dropout method is used in the fully connected layer which is shown in Figure 4. Some neurons are stochastically deactivated at each epoch. Dropout decreases the dependencies between nodes and reduce overfitting by turning the CNN into an ensemble classifier of many weak classifiers. The dropout parameter is set to 0.5 in our model.

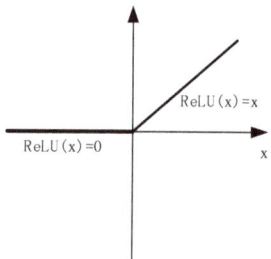

Figure 3. Schematic diagram of the ReLU function.

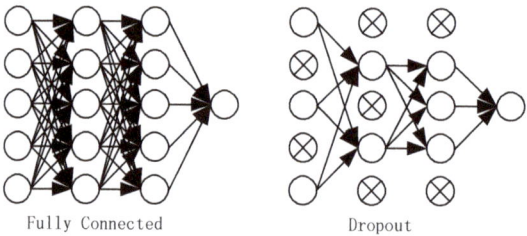

Figure 4. Schematic diagram of the random Dropout method.

4. Test Design and Environment

4.1. Test Design

First of all, in order to help understand the mechanism of feature extraction and evolution of the algorithm, the operation results of each convolution and pool layer will be visualized. Secondly, in order to show the feasibility and generalization ability of CNN–LSTM algorithm, the training performance and testing performance will be compared. The size of the original image was converted into 32 × 32, 64 × 64, and 128 × 128 as the initial input. The contribution source of the feasibility of the algorithm was illustrated by the influence of different input sizes on the composition algorithm. Among the tasks related to image feature extraction, CNN has been widely proved to be superior to traditional algorithms. Therefore, in order to fully reflect the superiority of the algorithm and illustrate the contribution sources of the superiority, performance comparison tests were conducted under the same hyper-parameters with the composition algorithm (CNN, LSTM) and CNN-3 (add a convolution and pooling layer, respectively). Finally, in order to illustrate the versatility of the CNN–LSTM algorithm, the performance of the algorithm was tested on the MNIST and FashionMNIST datasets in Appendix A.

In addition, in order to guarantee the fairness of the comparison test, other hyper-parameters such as convolution kernel size, pool window size, stride, network learning rate, activation function, optimizer, dropout value, LSTM's hidden layer unit number, etc., were set to be same. The algorithm was analyzed and compared using three criteria: recognition accuracy, convergence speed, and recognition time.

4.2. Test Environment

In terms of data sources, this paper relies on the key laboratory of Robotics and Welding Technology of Guilin University of Aerospace Technology to carry out the CO_2 welding test. In the actual welding process, welding defects are caused by a variety of factors and have great uncertainty. Through pre-processing, a total of 500 molten pool images of the three most common types of welding including welding through, welding deviation, and normal welding were collected. The original size of the images were 768 × 768. There were 300 pictures per class in the training set and 100 pictures in each class in the validation set and testing set. The tail of the molten pool corresponding to the welding through defect will leak to the back of the base metal and appear as a shadow on the image (Figure 5a, yellow area). Welding through defects are mainly caused by the welding current being too large, welding speed being too slow, the base material too thin, the base material not uniform, and so on. The weld pool corresponding to the weld deviation defect will deviate from the predetermined weld seam. Welding deviation defects are mainly caused by the vibration of the walking mechanism, the low accuracy of the positioning, the instability of the arc, and so on. At this point, the molten pool will deviate from the predetermined weld, which is reflected in the image as a part missing from the molten pool (Figure 5b, yellow area). A normal molten pool has an elliptical shape. Figure 5 shows a partial picture of the sample set.

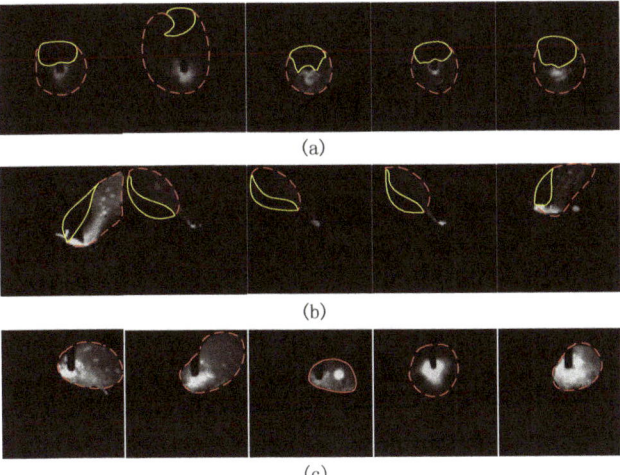

Figure 5. Part of the sample set images. (**a**) welding through; (**b**) welding deviation; (**c**) normal welding.

The algorithm performs performance tests under the ubuntu16.04 operating system, a GTX1080Ti graphics card, a hardware environment of 64 GB running, and the Tensorflow deep learning framework.

5. Test Results and Analysis

5.1. Visual Analysis

According to the details of the network framework, the algorithm has two layers of convolution and two layers of pooling for feature adaptive extraction of the molten pool image. Figure 6 shows the feature extraction and evolution mechanism of CNN.

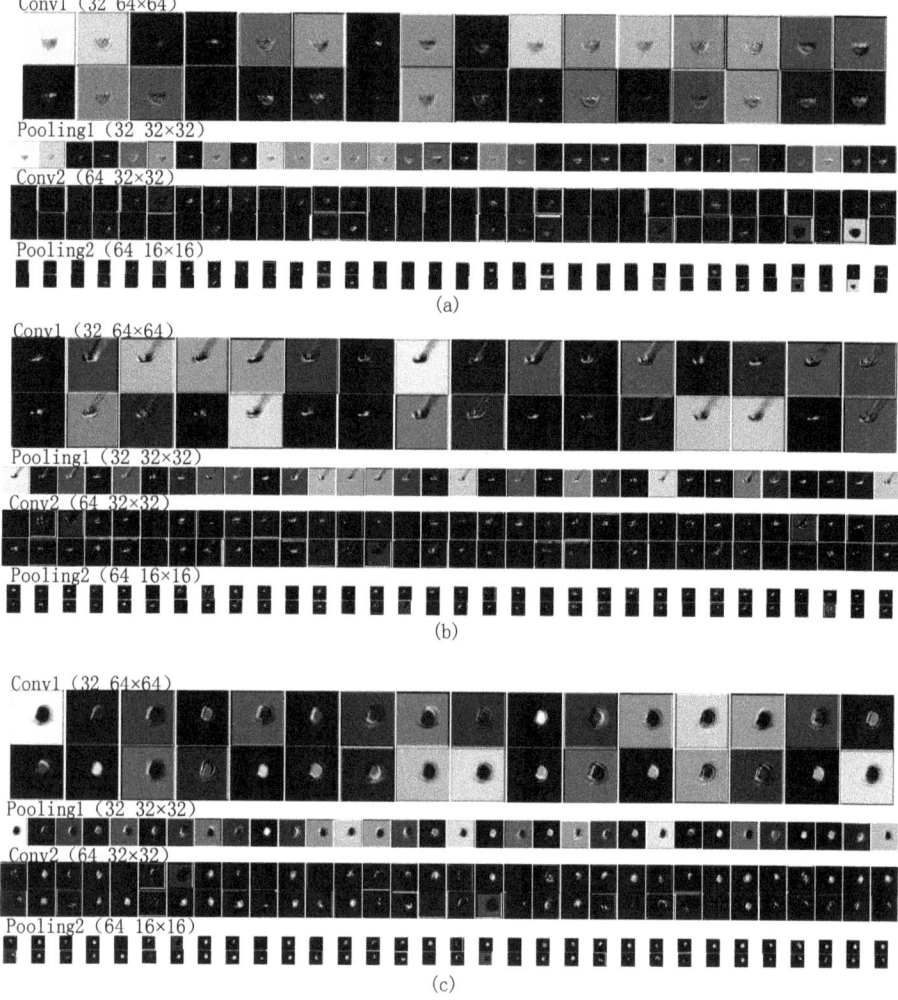

Figure 6. The feature images of the molten pool in three states: (**a**) welding through; (**b**) welding deviation; (**c**) normal welding.

As a whole, the comparison between the convolution results of the first layer and the convolution results of the second layer shows that the activation degree of the background of the molten pool decreases with the deepening of the convolution layer. The attention of convolution kernels gradually concentrates on the characteristic information of molten pool and ignores the background. The convolution kernels in the first layer are mainly used to detect low-order features, such as the edge, angle, and curve of the molten pool. The convolution kernels in the second layer are mainly used to detect the combined features of low-order features, such as the arc and shape of the molten pool. The result of the first convolution layer has a high spatial resolution, which is conducive to the accurate positioning of the target, such as the separation of the molten pool and the background but lacks robust feature representation. The spatial resolution of the second-layer convolution result is reduced due to the pooling operation, resulting in weaker positioning functions, but with deeper abstract features, and thus distortion tolerance [14,15]. The results of the second layer convolution are more ambiguous than the first layer, but the unique part of the category is highlighted. The maximum pooling operation can reduce the parameter calculation amount to prevent overfitting while better retaining the main texture features of the molten pool. It can be seen that maximum pooling of the first layer has a partial enhancement to the features extracted by the first convolution layer. However, after the maximum pooling of the second layer, some relatively abstract discrete blocks appeared, which are difficult to see by the naked eye. In addition, a small number of feature images are black or very similar to other feature images, which means that the convolution kernels failed to extract information or similar information were extracted by multiple convolution kernels. Therefore, it is necessary to select and combine the feature information by using LSTM network before the classification.

Specifically, in the same convolution layer, the convolution kernels extract different features in different molten pool states. Due to the excessive energy density of the weld, the molten pool will collapse in the middle, resulting in shadow behind the molten pool image (Figure 5a, yellow area). This feature is an irregular feature map similar to the elliptical gap in the convolution results. The welding deviation defects caused by improper groove angle, uneven assembly clearance or low precision of welding robot, resulting in the molten pool is only half (Figure 5b, yellow area), its own irregularity is very strong, resulting in a large deviation and irregularity of convolution results. Due to the regular shape of the molten pool in the normal state (Figure 5c, red area), the feature images extracted from the convolution kernels are also approximately elliptic, and the molten pool in the normal state is brighter than the other two states, so the convolution kernels have also extracted more bright features. In the same molten pool state, taking the normal state as an example, as mentioned above: the first layer of convolution kernels are more concerned with the characteristics of the approximate elliptical edge of the molten pool; the second layer of convolution kernels focuses on the overall morphological features of the approximate ellipse of the molten pool; the pooling layer improves computational efficiency while preserving texture features that can represent grayscale distributions of pixels and surrounding spatial neighborhoods.

5.2. Feasibility Analysis

The loss curve of CNN–LSTM algorithm in the training process is shown in Figure 7. On the training set, the CNN–LSTM network begins to converge after about 200 epochs and the convergence process is stable. On the validation set, CNN–LSTM begins to converge after about 300 epochs. Although the convergence process fluctuates slightly, the overall trend of loss is clearly decreasing. Although the accuracy of CNN–LSTM on the validation set is slightly lower than that on the training set, with the increase of the training epoch, the accuracy on the verification set does not decrease, and the recognition accuracy on the test set is 94%. Through the above analysis, we can find that the algorithm of maximum pooling, ReLU activation function, and random dropout can effectively suppress the overfitting due to the sparse set of samples, thus achieving excellent performance on both the training and testing stage, which embodies the effectiveness and strong generalization ability of the algorithm.

The effect of different input sizes on each algorithm is shown in Figure 8. It can be seen that with the increase of input sequence size, the convergence speed and recognition accuracy of the CNN network increase significantly. This is because with the increase of input sequence size, the details and features of molten pool image are more abundant, and the strong feature adaptive extraction ability of CNN network comes into play. But for the LSTM algorithm, the recognition accuracy is the highest when the size of the input image is 64 × 64. When the input size is 128 × 128, the recognition accuracy is lower than that input with smaller size. This is likely because that when the size of the input sequence is small, the details and features of the molten pool image are small and not obvious, and LSTM cannot extract too much effective characteristic sequence information of the molten pool. When the input sequence is long, LSTM can extract more effective sequence information, and at the same time, it can exert its strong gradient retention and long-term dependence ability. But when faced with a particularly long sequence, because the traditional neural network model uses an encoder-decoder structure, the model of this structure usually encodes the input sequence into a fixed-length vector representation. For short input sequences, the model can learn a reasonable vector representation. However, the problem with this model is that when the input sequence is very long, it is difficult for the model to learn a reasonable vector representation, so it is difficult to retain all necessary information [33–35]. However, the CNN–LSTM network proposed in this paper increases the convergence speed and recognition accuracy as the input sequence size increases. This indicates that the method of automatic feature extraction by CNN has the largest overall contribution to the feasibility of the CNN–LSTM algorithm, and the method of intelligent fusion of feature tensor by LSTM's long term dependence has the second largest contribution to the feasibility of the CNN–LSTM algorithm.

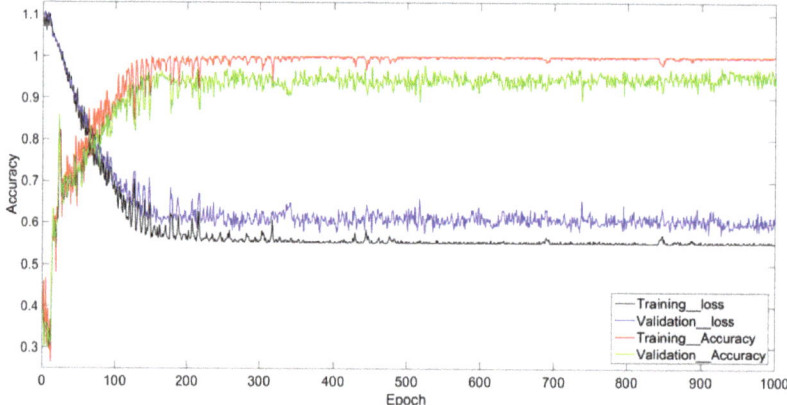

Figure 7. The recognition accuracy of this network training and testing process.

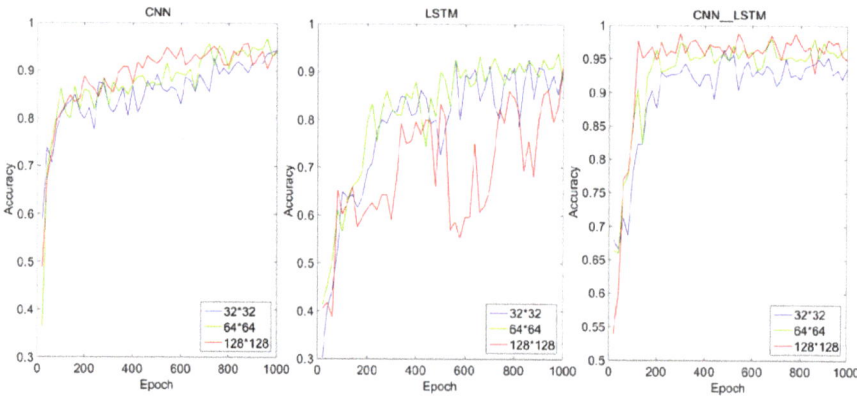

Figure 8. Influence of different input sizes on each algorithm.

5.3. Performance Analysis

The training performance comparison between the CNN–LSTM algorithm and each component algorithm under the same hyper-parameter is shown in Figure 9. Taking the input image size 64 × 64 as an example, in terms of convergence speed, the LSTM network starts to converge from about 700 epochs and eventually converges to about 90% accuracy. The CNN network converges from about 600 epochs and eventually converges to about 93% accuracy. The CNN-3 network converges from about 400 epochs and eventually converges to about 94% accuracy. The CNN–LSTM algorithm starts to converge from about 300 epochs and eventually converges to about 96% accuracy. In terms of defect recognition accuracy on the test set, the recognition accuracy of LSTM network is 88%, the recognition accuracy of CNN network is 89%, the recognition accuracy of CNN-3 network is 91%, the recognition accuracy of CNN–LSTM network is 94%. In terms of defect recognition speed, it can be found from Table 1 that LSTM algorithm has the fastest recognition speed under any input image size, followed by CNN. The CNN–LSTM has more training parameters due to the hybrid of CNN and LSTM, so it takes more time, but it is still faster than the CNN-3 which contains the three-layer convolution and pooling. In addition, when the input image size is 128 × 128, although CNN-3 can get a recognition accuracy of 94% on the test set, the recognition time of a single image is five times longer than that of the CNN–LSTM network when the input image size is 64 × 64. In the process of online monitoring of the molten pool state, the most important thing is to ensure the recognition accuracy of the algorithm, followed by the recognition time of a single image. Therefore, the CNN–LSTM network can guarantee high-recognition accuracy by adjusting the input size in the shallow network under the requirement of high-frequency molten pool monitoring while CNN needs to stack more convolution layers to improve the recognition accuracy, but it undoubtedly brings huge real-time computational cost. Therefore, considering all algorithms' recognition accuracy, convergence speed and single image recognition speed comprehensively, the CNN–LSTM algorithm is superior to all the other competitors in real-time welding applications. This superiority is a result of using LSTM in the feature fusion stage, which filters and hybridizes the feature tensor extracted by CNN with rows as the unit. This method can consume a shorter recognition time with the guarantee of reaching a very high accuracy.

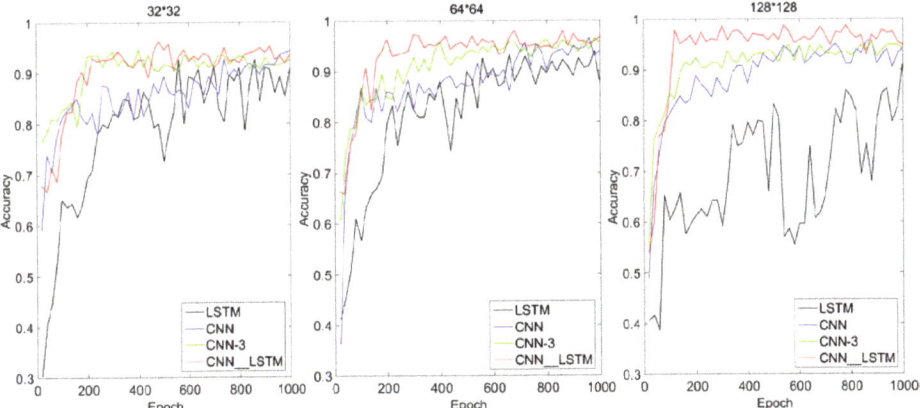

Figure 9. Performance comparison of three algorithms under different input dimensions.

Table 1. The recognition accuracy and recognition time of different algorithms under different input sizes.

Algorithm Type		Input Size		
		32 × 32	64 × 64	128 × 128
Accuracy (Recognition Time (t/ms))	LSTM	0.85 (0.017)	0.88 (0.033)	0.8 (0.167)
	CNN	0.88 (0.02)	0.89 (0.06)	0.92 (0.233)
	CNN-3	0.90 (0.04)	0.91 (0.099)	0.94 (0.33)
	CNN–LSTM	0.92 (0.033)	0.94 (0.067)	0.95 (0.2667)

6. Discussion

In order to meet the engineering requirements of high accuracy and real time in welding in an online monitoring process, a CNN–LSTM algorithm was proposed based on the traditional deep learning method. The original input of the model is a single image, and the LSTM network processes the feature map extracted by CNN instead of the original sequence, as in the literature. The motivation for using LSTM in this paper was to intelligently fuse the feature information that CNN has extracted, rather than to extract the dependencies between each individual in the sequence. The feasibility of the algorithm is based on using multiple convolution kernels to scan the whole image to obtain the redundant features of the molten pool. The hybrid algorithm was designed in such a way that the rows of the feature matrix extracted by CNN are considered as the basic units and put into the LSTM network for a feature hybrid. The algorithm has high accuracy and short time to identify defects in the molten pool, which completely meets the need of online monitoring in the molten pool. The experiment on the self-made molten pool image dataset shows that the contribution of the feasibility of the algorithm is more derived from the CNN's feature adaptive extraction capability. However, the superiority of the algorithm is derived from using LSTM in the feature hybrid stage, which filtered and hybridized the feature tensor extracted by CNN in rows. The successful application of the CNN–LSTM algorithm on the MNIST and FashionMNIST datasets show that the motivation of this algorithm is universal when dealing with similar non-strict sequential image data.

Although the feature hybrid method in the CNN–LSTM algorithm is superior to the traditional methods, there are still some shortcomings. In future research work, we should first consider obtaining more defect types and sample sets of molten pool. Secondly, the choice of hyper-parameters should be fully studied in the process of network construction. Thirdly, welding quality should be used as a bridge to establish a corresponding model between the welding process and the weld pool defects. Finally, a feedback control model should be established between the monitoring results of the molten pool and the welding process to realize online monitoring of the welding process based on the molten pool.

Author Contributions: Conceptualization, T.L. and J.B.; Data curation, T.L.; Formal analysis, T.L., J.B. and J.W.; Funding acquisition, J.B.; Investigation, T.L.; Methodology, T.L.; Project administration, J.B.; Software, T.L.; Supervision, J.B. and J.W.; Validation, T.L.; Visualization, T.L.; Writing—original draft, T.L.; Writing—review & editing, T.L., J.W. and Y.Z.

Funding: The present work has been supported by the National Natural Science Foundation of China (51475301), the Science and Technology Commission of Shanghai Municipality, China (17511109202), the Fundamental Research Funds for the Central Universities (16D110309).

Acknowledgments: Thanks to the welding test provided by Xiaogang Liu and Ke Qin from the key laboratory of Robotics and Welding Technology of Guilin University of Aerospace Technology.

Conflicts of Interest: The authors declare no conflict of interest.

Appendix A

The motivation of the CNN–LSTM algorithm is that each row of the feature matrix is regarded as a basic input unit of the LSTM network, and the number of rows of the feature matrix is regarded as the length of the sequence, so that the feature matrix can be filtered and hybridized by using the long-term dependence ability of the LSTM network. In order to illustrate the universality of the CNN–LSTM algorithm, this algorithm was tested on MNIST and FashionMNIST datasets which have no strict timing dependence and then compared with basic algorithms. The MNIST dataset includes 10 categories of handwritten numerals from 0 to 9, and FashionMNIST includes 10 categories such as T-shirt, Coat, Sandal, Bag, etc. The original size of the images in both datasets is 28 × 28. Figure A1 shows the variation of accuracy of different algorithms on these two public datasets during training. It can be seen that on the relatively simple dataset of MNIST, all four algorithms are very stable during training. Although the improvement accuracy of CNN–LSTM is not obvious compared with that of CNN-3, the CNN–LSTM algorithm converges faster. Both CNN-3 and CNN–LSTM converge rapidly on the more complex dataset of FashionMNIST, but with the increase of training epoch, CNN-3 is unable to improve the accuracy, and the accuracy of CNN–LSTM algorithm is still rising. It can also be seen from Figure A1 that CNN–LSTM has a higher accuracy than other algorithms in the early stages of training, which shows that the CNN–LSTM algorithm can quickly adapt to different image scenarios. Table A1 shows the recognition accuracy of each algorithm on the test set and the recognition time of a single image. It can be seen that the CNN–LSTM algorithm takes less time than the CNN-3 algorithm with similar recognition accuracy. As mentioned earlier, although the feature extraction ability of CNN is very strong, the long-term dependency ability of LSTM network is stronger than that of stacked convolution layer for fusing ability of feature tensor extracted by CNN. The successful application of the CNN–LSTM algorithm on MNIST and FashionMNIST datasets show that our innovative model of combining CNN and LSTM is generic when dealing with similar non-strict sequential image data.

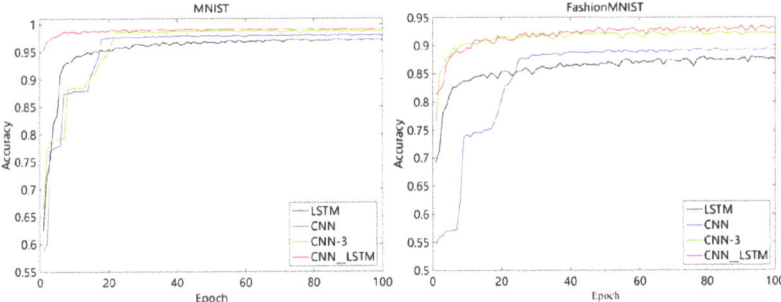

Figure A1. The representation of different algorithms on MNIST and FashionMNIST datasets.

Table A1. The recognition accuracy of each algorithm on the test sets and the recognition time of a single image.

		Dataset	
	Algorithm	MNIST	FashionMNIST
Accuracy (Recognition time (t/ms))	LSTM	0.9797 (0.008)	0.8671 (0.01)
	CNN	0.9911 (0.01)	0.8885 (0.012)
	CNN-3	0.992 (0.016)	0.9122 (0.018)
	CNN-LSTM	0.9922 (0.014)	0.9128 (0.016)

References

1. Zhao, W.; Li, S.; Zhang, B. Research status and Prospect of welding robot intelligent technology. *Dev. Appl. Mater.* **2016**, *31*, 108–114.
2. Zhang, G.; Li, Y. Towards Intelligent Welding in the Context of Industry 4.0. *Aeronaut. Manuf. Technol.* **2016**, *506*, 28–33.
3. Yong, Y.; Fu, W.; Deng, Q.; Chen, D. A comparative study of vision detection and numerical simulation for laser cladding of nickel-based alloy. *J. Manuf. Process.* **2017**, *28*, 364–372. [CrossRef]
4. Hong, Y.; Chang, B.; Peng, G.; Yuan, Z.; Hou, X.; Xue, B.; Du, D. In-Process Monitoring of Lack of Fusion in Ultra-Thin Sheets Edge Welding Using Machine Vision. *Sensors* **2018**, *18*, 2411. [CrossRef]
5. Liu, Z.; Wu, C.S.; Gao, J. Vision-based observation of keyhole geometry in plasma arc welding. *Int. J. Therm. Sci.* **2013**, *63*, 38–45. [CrossRef]
6. Garcia-Allende, P.B.; Mirapeix, J.; Conde, O.M.; Cobo, A.; Lopez-Higuera, J.M. Arc-Welding Spectroscopic Monitoring based on Feature Selection and Neural Networks. *Sensors* **2008**, *8*, 6496–6506. [CrossRef]
7. Luo, M.; Shin, Y.C. Vision-based weld pool boundary extraction and width measurement during keyhole fiber laser welding. *Opt. Laser. Eng.* **2015**, *64*, 59–70. [CrossRef]
8. Gao, X.; Zhang, Y. Monitoring of welding status by molten pool morphology during high-power disk laser welding. *Optik* **2015**, *126*, 1797–1802. [CrossRef]
9. Malarvel, M.; Sethumadhavan, G.; Bhagi, P.C.R.; Kar, S.; Saravanan, T.; Krishnan, A. Anisotropic diffusion based denoising on X-radiography images to detect weld defects. *Digit. Signal Process.* **2017**, *68*, 112–126. [CrossRef]
10. Ranjan, R.; Khan, A.R.; Parikh, C.; Jain, R.; Mahto, R.P.; Pal, S.; Pal, S.K.; Chakravarty, D. Classification and identification of surface defects in friction stir welding: An image processing approach. *J. Manuf. Process.* **2016**, *22*, 237–253. [CrossRef]
11. Boaretto, N.; Centeno, T.M. Automated detection of welding defects in pipelines from radiographic images DWDI. *NDT E Int.* **2016**, *86*, 7–13. [CrossRef]
12. Malarvel, M.; Sethumadhavan, G.; Bhagi, P.C.R.; Kar, S.; Thangavel, S. An improved version of Otsu's method for segmentation of weld defects on X-radiography images. *Optik* **2017**, *109*, 109–118. [CrossRef]
13. Rodríguez-Martín, M.; Lagüela, S.; González-Aguilera, D.; Martínez, J. Thermographic test for the geometric characterization of cracks in welding using IR image rectification. *Autom. Constr.* **2016**, *61*, 58–65. [CrossRef]
14. Yu, Y.; Yin, G.; Yin, Y. Defect recognition for radiographic image based on deep learning network. *Chin. J. Sci. Instrum.* **2014**, *35*, 2012–2019.
15. Zhou, Y.; Zhou, W.; Chen, X. Research of Laser Vision Seam Detection and Tracking System Based on Depth Hierarchical Feature. *Chin. J. Lasers* **2017**, *44*, 89–100.
16. Qin, K.; Liu, X.; Ding, L. Recognition of molten pool morphology in CO_2 welding based on convolution neural network. *Weld. Join.* **2017**, *6*, 21–26.
17. Wang, T.; Chen, J.; Gao, X.; Qin, Y. Real-time Monitoring for Disk Laser Welding Based on Feature Selection and SVM. *Appl. Sci.* **2017**, *7*, 884. [CrossRef]
18. Gao, X.; Li, G.; Xiao, Z. Detection and classification of welded defects by magneto-optical imaging based on multi-scale wavelet. *Opt. Precis. Eng.* **2016**, *24*, 930–936.
19. Cai, X.; Mu, X.; Gao, W. Application of PCA-Bayesian classification technology to recognition of weld defects. *Weld. Join.* **2014**, *3*, 31–35.
20. Vaithiyanathan, V.; Raj, M.M.A.; Venkatraman, B. PCA and clustering based weld flaw detection from radiographic weld images. *IJET* **2013**, *5*, 2879–2883.
21. Zhan, L. Study on Algebraic Feature Extraction of Arc Welding Pool Image Based on 2DPCA Method. Master's Thesis, Nanjing University of Science and Technology, Nanjing, China, 2013.
22. Greff, K.; Srivastava, R.K.; Koutnik, J.; Steunebrink, B.R.; Schmidhuber, J. LSTM: A Search Space Odyssey. *IEEE Trans. Neural Netw. Learn. Syst.* **2016**, *28*, 2222–2232. [CrossRef]
23. Wang, J.; Zhang, J.; Wang, X. Bilateral LSTM: A Two-Dimensional Long Short-Term Memory Model With Multiply Memory Units for Short-Term Cycle Time Forecasting in Re-entrant Manufacturing Systems. *IEEE Trans. Ind. Inform.* **2018**, *14*, 748–758. [CrossRef]
24. Tsironi, E.; Barros, P.; Weber, C.; Wermter, S. An Analysis of Convolutional Long-Short Term Memory Recurrent Neural Networks for Gesture Recognition. *Neurocomputing* **2017**, *268*, 76–86. [CrossRef]

25. Oehmcke, S.; Zielinski, O.; Kramer, O. Input Quality Aware Convolutional LSTM Networks for Virtual Marine Sensors. *Neurocomputing* **2018**, *275*, 2603–2615. [CrossRef]
26. Zhou, X.; Hu, B.; Chen, Q.; Wang, X. Recurrent Convolutional Neural Network for Answer Selection in Community Question Answering. *Neurocomputing* **2018**, *274*, 8–18. [CrossRef]
27. Sainath, T.N.; Vinyals, O.; Senior, A.; Sak, H. Convolutional, Long Short-Term Memory, fully connected Deep Neural Networks. In Proceedings of the IEEE International Conference on Acoustics, Speech and Signal Processing, South Brisbane, Australia, 19–24 April 2015.
28. Zhao, R.; Yan, R.; Wang, J.; Mao, K. Learning to Monitor Machine Health with Convolutional Bi-Directional LSTM Networks. *Sensors* **2017**, *17*, 273. [CrossRef]
29. Shi, X.; Chen, Z.; Wang, H.; Yeung, D.-Y.; Wong, W.-K.; Woo, W.-C. Convolutional LSTM Network: A Machine Learning Approach for Precipitation Nowcasting. Available online: https://arxiv.org/abs/1506.04214 (accessed on 18 May 2018).
30. Juan, C.N.; Raúl, C.; Juan, J.P.; Antonio, S.M.; José, F.V. Convolutional Neural Networks and Long Short-Term Memory for skeleton-based human activity and hand gesture recognition. *Pattern Recognit.* **2018**, *76*, 80–94.
31. Dahl, G.E.; Sainath, T.N.; Hinton, G.E. Improving deep neural networks for LVCSR using rectified linear units and dropout. In Proceedings of the IEEE International Conference on Acoustics, Speech and Signal Processing, Vancouver, BC, Canada, 26–30 May 2013.
32. Nair, V.; Hinton, G.E. Rectified linear units improve restricted boltzmann machines. In Proceedings of the International Conference on International Conference on Machine Learning, Haifa, Israel, 21–24 June 2010.
33. Sutskever, I.; Vinyals, O.; Le, Q.V. Sequence to Sequence Learning with Neural Networks. Available online: https://arxiv.org/abs/1409.3215 (accessed on 18 May 2018).
34. Cho, K.; Merrienboer, B.V.; Gulcehre, C.; Bahdanau, D.; Bougares, F.; Schwenk, H.; Bengio, Y. Learning Phrase Representations using RNN Encoder-Decoder for Statistical Machine Translation. Available online: https://arxiv.org/abs/1406.1078 (accessed on 18 May 2018).
35. Bahdanau, D.; Cho, K.; Bengio, Y. Neural Machine Translation by Jointly Learning to Align and Translate. Available online: https://arxiv.org/abs/1409.0473 (accessed on 18 May 2018).

© 2018 by the authors. Licensee MDPI, Basel, Switzerland. This article is an open access article distributed under the terms and conditions of the Creative Commons Attribution (CC BY) license (http://creativecommons.org/licenses/by/4.0/).

Article

Design of a Purely Mechanical Sensor-Controller Integrated System for Walking Assistance on an Ankle-Foot Exoskeleton

Xiangyang Wang [1], Sheng Guo [1,2,*], Haibo Qu [1,2] and Majun Song [1]

1 Robotics Research Center, School of Mechanical, Electronic and Control Engineering, Beijing Jiaotong University, Beijing 100044, China
2 Key Laboratory of Vehicle Advanced Manufacturing, Measuring and Control Technology, Ministry of Education, Beijing Jiaotong University, Beijing 100044, China
* Correspondence: shguo@bjtu.edu.cn; Tel.: +86-010-5168-8224

Received: 5 June 2019; Accepted: 16 July 2019; Published: 19 July 2019

Abstract: Propulsion during push-off (PO) is a key factor to realize human locomotion. Through the detection of real-time gait stage, assistance could be provided to the human body at the proper time. In most cases, ankle-foot exoskeletons consist of electronic sensors, microprocessors, and actuators. Although these three essential elements contribute to fulfilling the function of the detection, control, and energy injection, they result in a huge system that reduces the wearing comfort. To simplify the sensor-controller system and reduce the mass of the exoskeleton, we designed a smart clutch in this paper, which is a sensor-controller integrated system that comprises a sensing part and an executing part. With a spring functioning as an actuator, the whole exoskeleton system is completely made up of mechanical parts and has no external power source. By controlling the engagement of the actuator based on the signal acquired from the sensing part, the proposed clutch enables the ankle-foot exoskeleton (AFE) to provide additional ankle torque during PO, and allows free rotation of the ankle joint during swing phase, thus reducing the metabolic cost of the human body. There are two striking advantages of the designed clutch. On the one hand, the clutch is lightweight and reliable—it resists the possible shock during walking since there is no circuit connection or power in the system. On the other hand, the detection of gait relies on the contact states between human feet and the ground, so the clutch is universal and does not need to be customized for individuals.

Keywords: mechanical sensor; self-adaptiveness; ankle-foot exoskeleton; walking assistance

1. Introduction

Walking efficiency significantly influences the walking duration and metabolic cost of the human body. Humans have evolved a way of walking under natural selection, which is well-tuned and might be the most energetically efficient [1]. Since half of the positive work is done by the ankle during push-off [2], the design of ankle-foot exoskeleton (AFE) has become a hot research topic in the past years, and many devices have been developed to reduce the metabolic cost when normally walking [3–7]. The assistance or the torque provided by actuators rely on the recognition of people's real-time gait data and the adjustment of the control law promptly [8–11]. As a result, many gait-phase detection methods and algorithms have been proposed to provide accurate status feedback on the current gait stage.

Existing wearable sensor systems can be divided into the following types, including Electromyography (EMG) sensors [12], footswitches [13–15], foot pressure insoles [16], joint encoders, and inertial sensors [17].

The electrical activity in the gastrocnemius or soleus is an intuitive prediction of movement, which is employed by the control system of many AFEs [18–20]. There is no detection delay, and the voltage message can be detected as early as 50 ms before the contraction of muscles [21]. However, the signal varies when physical or physiological changes happen [22], for example, the magnitude changes due to the location shift of the electrodes [23], or the frequency drifts due to muscle fatigue [24]. Some complex processing approaches are also required since the sensing signal is too weak [25]. Force sensors are widely used in the detection of gaits, such as the foot pressure insoles which should be customized for different individuals, and the footswitches that have comparatively higher robustness by setting threshold rules [26], even though the number of the detectable gait phase is limited [27]. Sawicki et al. [28] have designed an AFE by controlling the contraction of the artificial muscle according to the contact states of the forefoot and effectively reduced the metabolic cost during walking.

The inertial sensors and the joint encoders can record and provide the kinematic information on the limb, and the system states of the controller [29–31]. However, the signal lags behind the EMG with respect to the generation time [32]. Recently, wearable capacitive sensors [33,34] and force myography-based sensors [35,36] have also been utilized for gait event detection and a satisfactory accuracy is achieved, but they have never been used for AFE control.

In order to increase the detecting accuracy of the gait phase, multi-sensors are usually integrated to achieve fusion-based recognition [37]. Together with the controller and actuators, a device with the bulk of the electronic control system is worn at the distal limb, resulting in an additional metabolic cost to the human body. It has been shown that the metabolic increase associated with adding mass to the foot is more than four times greater than the same mass attached to the waist [5,38]. The actuator and sensor system must be light so that the benefit from the energy input will not be counteracted by the additional metabolic cost caused by adding mass. Moreover, nearly all the sensors mentioned above are circuit-based. Shock during walking is an adverse factor towards the sensing accuracy and can destroy the circuit. When people walk fast, the delay in signal transmission, processing, and calculation may also arouse some problems.

Purely mechanical sensors are reliable when compared with the circuit-based sensors mentioned above because the signal transmission procedure relies on the motion of components under mechanical constraints. Also, the forces exerted by the environment can be transmitted by the rigid body without any delay. Collins [39,40] has designed a clutch based on the ratchet-pawl mechanism so that the spring linkage could be controlled by setting the timing of the pawl latch and release, which actually could be regarded as a mechanical sensor that is capable of detecting the key gait event during walking. However, the device must be customized to fit for different gait characteristics. Yandell et al. [41] presented an unpowered ankle exoskeleton with a slider under the shoes. The slider, together with a top and a bottom gripper, can identify a specific gait stage and control the transformation of energy so that the walking efficiency could be hugely improved.

The purpose of this paper is to design a purely mechanical sensor-controller integrated system to achieve the goal of gait identification and to reduce the metabolic cost during human walking. This system consists of a sensing part and a mechanical executing part. The sensing part acts as a sensor that possesses two input channels, while the executing part works like a controller that controls the engagement of a suspended spring behind the calf muscles. Unlike other electronic sensors that can record and transmit the data to the microprocessor unit for post-processing purposes, the sensing part must work together with the executing part, so that the motion signal could be "read" and contribute to the control of the spring in a mechanical way. Since the system is connected directly to the actuator (spring), in a purely mechanical system, the mechanical sensor should be analyzed together with the controller. That is the main difference between mechanical sensors and the circuit-based electronic sensors.

We briefly introduce the biomechanics method and goals of the design in Sections 2.1 and 2.2 and describe the structure and working process of the clutch in Sections 2.3 and 2.4. In Section 3, we simulate to evaluate the metabolic cost of the plantar muscles when the passive AFE is worn on the human body and conclude in Section 4.

2. Methods

2.1. Biomechanics and Energetics during Human Walking

In order to study the movement mechanism of the ankle joint and find the nature of energetics during human walking, we carried out a gait experiment. A 3D motion capture system (Cortex, Motion Analysis Co., Santa Rosa, CA, USA) was used to track the body movement. The Helen Hayes marker set was used with 15 markers worn on the lower limbs and six markers on the torso (including arms), as shown in Figure 1a.

Testers walked through an irradiation area formed by six cameras. The ground reaction force (GRF) was measured during walking along a 2.4 m walkway formed by four force plates (JP4060, Bioforcen Intelligent Tec. Ltd., Hefei, China). The captured gait cycle could be divided into four stages according to the contact state between the shoe sole and the ground. The cycle begins with the heel strike (HS) and experiences Flat foot (FF), Heel off (HO), and the swing phase. The ankle angle and moment were respectively calculated based on the kinematic data and the GRF data in the software cortex (see Figure 1b). As shown in Figure 1c, dorsiflexion happens almost during the FF, whereas plantarflexion is with the process of PO. By multiplying the joint velocity and the ankle moment, the instantaneous power of the ankle joint is derived. The ankle joint output work with completely different properties at different gait stage, i.e., negative work during FF, and positive work during PO. Although both positive and negative work cost energy to the human body [42], the positive work is usually performed to move body segments and increase the potential energy of body center of mass (COM), whereas the negative work is done with the mechanical energy (ME) transferred into other forms of energy. The reason for the negative part could be the energy stored in the tendons, and the energy dissipated due to the damped motion of fat, viscera, and muscle [43], which means only a part of the energy could be recycled, with most of the energy wasted.

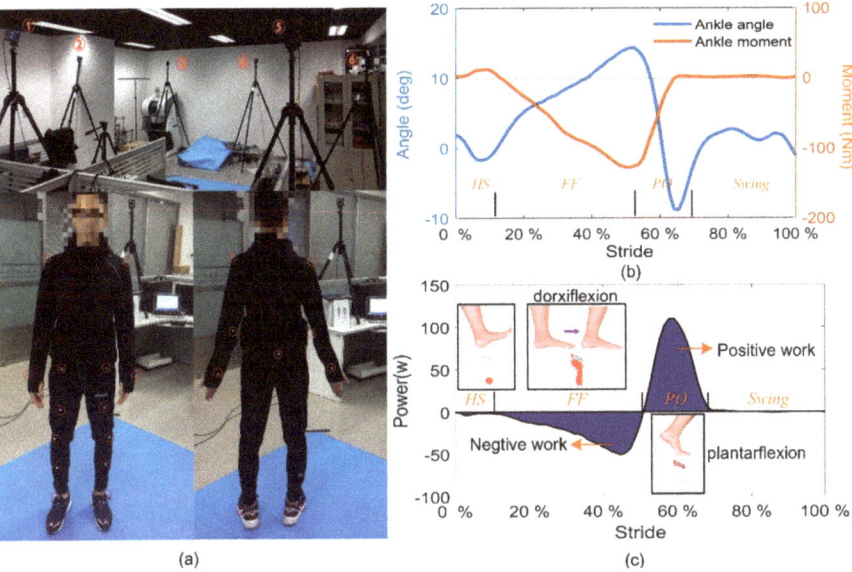

Figure 1. Kinematic data collection during the normal walking process. (**a**) The Helen Hayes marker set placement. (**b**) The obtained ankle angle and moment data within one gait cycle. (**c**) Instantaneous power of the ankle joint within one gait cycle.

Man consumes his biomass energy continuously to maintain the walking speed and compensate for the energy loss due to physiological activities in his body. Since energy is recycled with an elastic component almost without any loss, the increasing proportion of the energy transition into the elastic components could greatly reduce the energy consumed by the human tissue, increasing the walking efficiency, and achieving better walking economy.

2.2. Bio-Inspired Passive AFE and Its Description

Tendon-muscle units play an important role in the realization of human movements. Soleus and gastrocnemius (see Figure 2) are the most important plantar muscles. When a person begins to raise his heel, the calf muscles contract and generate a linear force F that causes a clockwise moment M about the rotation center of the ankle joint, outputting quantities of mechanical work.

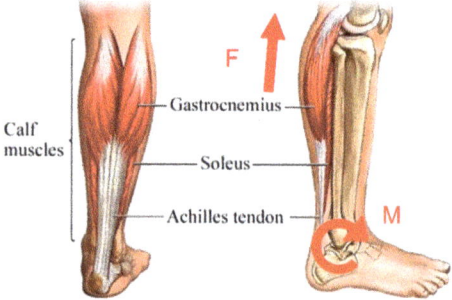

Figure 2. The composition of the calf muscles, which plays a key role in propelling human body forward [44].

Inspired by this, we developed a passive AFE that has a structure similar to a tendon-muscle unit (see Figure 3). An elastomer spring is suspended behind the calf muscles aiming to generate a linear force parallel to the muscle group during plantarflexion. The spring should be stretched at the beginning of FF, storing energy during the pendular motion of the human body, and transfers the energy into strain energy in the process of dorsiflexion. At the end of FF, the stored energy reaches its maximum and could be recycled to provide assistance. During the swing phase, the suspended spring is supposed to be disengaged, without impeding the free movement of the ankle joint.

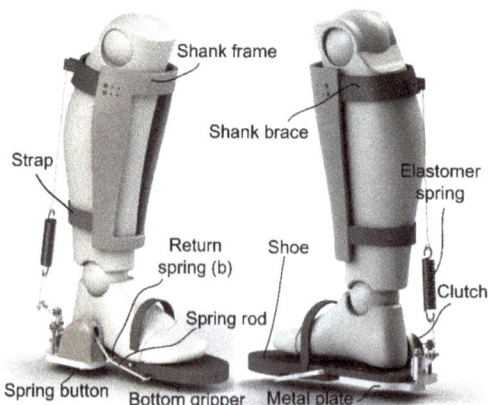

Figure 3. The proposed passive ankle-foot exoskeleton (AFE) and its components. The core and fundamental part is the clever clutch that ensures the implementation of the function.

In order to achieve the goal proposed above, a clever sensor-controller integrated system was designed to identify the key event of gait and control the engagement of the spring. A shoe is cut off from its back, with the upper heel kept to serve as a base, to which a metal plate is connected, so that the wearer could keep their foot completely flat when normally standing. The clutch is placed on the metal plate.

The interface between the shank and the exoskeleton consists of a shank brace and a shank frame. Since the tension of the spring can be up to hundreds of newtons, the possible slipping of the shank brace should be avoided by adjusting the position of the shank frame according to the size of the user's shank. With a nylon strap attached to the lower end of the frame, the spring force is held by the whole shank frame, so that comfort during its use is guaranteed.

2.3. Sensor-Controller Integrated System and Its Design

2.3.1. Mechanical Structure of the Clutch

The clutch is the core component in the design of the passive AFE that ensures that self-adaptiveness can be achieved. Self-adaptiveness here is defined as the capability of identifying different gait stages and providing assistance at a proper time mechanically.

Fixed at the outer side of the shoe sole, the clutch is composed of several tiny components. The overall size is small enough to be put into a spherical ball with a diameter of 42 mm. Figure 4a shows the mechanical structure of the clutch in a zoomed view. The whole clutch is divided into two parts, namely, the sensing part and the executing part. The executing part comprises a felly, a pulley, a pre-stressed spring, and a cord with one end attached to the pulley. The remaining components inside the clutch constitute the sensing part. The felly and the pulley are bolted together, concentric with a shaft passing through their center holes. Hence, the cord can be dragged out when the felly rotates clockwise (CW) and is dragged in when the felly rotates in the opposite direction. Routed to the back of the foot, the other end of the cord passes through three bearings and is connected to the lower end of the elastomer spring, the upper end of which is fixed at the shank brace. When the clutch is clutched (see Figure 4b), the rotation of the felly is restricted. As a result, the elastomer spring can only be stretched during the dorsiflexion process with elastic energy accumulated.

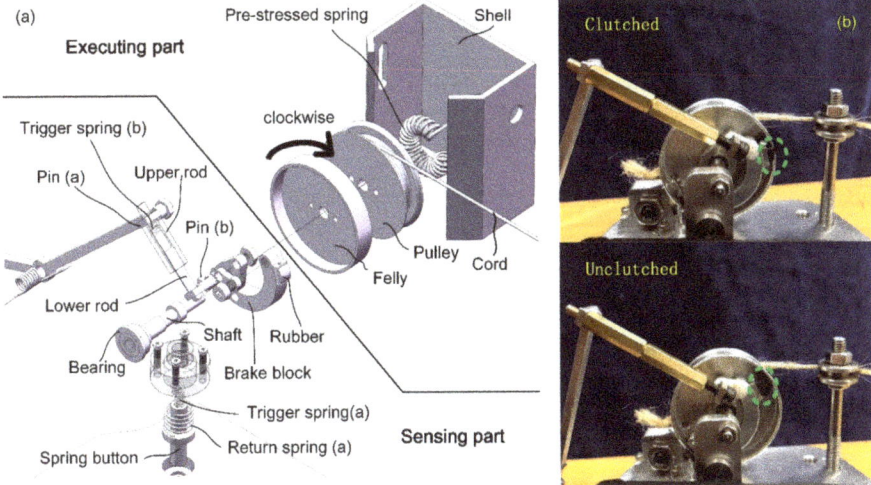

Figure 4. (**a**) The structure of the mechanical clutch. This sensor-controller integrated system consists of a sensing part and an executing part; (**b**) The clutched/unclutched state of the prototype, which relies on the contact state between the rubber and the wall of the felly flange.

In order to control the states switching of the clutch, a spring rod and a spring button are separately distributed at the fore shoe sole and the rear sole. They are pushed under the GRF. The forced motion of the spring button results in the compression of the trigger spring (a) (see Figure 5a) so that one end of the lever arm moves upward under the spring force f_1 (see Figure 5b), which results in the contact between the rubber and the flange of the felly with normal pressure f_p, causing a friction force. Similarly, when the spring rod is pushed under GRF, the trigger spring (b) is compressed (see Figure 5c). The spring force f_2 is generated and transmitted by a four-bar mechanism, pushing the rubber to contact the flange of the felly (see Figure 5d).

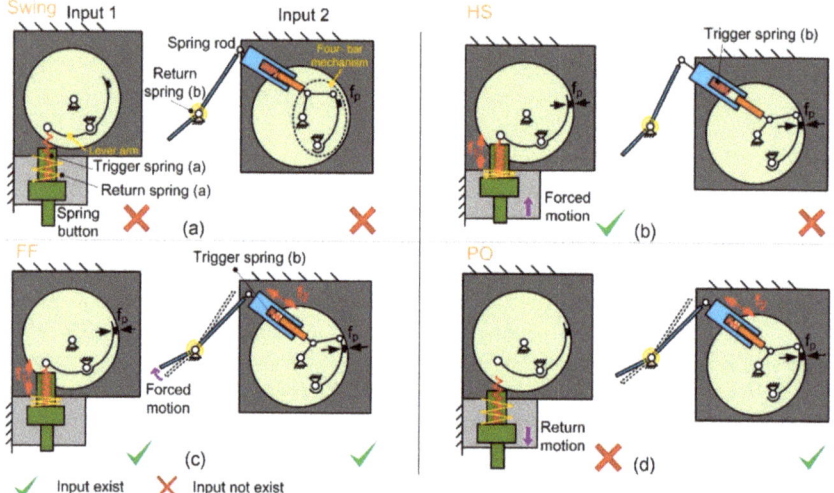

Figure 5. State of the executing part and the corresponding gait phases. (**a**) Swing: neither of trigger springs (a) and (b) is compressed; (**b**) heel strike (HS): only trigger spring (a) is compressed; (**c**) flat foot (FF): both trigger springs (a) and (b) are compressed; (**d**) push-off (PO): only trigger spring (b) is compressed.

When the rubber contacts the felly, the clutch is clutched due to the mechanical constraints. As shown in Figure 6, point M has different velocities with points A and B as the center of rotation. There is no doubt that point M cannot move along two directions at the same time. The pulling force F_d will cause deformation of the rubber, resulting in a larger normal pressure f_p and a corresponding friction force that stops the CW rotation of the felly. The greater the pulling force on the cord, the greater the resistance to rotation. Consequently, the clockwise rotation of the pulley is completely restricted.

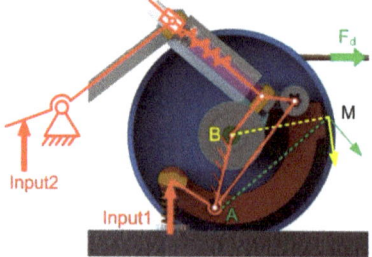

Figure 6. Two-inputs mechanical sensing system. For Input 1, the force is transmitted by a leveling mechanism. For Input 2, the force is transmitted by a four-bar mechanism.

In order to keep the cord taut all the time, a pre-stressed spring is placed inside the pulley with one of its ends fixed to the shell and the other end attached to the pulley. The pre-stressed spring could produce a counterclockwise torque. When the cord is slack, or the distance between point A and C is reduced (A and C are defined in Figure 7), excess cord length is dragged into the pulley immediately. Consequently, the cord always keeps tensioned.

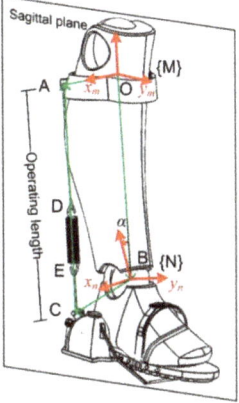

Figure 7. The sagittal plane and the definition of the operating length. x_n and x_m are all perpendicular to the sagittal plane.

Since the friction moment caused by the contact force f_p decreases as the felly counterclockwise (CCW) rotates, CCW rotation could still be achieved under the action of the pre-stressed spring. During the swing phase, the brake block leaves the flange due to the auto-return characteristics of the spring-linkage structure so that the clutch is unclutched. Two-way rotation is, therefore, allowed with the elastomer spring disengaged, which guarantees the full range movement of the ankle joint.

Each trigger spring is an independent input signal that controls the position of the brake block and the rubber attached to it. The state of the clutch under different conditions is shown in Table 1.

Table 1. State of the clutch during different gait phases.

Gait Phase	Forced Motion (Spring Button)	Forced Motion (Spring Rod)	State of the Executing Part	Rotation Restricted
Swing	No	No	unclutched	-
HS	Yes	No	clutched	CW
FF	Yes	Yes	clutched	CW
PO	No	Yes	clutched	CW

It could be concluded that when at least one input exists, the executing part is clutched and only one-way rotation of the pulley is allowed, which means the cord could not be dragged out from the pulley as the foot dorsiflexes. When no inputs exist, the clutch is unclutched, and two-way rotation is allowed, without impeding the free movement of the ankle joint. The return springs (a) and (b) ensure the spring button and the spring rod reverts to their original state when there is no GRF.

2.3.2. Working Principle

In the design of the passive AFE, the increment of the operating length corresponds to the extent of the energy stored. In the sagittal plane, the length could be derived through kinematic calculations.

The coordinates systems {M} and {N} are attached to the shank and the foot respectively, with z_m pointing along BO (B is the center of the ankle joint, O is attached to the shank as shown in Figure 7).

{N} is obtained by rthe otation about x_m with an angle α, as shown in Figure 7, a vector loop equation is obtained

$$CA = OA - OB - BC \tag{1}$$

Solving Equation (1) for CA, we obtain

$$CA = \begin{bmatrix} OA_x \\ OA_y \\ OA_z \end{bmatrix} - \begin{bmatrix} 0 \\ 0 \\ -l \end{bmatrix} - {}^M_N R \begin{bmatrix} BC_x \\ BC_y \\ BC_z \end{bmatrix} \tag{2}$$

And the coordinate-transformation matrix from coordinate {M} to {N} is

$${}^M_N R = \begin{bmatrix} 1 & 0 & 0 \\ 0 & \cos\alpha & -\sin\alpha \\ 0 & \sin\alpha & \cos\alpha \end{bmatrix} \tag{3}$$

where, angle α is obtained via inverse kinematics based on movement information collected by the motion capture system.

The operating length ranges from 370 mm to 395 mm in one gait cycle when normally worn, which means the spring could be stretched up to 25 mm with the corresponding energy stored. The maximum moment arm is about 98.2 mm. The states of the clutch and the contact status between the shoe and the ground, together with the variation of the operating length are presented in Figure 8.

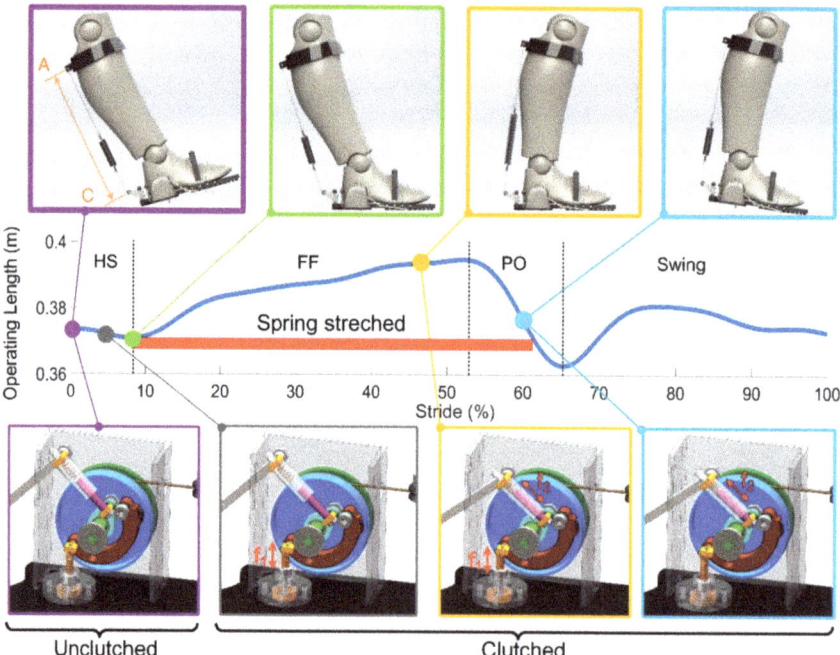

Figure 8. Schematics showing the working mode with respect to stride and the key event of the clever clutch in different gait phases.

In the beginning, the human foot is in the swing phase with the elastomer spring disengaged (in purple). The clutch allows the two-way rotation of the pulley without hindrance applied to the ankle joint. When the spring button is moved under GRF during HS, the operating length decreases a little bit as the ankle angle decreases, with the excess cord immediately pulled into the pulley by the pre-stressed spring (in grey). At maximum plantarflexion (in green), the foot is completely on the ground with the operating length reaching its minimum and preparing to perk up. Both the spring rod and the spring button are simultaneously pushed, resulting in the clutched state of the executing part until the end of the FF, which prevents the cord from being pulled out as the shank rotates. As a result, the elastomer spring is engaged and stretched as the ankle begins to dorsiflex into the terminal stance (in yellow).

The spring rod keeps the clutch clutched as the heel of the sole is raised with the spring button coming back to its original position (in blue), which ensures the steady release of the energy is achieved without any loss. The required force from the muscles is hugely reduced with the assistance provided by the spring.

Following the push-off, the clutch switches to be unclutched as the foot is off the ground.

2.3.3. Clutch Prototype Specification and Testing

The implementation of the state switching of the clutch significantly affects the timing when the assistance should be provided and ensures that free rotation of the ankle is achieved at the interval when no assistance is in need. In order to test the reliability of the switching states, a prototype is produced and tested.

In our prototype, the felly may be deformed under normal pressure f_p, which is mainly affected by two factors, namely, the pulling force on the cord and the GRF transmitted by the two mechanical inputs. Too high normal pressure could lead to mechanical failure. Considering the mass of the device, we first used 3D printed material (ABS) to build our prototype. For the convenience of the strength test, we used two shaft seats to hold the clutch mechanism. A torsional spring is placed on the shaft with its two arms attached to the brake block and the shaft seat respectively, ensuring that the clutch was unclutched when no inputs trigger the clutch. The flange wall has a thickness of 1 mm with a textured surface. When the cord is stretched and the clutch is set to be the clutched state, the resistance increases as the felly slightly rotates. The rubber is compressed as the pulling force increases, and the larger the pulling force, the larger the normal pressure f_p on the contacting area between the rubber and the wall of the felly flange, thus generating a larger friction force to stop the rotation of the felly. When the spring force was larger than 35N, the felly generated a visible deformation due to the characteristic of the plastic material. In practice, the pulling force could be up to a few hundred newtons when stiff spring is used. Hence, metal should be used instead. To achieve a better mass stiffness ratio, Al-alloy is employed as the base material in our test, as shown in Figure 9b. In our testing with the second prototype, the felly remains undeformed after a pull is exerted.

Assuming that the components in the two mechanical channels are all rigidly connected, which happens as the trigger spring (a) and (b) are fully compressed. GRF causes larger pressure than that caused by pulling the cord. In order to evaluate the structural limits of our prototype with the material Al-alloy, we carried out a finite element analysis based on the software Adams. When a person normally walks, the distribution of the pressure at the fore sole and the rear sole could be measured. In our case, one of the human subjects (24 years old, 180 cm tall, 74.2 kg) walked on a treadmill (Zebris FDM-T, Zebris Medical GmbH, Isny, Germany) at a speed of 4 km/h, the maximum load (% of body load) at the forefoot and the rear foot were respectively recorded, as shown in Figure 9e. The result showed that the forefoot carries 105% of the body weight, whereas the heel takes up only 76%, which implies the maximum GRF applied to the bottom grimmer and the spring rod is 559 N and 764 N, respectively.

Based on the above loading conditions, the pulley was meshed into a tetrahedral shape with the average element size of 0.2 mm in the simulation environment. The coefficient of friction between the rubber and the Al-alloy is 0.4. Assuming all the parts are rigid bodies except for the felly, we added the force directly to the pin (b) or the bottom gripper independently to compare the stress results.

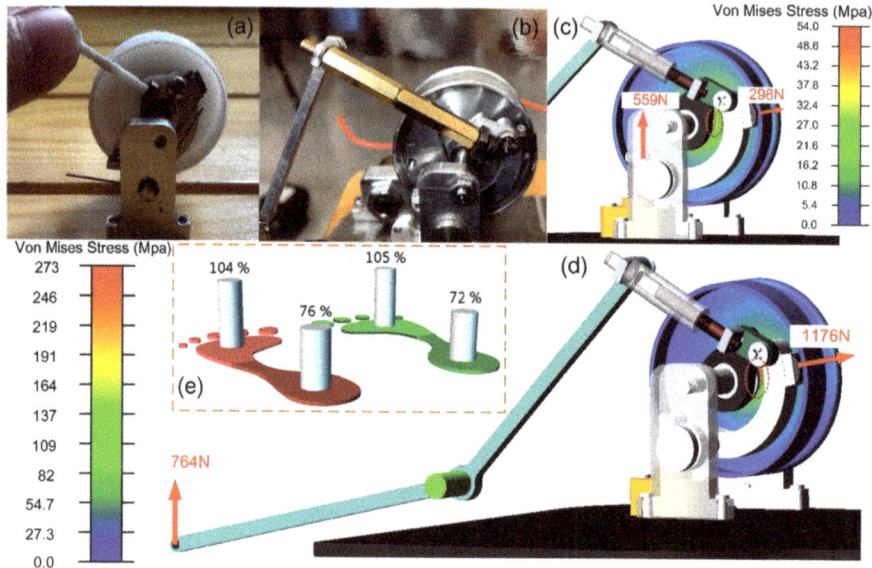

Figure 9. The prototype of the proposed sensor-control integrated clutch. (**a**) Original prototype with 3D printed pulley, felly, and connecting rod; (**b**) Second prototype with all components made of Al-alloy. (**c**) Finite element model when the bottom gripper suffers from the ground reaction force (GRF); (**d**) The finite element model when the bottom gripper suffers from the GRF; (**e**) The distribution of the body weight at the fore and rear foot.

The results are obtained and shown in Figure 9c,d. In both cases, the stress concentration points are mainly distributed near the contacting line between the felly and the bearing on the shaft (a). For case (b), the maximum Von Mises Stress is about 54 Mpa. As for the case (d), the maximum stress is approximately 273 Mpa. Since the elastic limiting pressure of the 6061-T6 Al-alloy is 333.75 Mpa in room temperature, the clutched clutch will generate invisible deformation during normal walking when the body weight of the users is less than 90 kg (corresponding to the elastic limit). The initial friction force ranges from 120 N to 470 N. When the felly clockwise rotates under the action of the pulling force on the wire rope, the normal pressure increases rapidly when the rubber block is deformed and compressed. Due to the mechanical constraint, the increased friction force prevents the pulley from being rotated and ensures the clutch is completely clutched.

2.4. Performance Evaluation

Unlike other circuit-based electronic sensors usually evaluated by the comparison between the detection results and the actual value, mechanical sensors only provide state feedback in the form of partial motion that can be directly "read" by the controller and the actuators. Hence, it works better to focus on the evaluation of the performance of the whole system.

With the help of the software OpenSim [45], we carried out a musculoskeletal simulation with a human skeleton model wearing the AFE. The model weights 72 kg and is 1.8 m tall (see Figure 10). In our case, the stance phase lasts 0.8 s. Since the assistance is provided only during the PO, the stance phase is selected as the main focus. Based on the force-displacement relation of spring, we defined two path actuators to mimic the force output of the elastomer spring. A set of muscle forces that drive a dynamic musculoskeletal model to track the walking behavior has been calculated by the method of Thelen [46].

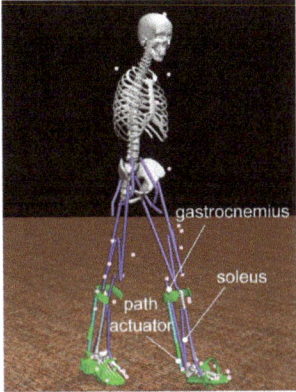

Figure 10. A Human musculoskeletal model with the passive AFE worn on both legs in the environment in OpenSim.

3. Simulation Results and Discussion

As shown in Figure 11, we compared the forces and power of the plantar flexor muscles (i.e., gastrocnemius and soleus) of the right leg by adjusting the spring stiffness, because these two muscles contribute a major part of the positive work during PO to propel human body forward.

Data of the force and power are listed in Table 2. The range of power is defined as:

$$P_r = P_{max} - P_{min} \qquad (4)$$

As the stiffness increases, the required muscle forces decrease consequently, especially for the soleus. For every 10 N/mm increase in stiffness, the force decreases nearly by one third, as does the change in power. The changes in gastrocnemius force and power are not as obvious as soleus, but the decline is also noticeable. It is worth noting that there is little difference between the case without assistance ($k = 0$ N/mm) and the case of normal walking since the weight of the exoskeleton is directly transferred to the ground during the stance phase. Therefore, there is almost no effect on the muscles.

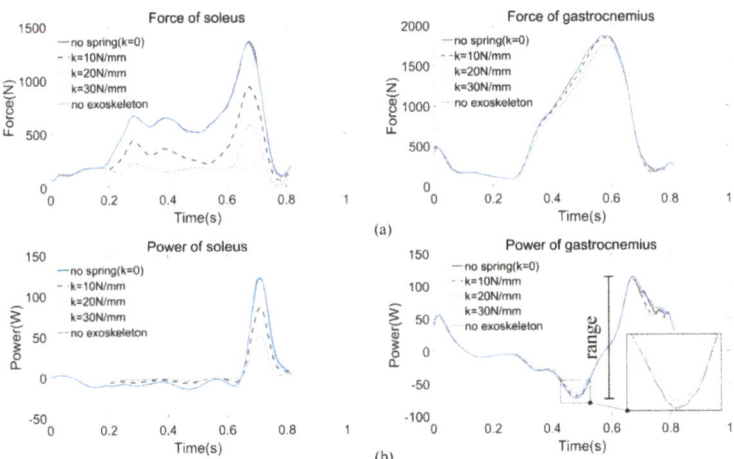

Figure 11. (**a**) Forces of soleus and gastrocnemius under different stiffness conditions. (**b**) Power comparison during the stance phase.

Table 2. Comparison of forces and power of plantar flexor muscles.

Reduction (in brackets)	Soleus		Gastrocnemius	
	Maximum Force (N)	Range of Power (W)	Maximum Force (N)	Range of Power (W)
no exos	1341.8	122.0	1865.1	186.8
no spring	1356.2 (−1.1%)	120.6 (−1.1%)	1866.4 (−0.07%)	187.42 (−0.33%)
10 N/mm	934.3 (30.4%)	84.9 (30.4%)	1844.9 (1.1%)	180.7 (3.3%)
20 N/mm	578.3 (56.9%)	50.0 (59.0%)	1745.9 (6.4%)	173.2 (7.3%)
30 N/mm	234.2 (82.5%)	18.8 (84.6%)	1657.7 (11.12%)	172.4 (7.7%)

Metabolic probes were used in the simulation to evaluate the metabolic consuming based on the model proposed by Umberger [47,48]. Since the whole-body metabolic rate is difficult to obtain by simulation, we then estimated the metabolism of the soleus muscle as an alternative. As shown in Figure 12, the results obtained from the probes have almost the same tendency and magnitude as the work rate obtained by computing the dot product of the muscle force vector and the contraction velocity (see Figure 11b). It is difficult to attribute changes in whole-body metabolic rate to a particular change of muscle mechanics [49]; however, simply reducing muscle force can save metabolic energy [39].

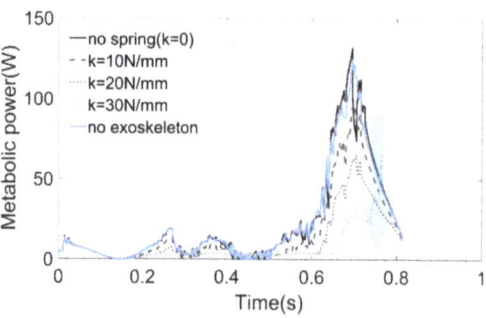

Figure 12. Metabolic cost of soleus under different stiffness conditions.

In order to facilitate the design, an upper limit for the stiffness (k = 62 N/mm) is found where the peak ankle moment during normal walking can be achieved. The moment provided by the AFE is then calculated and compared with the actual ankle moment of the musculoskeletal model (see Figure 13). It can be seen that during most of the stance phase (0–60% stride cycle), the AFE moment is greater than the required value. As a result, muscles antagonizing soleus must output additional positive work, so that the desired joint trajectories can be realized. Additional energy must be provided by the human body to compensate for the energy required to be stored in the spring as it is stretched. This can leads to wearing discomfort and even hinder the normal movement of the human body. In addition, 62 N/mm is an incredibly stiff spring and hard to be manufactured.

Even within a certain range, the higher the stiffness, the more energy can be recycled, but some tradeoffs must be made. Due to the individual weight differences, there is no suitable spring stiffness. On the one hand, excessive stiffness may change one's walking habit; on the other hand, it is possibly slower and could even hinder the pendular motion of the human body. The stiffness k = 25 N/mm is a recommended value based on the simulation results (see Figure 11a), where the soleus force approaches to zero during push-off when k = 20 N/mm and decreases to zero when k = 30 N/mm. Hence, we took an intermediate value. Since the exoskeleton plays a role in assisting, rather than a role that replaces the human muscles, we hoped the force of soleus was larger than zero all the time.

In our case, the maximum moment arm was 98.2 mm, and the maximum moment provided by AFE was 52.4 Nm. Compared with the peak value of the torque during normal walking (130.3 Nm), only 59.8% of the moment is required from the human body. Even additional energy is still required at the beginning of FF (12–50%), but this can be overcome easily without too many hindrances applied to the human body.

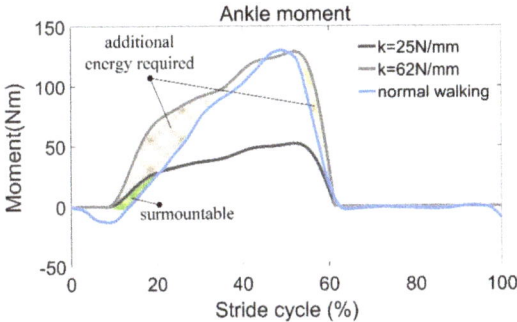

Figure 13. Comparison between the required joint moment and the calculated moment contributed by the AFE.

4. Conclusions

Based on the analysis of the movement mechanisms and energetics in different gait phases, we designed a novel passive AFE in analogy with the muscle-tendon unit to mimic the muscle force. Such a device is capable of storing energy and providing assistance during human walking. In order to detect the gait phases mechanically, a sensor-controller integrated system, or the clutch, was designed and introduced. The sensing part has two-inputs that are used to detect the contact status between the shoe sole and the ground, and each of the inputs could control the state of the executing part independently.

The energy storage and release process rely on the state of the clutch. When the clutch is clutched, the spring suspended behind the calf muscles is engaged. The energy is stored in the process of plantarflexion and released during the PO to provide walking assistance. When the clutch is unclutched, the spring is disengaged, without impeding the free rotation of the ankle joint.

Since the mechanical sensor is incorporated into the system, the assistance of the AFE under the control of the clutch is evaluated by the simulation. By comparing the force and power of plantar muscles under different stiffness conditions, the relationship between the spring stiffness and the assistance provided is obtained. The metabolic cost of the soleus is also estimated to show the walking economy.

The system is entirely passive and user-friendly with high reliability and can resist the disturbance of shocking during human walking. In our future work, we plan to make a prototype and do a more related test with the AFE worn on the human body.

Author Contributions: Conceptualization—X.W. and S.G.; Formal analysis—H.Q.; Funding acquisition—S.G.; Investigation—X.W.; Methodology—X.W.; Software—M.S.; Writing - original draft—X.W.; Writing – review & editing—S.G.

Funding: This research was supported by the Beijing Natural Science Foundation (grant number L172021), National Natural Science Foundation of China (grant numbers 51875033) and the Fundamental Research Funds for the Central Universities (grant number 2019YJS164).

Acknowledgments: Xiangyang Wang would like to thank An Du and Bojian Qu for their contribution to data collection and valuable suggestions. Sheng Guo would like to thank Yang Du for providing equipment support (Zebris FDM-T, Zebris Medical GmbH, Germany).

Conflicts of Interest: The authors declare no conflict of interest.

References

1. Alexander, R.M. *Principles of Animal Locomotion*; Princeton University Press: Princeton, NJ, USA, 2003.
2. Winter, D.A. Energy generation and absorption at the ankle and knee during fast, natural, and slow cadences. *Clin. Orthop. Relat. R.* **1983**, *175*, 147–154. [CrossRef]
3. Blaya, J.A.; Herr, H. Adaptive control of a variable-impedance ankle-foot orthosis to assist drop-foot gait. *IEEE Trans. Neural Syst. Rehabil. Eng.* **2004**, *12*, 24–31. [CrossRef] [PubMed]
4. Mooney, L.M.; Rouse, E.J.; Herr, H.M. Autonomous exoskeleton reduces metabolic cost of human walking during load carriage. *J. NeuroEng. Rehabil.* **2014**, *11*, 80. [CrossRef] [PubMed]
5. Mooney, L.M.; Herr, H.M. Biomechanical walking mechanisms underlying the metabolic reduction caused by an autonomous exoskeleton. *J. NeuroEng. Rehabil.* **2016**, *13*, 4. [CrossRef] [PubMed]
6. Jackson, R.W.; Collins, S.H. An experimental comparison of the relative benefits of work and torque assistance in ankle exoskeletons. *Am. J. Physiol. Heart C* **2015**, *119*, 541–557. [CrossRef] [PubMed]
7. Witte, K.A.; Zhang, J.; Jackson, R.W.; Collins, S.H. Design of two lightweight, high-bandwidth torque-controlled ankle exoskeletons. In Proceedings of the IEEE International Conference on Robotics and Automation (ICRA), Seattle, WA, USA, 26–30 May 2015; pp. 1223–1228.
8. Au, S.; Berniker, M.; Herr, H. Powered ankle-foot prosthesis to assist level-ground and stair-descent gaits. *Neural Netw.* **2008**, *21*, 654–666. [CrossRef] [PubMed]
9. Sup, F.; Varol, H.A.; Goldfarb, M. Upslope walking with a powered knee and ankle prosthesis: Initial results with an amputee subject. *IEEE Trans. Neural Syst. Rehabil. Eng.* **2011**, *19*, 71–78. [CrossRef]
10. Li, D.Y.; Becker, A.; Shorter, K.A.; Bretl, T.; Hsiao-Wecksler, E.T. Estimating system state during human walking with a powered ankle-foot orthosis. *IEEE/ASME Trans. Mechatron.* **2011**, *16*, 835–844. [CrossRef]
11. Varol, H.A.; Frank, S.; Michael, G. Multiclass real-time intent recognition of a powered lower limb prosthesis. *IEEE Trans. Biomed. Eng.* **2010**, *57*, 542–551. [CrossRef]
12. Huang, H.; Kuiken, T.A.; Lipschutz, R.D. A strategy for identifying locomotion modes using surface electromyography. *IEEE Trans. Biomed. Eng.* **2009**, *56*, 65–73. [CrossRef]
13. Yu, F.; Zheng, J.; Yu, L.; Zhang, R.; He, H.; Zhu, Z.; Zhang, Y. Adjustable method for real-time gait pattern detection based on ground reaction forces using force sensitive resistors and statistical analysis of constant false alarm rate. *Sensors* **2018**, *18*, 3764. [CrossRef]
14. Agostini, V.; Balestra, G.; Knaflitz, M. Segmentation and classification of gait cycles. *IEEE Trans. Neural Syst. Rehabil. Eng.* **2014**, *22*, 946–952. [CrossRef]
15. Tang, J.; Zheng, J.; Wang, Y.; Yu, L.; Zhan, E.; Song, Q. Self-Tuning Threshold Method for Real-Time Gait Phase Detection Based on Ground Contact Forces Using FSRs. *Sensors* **2018**, *18*, 481. [CrossRef]
16. González, I.; Fontecha, J.; Hervás, R.; Bravo, J. An ambulatory system for gait monitoring based on wireless sensorized insoles. *Sensors* **2015**, *15*, 16589–16613. [CrossRef]
17. Atallah, L.; Lo, B.; Ali, R.; King, R.; Yang, G.Z. Real-time activity classification using ambient and wearable sensors. *IEEE Trans. Inf. Technol. Biomed.* **2009**, *13*, 1031–1039. [CrossRef]
18. Ferris, D.P.; Gordon, K.E.; Sawicki, G.S.; Peethambaran, A. An improved powered ankle–foot orthosis using proportional myoelectric control. *Gait Posture* **2006**, *23*, 425–428. [CrossRef]
19. Kao, P.C.; Lewis, C.L.; Ferris, D.P. Invariant ankle moment patterns when walking with and without a robotic ankle exoskeleton. *J. Biomech.* **2010**, *43*, 203–209. [CrossRef]
20. Zhang, J.; Fiers, P.; Witte, K.A.; Jackson, R.W.; Poggensee, K.L.; Atkeson, C.G.; Collins, S.H. Human-in-the-loop optimization of exoskeleton assistance during walking. *Science* **2017**, *356*, 1280–1284. [CrossRef]
21. Cavanagh, P.R.; Komi, P.V. Electromechanical delay in human skeletal muscle under concentric and eccentric contractions. *J. Electromyogr. Kinesiol.* **1979**, *42*, 159–163. [CrossRef]
22. Liu, M.; Zhang, F.; Huang, H. An Adaptive Classification Strategy for Reliable Locomotion Mode Recognition. *Sensors* **2017**, *17*, 2020.
23. Campanini, I.; Merlo, A.; Degola, P.; Merletti, R.; Vezzosi, G.; Farina, D. Effect of electrode location on EMG signal envelope in leg muscles during gait. *J. Electromyogr. Kinesiol.* **2007**, *17*, 515–526. [CrossRef]
24. Tkach, D.; Huang, H.; Kuiken, T.A. Study of stability of time-domain features for electromyographic pattern recognition. *J. NeuroEng. Rehabil.* **2010**, *7*, 21. [CrossRef]
25. Bunderson, N.E.; Kuiken, T.A. Quantification of feature space changes with experience during electromyogram pattern recognition control. *IEEE Trans. Neural Syst. Rehabil. Eng.* **2012**, *20*, 239–246. [CrossRef]

26. Goršič, M.; Kamnik, R.; Ambrožič, L.; Vitiello, N.; Lefeber, D.; Pasquini, G.; Munih, M. Online phase detection using wearable sensors for walking with a robotic prosthesis. *Sensors* **2014**, *14*, 2776–2794. [CrossRef]
27. Taborri, J.; Palermo, E.; Rossi, S.; Cappa, P. Gait partitioning methods: A systematic review. *Sensors* **2016**, *16*, 66. [CrossRef]
28. Sawicki, G.S.; Gordon, K.E.; Ferris, D.P. Powered lower limb orthoses: Applications in motor adaptation and rehabilitation. In Proceedings of the 9th International Conference on Rehabilitation Robotics, Chicago, IL, USA, 28 June–1 July 2005; pp. 206–211.
29. Zhang, J.; Cheah, C.C.; Collins, S.H. Experimental comparison of torque control methods on an ankle exoskeleton during human walking. In Proceedings of the IEEE International Conference on Robotics and Automation (ICRA), Seattle, WA, USA, 26–30 May 2015; pp. 5584–5589.
30. Au, S.K.; Dilworth, P.; Herr, H. An ankle-foot emulation system for the study of human walking biomechanics. In Proceedings of the IEEE International Conference on Robotics and Automation (ICRA), Orlando, FL, USA, 15–19 May 2006; pp. 2939–2945.
31. Caputo, J.M.; Collins, S.H. A universal ankle–foot prosthesis emulator for human locomotion experiments. *J. Biomech. Eng.* **2014**, *136*, 035002. [CrossRef]
32. Godiyal, A.K.; Mondal, M.; Joshi, S.D.; Joshi, D. Force Myography Based Novel Strategy for Locomotion Classification. *IEEE Trans. Hum. Mach. Syst.* **2018**, *99*, 1–10. [CrossRef]
33. Chen, B.; Zheng, E.; Fan, X.; Liang, T.; Wang, Q.; Wei, K.; Wang, L. Locomotion mode classification using a wearable capacitive sensing system. *IEEE Trans. Neural Syst. Rehabil. Eng.* **2013**, *21*, 744–755. [CrossRef]
34. Zheng, E.; Chen, B.; Wei, K.; Wang, Q. Lower limb wearable capacitive sensing and its applications to recognizing human gaits. *Sensors* **2013**, *13*, 13334–13355. [CrossRef]
35. Godiyal, A.K.; Verma, H.K.; Khanna, N.; Joshi, D. A force myography-based system for gait event detection in overground and ramp walking. *IEEE Trans. Instrum. Meas.* **2018**, *99*, 1–10. [CrossRef]
36. Jiang, X.; Chu, K.; Khoshnam, M.; Menon, C. A Wearable Gait Phase Detection System Based on Force Myography Techniques. *Sensors* **2018**, *18*, 1279. [CrossRef]
37. Huang, H.; Zhang, F.; Hargrove, L.J.; Dou, Z.; Rogers, D.R.; Englehart, K.B. Continuous locomotion-mode identification for prosthetic legs based on neuromuscular–mechanical fusion. *IEEE Trans. Biomed. Eng.* **2011**, *58*, 2867–2875. [CrossRef]
38. Browning, R.C.; Modica, J.R.; Kram, R.; Goswami, A. The effects of adding mass to the legs on the energetics and biomechanics of walking. *Med. Sci. Sports Exerc.* **2007**, *39*, 515–525. [CrossRef]
39. Collins, S.H.; Wiggin, M.B.; Sawicki, G.S. Reducing the energy cost of human walking using an unpowered exoskeleton. *Nature* **2015**, *522*, 212. [CrossRef]
40. Wiggin, M.B.; Sawicki, G.S.; Collins, S.H. An exoskeleton using controlled energy storage and release to aid ankle propulsion. In Proceedings of the IEEE International Conference on Rehabilitation Robotics (ICRR), Zurich, Switzerland, 29 June–1 July 2011; pp. 1–5.
41. Yandell, M.B.; Tacca, J.R.; Zelik, K.E. Design of a Low Profile, Unpowered Ankle Exoskeleton That Fits Under Clothes: Overcoming Practical Barriers to Widespread Societal Adoption. *IEEE Trans. Neural Syst. Rehabil. Eng.* **2019**, *27*, 712–723. [CrossRef]
42. Tzu-wei, P.H.; Kuo, A.D. Mechanics and energetics of load carriage during human walking. *J. Exp. Biol.* **2014**, *217*, 605–613.
43. Kuo, A.D.; Donelan, J.M.; Ruina, A. Energetic Consequences of Walking Like an Inverted Pendulum: Step-to-Step Transitions. *Exerc. Sport Sci. Rev.* **2005**, *33*, 88–97. [CrossRef]
44. The Holistic Clinic. Available online: http://almawiclinic.com/2017/03/13/calf-pains-symptoms-causes-treatment (accessed on 24 April 2019).
45. Seth, A.; Hicks, J.L.; Uchida, T.K.; Habib, A.; Dembia, C.L.; Dunne, J.J.; Ong, C.F.; DeMers, M.S.; Rajagopal, A.; Millard, M.; et al. OpenSim: Simulating musculoskeletal dynamics and neuromuscular control to study human and animal movement. *Plos Comput. Biol.* **2018**, *14*, 1006223. [CrossRef]
46. Thelen, D.G.; Anderson, F.C. Using computed muscle control to generate forward dynamic simulations of human walking from experimental data. *J. Biomech.* **2006**, *39*, 1107–1115. [CrossRef]
47. Umberger, B.R.; Gerritsen, K.G.; Martin, P.E. A model of human muscle energy expenditure. *Comput. Methods Biomech. Biomed. Eng.* **2003**, *6*, 99–111.

48. Umberger, B.R. Stance and swing phase costs in human walking. *J. R. Soc. Interface* **2010**, *7*, 1329–1340. [CrossRef]
49. Umberger, B.R.; Rubenson, J. Understanding muscle energetics in locomotion: New modeling and experimental approaches. *Exerc. Sport. Sci. Rev.* **2011**, *39*, 59–67. [CrossRef]

 © 2019 by the authors. Licensee MDPI, Basel, Switzerland. This article is an open access article distributed under the terms and conditions of the Creative Commons Attribution (CC BY) license (http://creativecommons.org/licenses/by/4.0/).

Article

Robust Visual Tracking Using Structural Patch Response Map Fusion Based on Complementary Correlation Filter and Color Histogram

Zhaohui Hao [1,2], Guixi Liu [1,2,*], Jiayu Gao [2,3] and Haoyang Zhang [1,2]

1. School of Mechano-Electronic Engineering, Xidian University, Xi'an 710071, Shaanxi, China; haozhaohui@stu.xidian.edu.cn (Z.H.); zhanghy@stu.xidian.edu.cn (H.Z.)
2. Shaanxi Key Laboratory of Integrated and Intelligent Navigation, Xi'an 710068, Shaanxi, China; gaojiayu_cool@126.com
3. Xi'an Research Institute of Navigation Technology, Xi'an 710068, Shaanxi, China
* Correspondence: gxliu@xidian.edu.cn; Tel.: +86-137-0029-6049

Received: 10 August 2019; Accepted: 23 September 2019; Published: 26 September 2019

Abstract: A part-based strategy has been applied to visual tracking with demonstrated success in recent years. Different from most existing part-based methods that only employ one type of tracking representation model, in this paper, we propose an effective complementary tracker based on structural patch response fusion under correlation filter and color histogram models. The proposed method includes two component trackers with complementary merits to adaptively handle illumination variation and deformation. To identify and take full advantage of reliable patches, we present an adaptive hedge algorithm to hedge the responses of patches into a more credible one in each component tracker. In addition, we design different loss metrics of tracked patches in two components to be applied in the proposed hedge algorithm. Finally, we selectively combine the two component trackers at the response maps level with different merging factors according to the confidence of each component tracker. Extensive experimental evaluations on OTB2013, OTB2015, and VOT2016 datasets show outstanding performance of the proposed algorithm contrasted with some state-of-the-art trackers.

Keywords: visual tracking; correlation filter; color histogram; adaptive hedge algorithm

1. Introduction

Visual object tracking is a fundamental research task and plays a crucial role in numerous computer vision applications including motion analysis, surveillance, segmentation, and autonomous driving and so forth [1]. Basically, the purpose of visual tracking is to estimate the motion trajectory of the target over successive video frames, only initializing its state at the first frame. Numerous robust tracking algorithms [2–4] have emerged and taken exciting progress gains in recent years. However, it is still a very challenging task to design a robust tracking algorithm due to significant target appearance variation caused by factors such as fast motion, shape deformation, partial occlusion, illumination change, background clutter, and so on. To overcome these issues, a more discriminative appearance representation which is a key part of successful tracking is needed.

Recently, tracking approaches based on discriminative correlation filters (DCFs) [5–9] have attracted considerable attentions and obtained excellent performances on several tracking benchmark datasets [10–12]. Benefited from the circular assumption of training samples, the DCFs-based algorithms can be learned and detected very efficiently in the Fourier domain by element-wise multiplication and, hence, is of significance for real-time tracking application. However, as traditional DCFs that use histogram of oriented gradients (HOG) features [13] strongly depend on the spatial layout of the tracked object, it is hard for them to handle deformation and rotation well.

To tackle the above shortcoming, an effective tracker termed as Staple [14] has been proposed to compensate for the deficiencies of both color histograms and DCFs via linearly combining their response maps, which successfully deals with deformation and illumination variation simultaneously. However, there emerge two principal lacks of the Staple tracker. Firstly, the Staple tracker only employs holistic appearance representations of color histogram and DCFs, ignoring the underlying spatial local structural information, thereby its component trackers $Staple_{ch}$ (only applying color histogram-based tracker) and $Staple_{cf}$ (only applying DCFs-based tracker), are likely to perform poorly alone in some challenging scenarios such as partial occlusion and drastic deformation. This always leads to failure due to the merged inaccurate response maps. Secondly, Staple tracker resorts to a fixed merging percentage factor (i.e., 0.3) for overall performance on datasets, which may cause tracking failure because of considering too much unreliable component trackers in some complex scenes. Figure 1 illustrates the tracking results on four sequences to explain the above findings of the Staple tracker. Due to the failures of both $Staple_{cf}$ and $Staple_{ch}$ at frame 176 and frame 88 in Surfer and Shaking sequences respectively, Staple which is the result of merging these two components fails at these instants as well. $Staple_{cf}$ fails at the 560th frame of the BlurCar1 sequence and $Staple_{ch}$ fails at the 506th frame of the Box sequence. These tracking failures also lead to the failure of Staple tracker since it has no emphasis on reliable component tracker. The LGCmF tracker [15], which is an improved method based on the Staple, performs well on both Surfer and Box sequences, while fails on the BlurCar1 and Shaking sequences.

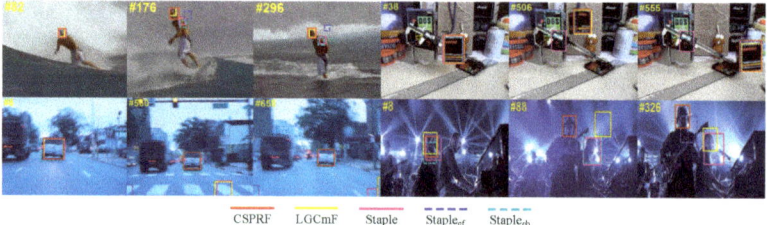

Figure 1. Tracking results of the Staple, its component trackers including $Staple_{cf}$ and $Staple_{ch}$, the LGCmF and our CSPRF tracker on four sequences. The Staple tracks failure on all four sequences and the LGCmF can track two of the four sequences, which illustrates the LGCmF indeed has some improvement. Our CSPRF can perform well on four sequences. From left to right and from top to bottom are Surfer, Box, BlurCar1, and Shaking sequences.

To alleviate the aforementioned deficiencies, in this work, we follow the research line of merging response maps between color histograms and DCFs [14] and embed a spatial local patch-structured framework to it for visual tracking. We first construct two part-based component trackers: correlation filter-based structural patch tracker (CFSP) and color histogram-based structural patch tracker (CHSP). In each of them, an adaptive hedge algorithm is introduced to determine weights of structural patches. The standard hedge algorithm [16] is an online decision theoretical method for multi-expert, which uses the difference between the loss of an expert and the weighted average loss of all experts to define the regret of this expert. This algorithm uses the cumulative regret corresponding to each expert to generate its weight in each frame. In this work, we treat each tracked patch as an expert and design a reliable loss metric for each expert by analyzing the similarity or the discrimination of these patches and the difference among displacement of each patch with the target. Then based on the tracking reliabilities of CFSP and CHSP, we selectively combine the response maps of them to formulate the final complementary structural patches response fusion tracker (CSPRF). Inspired by [15,17], we train and update a SVM detector for determining the confidences of the component trackers CFSP and CHSP, and implement a re-detect procedure when both of the component trackers are unreliable. From Figure 1, it can be seen that our proposed CSPRF tracker performs favorably when the Staple and LGCmF lose the target.

The main contributions of this work can be summarized as: (1) In contrast to existing Staple tracker which only uses holistic appearance representations of color histograms and DCFs, we use the local structural appearance information and propose an novel structural patch response map fusion tracking algorithm using complementary correlation filter and color histogram. (2) We develop an adaptive hedge algorithm for part-based tracking framework by adaptively considering the proportion of instantaneous and cumulative regrets of each expert over time. (3) We design two reliable loss measurement methods in correlation filter and color histogram models to provide credible inputs for the adaptive hedge algorithm, by which the correlation filter-based structural patch tracker (CFSP) and the color histogram-base structural patch tracker (CHSP) are proposed. (4) We execute the extensive experiments on three tracking benchmark datasets OTB2013 [10], OTB2015 [11], and VOT2016 [12] to demonstrate the efficiency and robustness of our proposed CSPRF tracker in comparison with the state-of-the-art trackers.

2. Related Work

2.1. Correlation Filter-Based Tracking

Since discriminative correlation filter-based tracking method was initially proposed by Bolme et al. [5], now it has been widely applied to the visual object tracking community and has been demonstrated very impressive performance on benchmark datasets. The work in [5] optimizes a minimum output sum of squared error filter (MOSSE) that uses simple grayscale features to represent the target appearance. According to the circular matrix structure and kernel trick, Henriques et al. [18] propose the circular structure with kernels tracking algorithm (CSK), and soon after extend this work to handle multi-channel features such as histograms of oriented gradients (HOG) [13], namely kernelized correlation filters (KCF) [6]. Danelljan et al. [19] introduce color names feature [20] to correlation filter for improving tracking performance. To resolve scale changing problem during tracking process, Danelljan et al. propose the DSST tracker [21] with a separate multi-scale correlation filter. To mitigate the boundary effects, Danelljan et al. propose the SRDCF tracker [7] by using a spatially regularized weight to penalize filter coefficients far away from the target center. Li et al. present spatial-temporal regularized correlation filters (STRCF) [9] by introducing temporal regularization to SRDCF. Additionally, Yang et al. present parallel correlation filters (PCF) [22] for visual tracking by constructing two parallel correlation filters. Zhang et al. propose a novel motion-aware correlation filters (MACF) tracking algorithm [23], which integrates instantaneous motion estimation Kalman filters into the correlation filters.

2.2. Color Histogram-Based Tracking

Color histograms [24–26] are a common method to model the object appearance representation among earlier tracking approaches. Compared to others features such as HOG or pixels, color histogram is robust to shape deformation and rotation, hence it is meaningful to track non-rigid objects. The early mean shift tracker [24] minimizes the Bhattacharyya distance of color histograms between the target object and the reference regions iteratively. Abdelai et al. [25] present an efficient accept–reject color histogram-based scheme embedding integral image into a Bhattacharyya kernel to find most similar area with target. Duffner et al. [26] construct a probabilistic segmentation using back-projection maps between foreground and background, where the target tracking process is accomplish by applying a generalized Hough transform with pixel-based descriptors. The distractor-aware tracker (DAT) [27] proposed by Possegger et al. formulates an efficient discriminative color histograms model to identify potentially distracters and significantly reduce the risk of drifting.

In recent years, there appear complementary learners [14,28] combining color histogram and correlation filter to represent the target, which are able to compensate each other in visual tracking. In [14], Bertinetto et al. linearly incorporate the output response maps of color histograms and correlation filters to achieve high tracking performance and speed. Fan et al. [29] present a dual color clustering and spatio-temporal regularized correlation regressions-based complementary tracker, where a color clustering-based histogram and a spatio-temporal regularized correlation filters are formulated as complementary learners to improve the tracking performance of [14]. Lukezic et al. [28]

construct a spatial reliability map to adjust the filter support to the part of the target object suitable for tracking by exploiting color histograms. Zhang et al. [15] propose a collaborative local-global layer visual tracking method (LGCmF), in which a block tracker (SLC) utilizing structural local color histograms feature and a global correlation filter tracker based on HOG feature are merged in the response map level. Inspired by [15], the block strategy also is adopted in this work. In contrast to [15] that only applies part-based tracking strategy in color histogram model, we employ more complete blocking strategy in both component trackers and more efficient block weighting method for each patch based on adaptive hedge algorithm.

2.3. Part-Based Tracking

Part-based tracking algorithms focus on the local parts of the target and, hence, they are very robust to handle partial occlusion and severe deformation. Commonly, the visible parts can still provide reliable cues for tracking when the target is partially occluded. Nejhum et al. [30] match the intensity histograms of foreground blocks by dividing the foreground shape as several rectangular blocks to update the target shape and adjust layout of them. Zhang et al. [31] propose a part matching tracker (PMT) based on a locality-constrained low-rank sparse learning method to optimize partial permutation matrices for image blocks among multiple frames. Yao et al. [32] present a latent structured learning method to model the unknown parts of target.

Several recent tracking methods have attempted to integrate the correlation filters into a part-based framework for improving the tracking performance [33,34]. Liu et al. [33] propose a part-based tracker with multiple adaptive correlation filters, where the Bayesian inference framework and a structural constraint mask are adopted to be robust to partial occlusion and deformation. Li et al. [34] identify the reliability of patches according to the motion trajectory and trackability of each patch. Sun et al. [35] present a shape-preserved kernelized correlation filter within a level set framework for deformable tracking of individual patches. Wang et al. [36] formulate an occlusion-aware part-based tracker that can convert between the global model and local model adaptively to avoid polluting target templates by background information.

2.4. Sparse-Based Tracking and Deep Learning-Based Tracking

In addition to correlation filter tracking and color tracking, popular tracking algorithms in recent years include sparse tracking [37–39] and deep learning tracking [40–43] as well. In sparse tracking, Zhang et al. [37] propose a novel sparse tracking method by matching framework for robust tracking based on basis matching. Zhang et al. [38] propose a tracker using a semi-supervised appearance dictionary learning method. Zhang et al. [39] develop a biologically inspired appearance model for robust visual tracking. As for deep learning based tracking, the work of [40] learns multi-level correlation filters with hierarchical convolutional features to integrate the correlation responses proportionally. Subsequently, Qi et al. [41] exploit an adaptive hedge algorithm to make a weighted decision of all weak correlation filter trackers. Zhang et al. [42] integrate the point-to-set distance metric learning (DML) into visual tracking tasks and take full advantage of all the training samples when determining the best target candidate. Danelljan et al. [43] introduce a novel tracking architecture consisting of two components designed exclusively for target estimation and classification. This method achieves a considerable performance gain against the previous tracking approach.

3. Proposed Algorithm

3.1. Overview

Following the Staple [14], our work also relies on the strengths of both correlation filters and color histograms. However, the Staple employing the holistic appearance information is likely to drift or fail in the scenes of severe deformation or partial occlusion. A part-based tracking strategy can achieve favorable tracking results for above challenging scenes, since reliable cues for tracking can be provided by remaining visible parts or undeformed parts. Therefore, in this work we take into account the structural local information of both correlation filters and color histograms, which show promising tracking performance improvement over Staple.

Since the trackability of individual patches is distinct in different scenes, it should be highlighted for these patches with high trackability. The LGCmF [15] calculates the discrimination value to determine the trackability of individual patch in its component tracker SLC by the foreground-background discrimination analysis, which only considers the appearance information of the individual patch. To fully utilize both appearance discrimination and spatial motion information, we do not only consider discrimination value, but also allow motion consistency of individual patch with the target. And according to them, we formulate the loss metric of each patch tracker in CHSP, which is used as input for adaptive hedge algorithm. Figure 2c illustrates that our component tracker CHSP allocates more desirable weights to individual patch trackers than the SLC [15]. For instance the weights of our CHSP are more uniform than them of SLC when all patches are clearly visible at frame 4. And when the target is partially occluded at frames 39 and 253, the remaining visible patches 1, 2, 3, and 4 still can provide reliable tracking cues, which mean these patches are more likely to be tracked correctly, hence these patches are given higher weights in our CHSP.

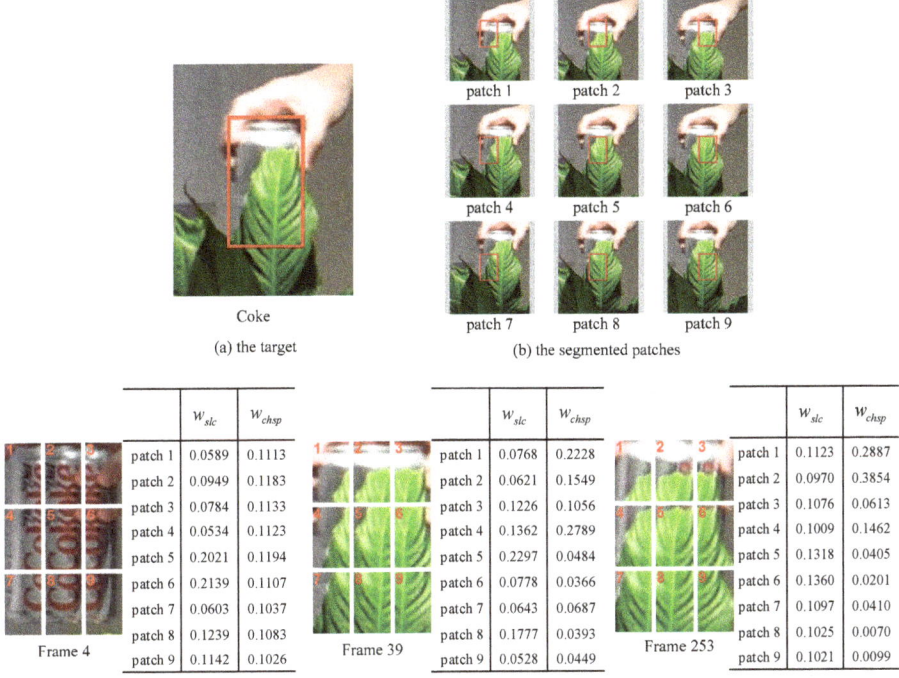

Figure 2. The segmented instance and the comparison of weights of patches in SLC and CHSP. (**a**) shows the tracking target with red bounding box in Coke sequence. (**b**) shows that the target is divided into nine overlapped patches using red bounding box. In (**c**), the tables list the patch's weights of SLC and CHSP for the frames 4, 39, and 253. w_{slc} and w_{chsp} represent the weights of corresponding patches of SLC and CHSP, respectively.

In CFSP, we calculate the loss of individual patch tracker according to similarity and motion consistency of individual patch. The motion consistency refers to the difference among displacement of individual patch with the displacement of predicted target. And the similarity of each patch is measured by employing the intensity and the smooth constraint of response map between patch expert trackers.

We describe the main steps of the proposed approach in Figure 3. In this section, we first present the adaptive hedge algorithm, and then describe the two component trackers in detail. Based on these component trackers, we formulate the final complementary structural patches response fusion tracker.

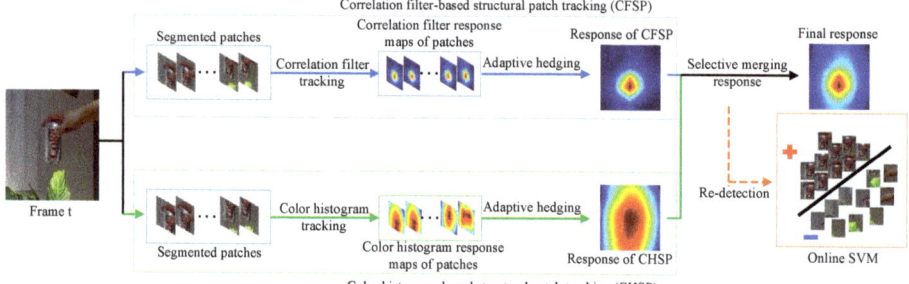

Figure 3. The flow chart of the proposed tracking algorithm. When a new frame t arrives, we first divide the target into several overlapped patches. The responses of correlation filter and color histogram of all these patches are computed. These correlation filter responses are combined together by the adaptive hedge algorithm to constitute the component tracker CFSP. With the same way, the component tracker CHSP is also constructed. Finally, responses of CFSP and CHSP are selectively fused to obtain final response and the new location of the target is estimated at its peak. When both of the combined responses are unreliable, an online SVM classifier is activated to re-detect the target.

3.2. Adaptive Hedge Algorithm

The standard hedge algorithm [16] for decision theoretic online learning problem generates a weight distribution w_t^i over all experts $i \in \{1, 2, \ldots, K\}$ at frame t, where K is the number of experts. Each expert i incurs a loss l_t^i, and the expected loss is calculated as:

$$l_t^A = \sum_{i=1}^{K} w_t^i l_t^i \qquad (1)$$

The standard hedge algorithm introduces a new notion of regret to generate a new weight distribution over all experts for next frame $t+1$. The instantaneous regret to expert i is defined as:

$$r_t^i = l_t^A - l_t^i \qquad (2)$$

Its cumulative regret to expert i for frame t is:

$$R_t^i = \sum_{\tau=1}^{t} r_\tau^i = R_{t-1}^i + r_t^i \qquad (3)$$

The purpose of the hedge algorithm is to minimize the cumulative regret R_t^i over all experts throughout the whole video frames.

Since the cumulative regret R_t^i is computed by simply summing the historical regret R_{t-1}^i and instantaneous regret r_t^i as shown in Equation (3), where R_{t-1}^i and r_t^i contribute equally in the loss function, the standard hedge algorithm [16] performs not well in real-world tracking tasks as it ignores two key factors. First, the target appearance is possible to change with irregular velocity throughout a video sequence, which means that the historical regret R_{t-1}^i should be considered with a varying proportion over time to better reflect the target state for visual tracking. Second, since each expert tracker captures a different part of the target in this work, it is less effective to utilize a fixed proportion for the historical regret over all expert trackers.

Similar to [41,44], to overcome the above two shortcomings, we propose an adaptive hedge algorithm, which is the use of an adaptive regret mechanism to determine the proportion of the historical as well as instantaneous regrets over time. Since the appearance variation of target occurs slowly in a short time period, we formulate the loss of each expert l^i during time period Δt via a Gaussian distribution with standard variance σ_t^i and mean μ_t^i:

$$\mu_t^i = \frac{1}{\Delta t} \sum_{\tau=t-\Delta t+1}^{t} l_\tau^i \tag{4}$$

$$\sigma_t^i = \sqrt{\frac{1}{\Delta t - 1} \sum_{\tau=t-\Delta t+1}^{t} \left(l_\tau^i - \mu_t^i\right)^2} \tag{5}$$

The stability of expert i at frame t is decided by:

$$s_t^i = \frac{|l_t^i - \mu_t^i|}{\sigma_t^i} \tag{6}$$

A large s_t^i means that this expert varies highly and, hence, its cumulative regret should mainly depend on its historical regret. In contrast, a small s_t^i means this expert tends to be more stable than the one with a larger s_t^i. Hence, its cumulative regret should take a large proportion on its instantaneous regret. Based on above rules, the adaptive cumulative regret for each expert is computed as follow:

$$\alpha_t^i = \exp\left(-\gamma s_t^i\right) \tag{7}$$

$$R_t^i = \left(1 - \alpha_t^i\right) R_{t-1}^i + \alpha_t^i r_t^i \tag{8}$$

where γ is a parameter to control the shape of the exponential in Equation (7).

Our adaptive hedge algorithm also has the same solution form with the standard one [16]. The weight of each expert is updated for the next frame as follow:

$$w_{t+1}^i \propto \frac{[R_t^i]_+}{c_t} \exp \frac{(R_t^i)^2}{2c_t} \tag{9}$$

Here $[R_t^i]_+$ denotes $max\{0, R_t^i\}$ and c_t is a scale parameter constrained by:

$$\frac{1}{K} \sum_{i=1}^{K} \exp\left(\frac{\left([R_t^i]_+\right)^2}{2c_t}\right) = e \tag{10}$$

In this work, we apply the proposed adaptive hedge algorithm to the following component trackers, respectively. In addition, different metrics used to calculate the loss of patch experts in this two component trackers are proposed.

3.3. Correlation Filter-Based Structural Patch Tracking (CFSP)

In CFSP, the target is split into multiple overlapped image patches p^i, $i \in \{1, 2, \ldots, K\}$, where K is the number of patches. The tracking task is then to locate these patches. During tracking, an image block z^i with the same size of appearance template x^i is extracted out at the location of patch p^i in the previous frame. After that, a kernelized correlation filter (KCF) [6], which can be considered as an expert, is applied on each patch to track its position. The response map of the ith patch is calculated as:

$$\mathcal{R}_{cf}^{i}(z^{i}) = F^{-1}\left(F(k^{x^{i}z^{i}}) \odot F(\alpha^{i})\right) \tag{11}$$

where the subscript cf represents the correlation filter operator. The patch p^i in current frame is localized according to the location where the peak of the response map $\mathcal{R}_{cf}^{i}$. The tracking details of KCF can be found in [6].

Based on the adaptive hedge algorithm proposed in the previous section, it is natural to fuse response maps of all patches at the frame t by:

$$\mathcal{R}_{cf,t} = \sum_{i=1}^{K} w_{cf,t}^{i} \mathcal{R}_{cf,t}^{i} \tag{12}$$

where $w_{cf,t}^{i}$ is the weight of patch p^i at frame t and $\sum_{i=1}^{K} w_{cf,t}^{i} = 1$. Then at frame t, the target is located by searching the peak of the fused response map $\mathcal{R}_{cf,t}$.

The loss of each expert tracker need to be computed and is used by the adaptive hedge algorithm described in the above section to update the weights of all expert trackers. In CFSP, we consider two aspects for calculating the loss of each expert tracker. First, we use intensity and the smooth constraint of each patch's response map to reflect the similarity of patch between current frame and previous frames. The peak-to-sidelobe ratio (PSR) [5] that quantifies the sharpness of the response map peak is used to estimate the intensity of response map. It is defined as:

$$PSR_{t}^{i} = \frac{\max(\mathcal{R}_{cf,t}^{i}) - \operatorname{mean}(\mathcal{R}_{cf,t}^{i})}{\operatorname{var}(\mathcal{R}_{cf,t}^{i})} \tag{13}$$

where $\operatorname{mean}(\mathcal{R}_{cf,t}^{i})$ and $\operatorname{var}(\mathcal{R}_{cf,t}^{i})$ are the mean and the standard variance of the ith patch's response map at frame t respectively. The smooth constraint of response map (SCRM) [33] is defined as:

$$SCRM_{t}^{i} = \|\mathcal{R}_{cf,t}^{i} - \mathcal{R}_{cf,t-1}^{i} \oplus \Delta\|_{2}^{2} \tag{14}$$

where $\oplus$ means a shift operation of the response map and Δ denotes the corresponding shift of maximum value in response maps from frame $t-1$ to t. Then the normalized similarity of patch p^i can be represented as:

$$S_{t}^{i} = \frac{(PSR_{t}^{i}/SCRM_{t}^{i})}{\sum_{i=1}^{K}(PSR_{t}^{i}/SCRM_{t}^{i})} \tag{15}$$

Second, we consider the displacement difference between each patch and the predicted target at frame t:

$$D_{cf,t}^{i} = \frac{\|dis_{cf,t}^{i} - dis_{cf,t}^{tar}\|_{2}^{2}}{\sum_{i=1}^{K} \|dis_{cf,t}^{i} - dis_{cf,t}^{tar}\|_{2}^{2}} \tag{16}$$

where $dis_{cf,t}^{i}$ and $dis_{cf,t}^{tar}$ denote the displacements of corresponding patch p^i and target with respect to frame t, respectively. The loss of the ith patch expert tracker at frame t is defined as

$$l_{cf,t}^{i} = (1-\beta)\left(1 - S_{t}^{i}\right) + \beta D_{cf,t}^{i} \tag{17}$$

where β is the trade-off between the similarity and the displacement difference. The loss calculated from Equation (17) is put into the adaptive hedge algorithm to update the weight of patch p^i for frame $t+1$ in CFSP. Figure 4a illustrates the weight distribution of the sequence Bolt generated by CFSP in some frames, in which different patches have different weights. Patch 8 lies in the leg area and undergoes sever deformation. Hence, the weights of patch 8 are relatively smaller. The tracking procedure of CFSP tracker is summarized in Algorithm 1.

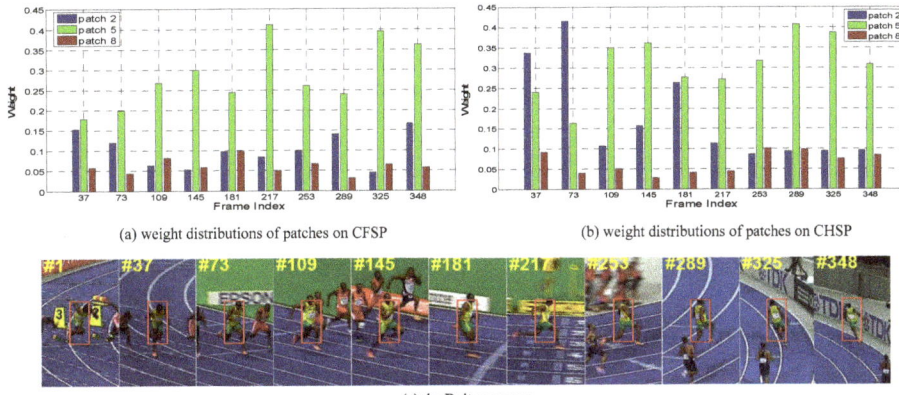

Figure 4. The weight distribution of component trackers in the Bolt sequence. For the sake of clarity, we only show the weight distribution of patches 2, 5, and 8. (**a**) and (**b**) are the weight distributions of the component trackers CFSP and CHSP at some frames, respectively. (**c**) shows the tracking target with red bounding box in Bolt sequence, in which the target suffers from severe deformation.

Algorithm 1: Correlation filter-based structural patch tracking

---

Inputs: current weight distribution $w_{cf,t}^1, \cdots, w_{cf,t}^K$; estimated target position pos_{t-1} in the previous frame;
Output: updated weight distribution $w_{cf,t+1}^1, \cdots, w_{cf,t+1}^K$; the response map $\mathfrak{R}_{cf,t}$ in the current frame.
Repeat:
1: compute correlation filter response of each patch using Equation (11);
2: compute the fused response map $\mathfrak{R}_{cf,t}$ using Equation (12);
3: compute the similarity and displacement difference of each patch using Equations (13–16);
4: compute loss of each patch tracker using Equation (17);
5: update stability models using Equations (4) and (5);
6: measure each patch tracker's stability using Equation (6);
7: update regret of each patch using Equations (1), (2), (7), and (8);
8: update weight distribution $w_{ch,t+1}^1, \cdots, w_{ch,t+1}^K$ for each patch tracker using Equation (9).

3.4. Color Histogram-Based Structural Patch Tracking (CHSP)

For the overlapped image patches $p^i, i \in \{1, 2, \ldots, K\}$, we apply the same color histogram tracking method as SLC [15] to track each of them. And each color patch tracker can be regarded as an expert. Let R_o^i, R_f^i and R_s^i represent the target region, foreground and surrounding background regions of patch p^i, respectively, where the foreground region R_f^i is slightly smaller than the target region R_o^i. Additionally, we denote y_u^i as the observation of pixel u within patch p^i, which is represented by the bin of u in the color histograms. The likelihood of pixel u belongs to the region R_o^i can be derived by applying Bayes rule like [27]:

$$P\left(u \in R_o^i \middle| R_f^i, R_s^i, y_u^i\right) \approx \frac{P\left(y_u^i \middle| u \in R_f^i\right) P\left(u \in R_f^i\right)}{\sum_{\psi \in \{R_f^i, R_s^i\}} P\left(y_u^i \middle| u \in \psi\right) P(u \in \psi)} \quad (18)$$

The likelihood terms can be derived from color histogram:

$$P\left(y_u^i \mid u \in R_f^i\right) \approx \frac{H_f^i\left(y_u^i\right)}{\left|R_f^i\right|} \text{ and } P\left(y_u^i \mid u \in R_s^i\right) \approx \frac{H_s^i\left(y_u^i\right)}{\left|R_s^i\right|} \quad (19)$$

where $\left|R_f^i\right|$ and $\left|R_s^i\right|$ denote the number of pixels in the foreground and surrounding background regions of patch p^i respectively. $H_f^i\left(y_u^i\right)$ and $H_s^i\left(y_u^i\right)$ denote the color histogram over foreground and surrounding background regions. The prior probability can be approximated as:

$$P\left(u \in R_f^i\right) \approx \frac{\left|R_f^i\right|}{\left|R_f^i\right|+\left|R_s^i\right|} \text{ and } P\left(u \in R_s^i\right) \approx \frac{\left|R_s^i\right|}{\left|R_f^i\right|+\left|R_s^i\right|} \quad (20)$$

Thus, the probability that pixel u belongs to the patch p^i can be simplified to:

$$P\left(u \in R_o^i\right) = P\left(u \in R_o^i \mid R_f^i, R_s^i, y_u^i\right) \approx \frac{H_f^i\left(y_u^i\right)}{H_f^i\left(y_u^i\right)+H_s^i\left(y_u^i\right)} \quad (21)$$

In the tracking stage, for patch p^i, we extract a rectangular searching region centered at its location in previous frame. And the response map of patch p^i can be evaluated by using its color histogram model. Using a dense sliding-window searching way over probability map $P\left(u \in R_o^i\right)$ derived from Equation (21), we can obtain the response map of patch p^i as follow:

$$\mathfrak{R}_{ch}^i\left(h_j\right) = \frac{\sum_{u \in h_j} P\left(u \in R_o^i\right)}{\left|h_j\right|} \quad (22)$$

Here $\left|h_j\right|$ represents the number of pixels in the jth sliding window h_j, the size of which is the same as patch p^i. The location of the ith patch at this frame is estimated by searching for the peak of the response map $\mathfrak{R}_{ch}^i$.

Similar as the above proposed CFSP, we also treat each patch tracker as an expert and apply the weights calculated from the adaptive hedge algorithm to fuse response maps of all patches at the frame t:

$$\mathfrak{R}_{ch,t} = \sum_{i=1}^{K} w_{ch,t}^i \mathfrak{R}_{ch,t}^i \quad (23)$$

where $w_{ch,t}^i$ is the weight of patch p^i at frame t and $\sum_{i=1}^{K} w_{ch,t}^i = 1$. The subscript ch denotes the color histogram operator. Then the target is located by searching the peak of the fused response map $\mathfrak{R}_{ch,t}$.

Different from SLC [15] only exploits appearance discrimination to determine the weight of each patch, we employ both the discrimination value and displacement difference to calculate the loss of each expert tracker and put this loss into adaptive hedge algorithm to update weight. Figure 2c illustrates that our weighted method has better performance. The discrimination values [15] of patches are calculated by considering their variance ratios (VR) [45] and histogram similarities between the foreground and surrounding background regions.

The variance ratio (VR) [35,45] is to measure the discriminative power of each patch against its surrounding background. The log likelihood of pixel u within patch p^i at frame t can be computed by using color histogram as follow:

$$L_t^i(u) = \log \frac{\max\{H_{f,t}^i(u), \delta\}}{\max\{H_{s,t}^i(u), \delta\}} \quad (24)$$

where δ is a small value to prevent dividing by zero. The log likelihood L_t^i maps the histogram into positive for colors associated with the foreground of the ith patch, and negative for colors associated with the surrounding background of the ith patch. Then the variance ratio (VR) of patch p^i at frame t can be computed as:

$$VR_t^i\left(L_t^i, H_{f,t}^i, H_{s,t}^i\right) = \frac{\text{var}\left(L_t^i; \left(H_{f,t}^i + H_{s,t}^i\right)/2\right)}{\text{var}\left(L_t^i; H_{f,t}^i\right) + \text{var}\left(L_t^i; H_{s,t}^i\right)} \quad (25)$$

where $\text{var}(L; H)$ defines the variance of $L(u)$ with respect to the color histogram $H(u)$ and is calculated as:

$$\text{var}(L; H) = \sum_u H(u)L^2(u) - \left[\sum_u H(u)L(u)\right]^2 \quad (26)$$

In Equation (25), the denominator is small when the log likelihood values of pixels in the patch and background classes are tightly clustered, while the numerator is large when the two clusters are widely separated. Thus, patches with large variance ratio show stronger discriminative power to separate the foreground and surrounding background.

Moreover, less similarity of histograms between foreground and surrounding background can readily distinguish the target from its surroundings. Therefore, the Bhattacharyya distance can be exploited:

$$\rho_t^i\left(H_{f,t}^i, H_{s,t}^i\right) = \sum_u \sqrt{H_{f,t}^i(u) H_{s,t}^i(u)} \quad (27)$$

Thus, the normalized discrimination of patch p^i can be defined as:

$$d_t^i = \frac{VR_t^i/\rho_t^i}{\sum_{i=1}^K (VR_t^i/\rho_t^i)} \quad (28)$$

Therefore, the loss of the ith patch expert at frame t is defined as:

$$l_{ch,t}^i = (1 - \beta)\left(1 - d_t^i\right) + \beta D_{ch,t}^i \quad (29)$$

where $D_{ch,t}^i$ denotes the displacement difference between the ith patch and the predicted target in CHSP at frame t:

$$D_{ch,t}^i = \frac{\|dis_{ch,t}^i - dis_{ch,t}^{tar}\|_2^2}{\sum_{i=1}^K \|dis_{ch,t}^i - dis_{ch,t}^{tar}\|_2^2} \quad (30)$$

Figure 4b displays the weight distribution of the sequence Bolt generated by CHSP at some frames. Similar as CFSP, different patches also have different weights and patches 5 and 8 have obvious distinction, of which the patch 5 is the middle part of the body whereas the patch 8 is the leg area. The leg area contains more background interference and has poor motion consistency with the body part. The tracking procedure of CHSP tracker is summarized in Algorithm 2.

Algorithm 2: Color histogram-based structural patch tracking

Inputs: current weight distribution $w^1_{ch,t}, \cdots, w^K_{ch,t}$; estimated target position pos_{t-1} in the previous frame;
Output: updated weight distribution $w^1_{ch,t+1}, \cdots, w^K_{ch,t+1}$; the response map $\mathfrak{R}_{ch,t}$ in the current frame.
Repeat:
1: compute color histogram response of each patch using Equation (22);
2: compute the response map $\mathfrak{R}_{ch,t}$ using Equation (23);
3: compute the discrimination and displacement difference of each patch using Equations (24)–(28) and (30);
4: compute loss of each patch tracker using Equation (29);
5: update stability models using Equations (4) and (5);
6: measure each patch tracker's stability using Equation (6);
7: update regret of each patch using Equations (1), (2), (7) and (8);
8: update weight distribution $w^1_{cf,t+1}, \cdots, w^K_{cf,t+1}$ for each patch tracker using Equations (9);

3.5. Response Maps Fusion between CFSP and CHSP

To complement the strengths of CFSP and CHSP, inspired by [15], we combine their response maps in a selective strategy as well. Different from LGCmF [15] using the peak value of response map in the global layer tracker to analyzing the confidence, we apply the online support vector machine (SVM) classifier on both the tracking results of CHSP and CFSP to evaluate their confidences. Specifically, we first use the SVM classifier on the tracking results of CFSP and CHSP to obtain the confidence scores C_{cfsp} and C_{chsp}. When C_{cfsp} or C_{chsp} are larger than the predefined thresholds T_{cfsp} or T_{chsp}, we consider that the CFSP or the CHSP tends to be credible. Therefore, the merging factor (η_{cfsp} or η_{chsp}) can be picked according to the credibility of the two component trackers:

$$\mathfrak{R} = \eta \mathfrak{R}_{ch} + (1-\eta)\mathfrak{R}_{cf} \qquad (31)$$

where $\mathfrak{R}_{ch}$ and $\mathfrak{R}_{cf}$ are the response maps of CHSP and CFSP, respectively. $\eta = \eta_{cfsp}$ or η_{chsp} is the merging factor that is chosen based on the confidences of CFSP and CHSP. If the confidence scores C_{cfsp} and C_{chsp} are both below the thresholds T_{cfsp} and T_{chsp}, we consider that the CFSP and CHSP are unreliable at this frame. Similar as [15,17], a re-detection process using the SVM classifier is performed by drawing dense candidates around the searching region. In this case the detected result of the SVM can be adopted only if its maximum detecting score $\max(C_{svm})$ is above a threshold T_{svm} to guarantee the accuracy. Once $\max(C_{svm}) < T_{svm}$, the re-detected result is given up and we select the η_{cfsp} as the merging factor in Equation (31). At this time the target usually suffers from partial occlusion or severe deformation, we trust the CFSP tracker more as its performance is more robust and accurate compared to the CHSP tracker, which is illustrated in experiment section. The tracking procedure of final CSPRF tracker is summarized in Algorithm 3.

Algorithm 3: Complementary structural patches response fusion tracking (CSPRF)

Inputs: the responses of the CFSP and CHSP $\mathfrak{R}_{cf,t}$, $\mathfrak{R}_{ch,t}$; estimated target position pos_{t-1} in the previous frame;
Output: estimated current target position pos_t.
Repeat:
1: obtain the confidence scores C_{cfsp} and C_{chsp} using the SVM classifier on the tracking results of CFSP and CHSP.
2: **if** $C_{chsp} \geq T_{chsp}$ **then**
3:    set $\eta = \eta_{chsp}$ and compute the current target position pos_t using Equation (31);
4: **else if** $C_{cfsp} \geq T_{cfsp}$ **then**
5:    set $\eta = \eta_{cfsp}$ and compute the current target position pos_t using Equation (31);
6: **else**
7:    use the online SVM classifier to draw dense candidates around pos_{t-1} and obtain the detecting scores C_{svm} of all candidate samples;
8:    **if** $\max(C_{svm}) \geq T_{svm}$ **then**
9:      current target position $pos_t = \mathrm{argmax}(C_{svm})$;
10:    **else**
11:      set $\eta = \eta_{cfsp}$ and compute the current target position pos_t using Equation (31);
12:    **end**
13: **end**
14: **end**

3.6. Update Scheme

To adapt to the target appearance variations, we need to update CFSP tracker, CHSP tracker and the SVM classifier. For CFSP tracker, we incrementally update the correlation filter of each patch when its response map peak $\max\left(\mathfrak{R}_{cf,t}^i\right)$ at frame t is above the threshold T_{peak}:

$$\widetilde{\alpha}_t^i = \begin{cases} (1-\xi)\widetilde{\alpha}_{t-1}^i + \xi\alpha_t^i, & if\ \max\left(\mathfrak{R}_{cf,t}^i\right) \geq T_{peak} \\ \widetilde{\alpha}_{t-1}^i, & otherwise \end{cases} \quad (32a)$$

$$\widetilde{x}_t^i = \begin{cases} (1-\xi)\widetilde{x}_{t-1}^i + \xi x_t^i, & if\ \max\left(\mathfrak{R}_{cf,t}^i\right) \geq T_{peak} \\ \widetilde{x}_{t-1}^i, & otherwise \end{cases} \quad (32b)$$

Here ξ is the learning rate. For CHSP tracker, the color histograms of each patch are update as follow:

$$\widetilde{H}_{c,t}^i = \begin{cases} (1-\tau)\widetilde{H}_{c,t-1}^i + \tau H_{c,t}^i, & if\ d_t^i \geq T_{dis} \\ \widetilde{H}_{c,t-1}^i, & otherwise \end{cases} \quad (33)$$

where τ is the learning rate and $H_{c,t}^i \in \left\{H_{f,t}^i, H_{s,t}^i\right\}$ indicates the learned color histograms of foreground and surrounding background regions of patch p^i at frame t. d_t^i is the discrimination value of patch p^i at frame t computed from Equation (28), and T_{dis} is the predefined threshold.

For the SVM classifier, it is updated only when $C_{cfsp} \geq T_{cfsp}$ or $C_{chsp} \geq T_{chsp}$, since at this time we consider the current tracking result is credible. We incrementally update the SVM classifier by applying the passive-aggressive algorithm [46] efficiently, which is similar to [17].

3.7. Scale Estimation

Similar to the DSST tracker [21], we first localize the target in a new frame and subsequently estimate scale variation. We train a one-dimensional correlation filter to perform scale estimation. A scaling set $S = \left\{a^n \middle| n \in \left\{\left\lfloor -\frac{N_s-1}{2} \right\rfloor, \ldots, \left\lfloor \frac{N_s-1}{2} \right\rfloor\right\}\right\}$ is built, where a and N_s denote the scale parameter and the number of scales respectively. Let $M \times N$ be the target size in the current frame and for each scale $s \in S$, an image patch z_s of size $sM \times sN$ centered at the target location is extracted to construct a feature pyramid. We exploit the correlation filter on these image patches z_s with corresponding to one dimensional Gaussian regression label y_s. The estimated scale is derived as:

$$s_{opt} = \text{argmax}\{f(z_s)|s \in S\} \qquad (34)$$

where s_{opt} is the maximum value of the scale correlation response. This implementation details can refer to [21].

4. Experimental Results

We first evaluate our complementary structural patches response fusion tracker (CSPRF) by comparing with others state-of-the-art trackers on OTB2013 and OTB2015. Then, the performance comparison of the LGCmF with our CSPRF is conducted. After that, to validate the effectiveness of two component trackers (CFSP and CHSP), we compare them with several relevant tracking algorithms, respectively. Finally, we conduct comparative experiments on VOT2016 [12].

4.1. Experimental Setup

We conducted our experiments on OTB2013 [10] and OTB2015 [11] benchmarks. All these sequences cover 11 challenging attributes: background clutters (BC), deformation (DEF), fast motion (FM), scale variation (SV), out-of-plane rotation (OPR), motion blur (MB), out-of-view (OV), in-plane rotation (IPR), illumination variation (IV), occlusion (OCC), and low resolution (LR). The tracking methods are evaluated by the following metrics: center location error (CLE), distance precision rate (DP), and overlap success rate (OS). The CLE is defined as the average Euclidean distance between the ground truth and the estimated center location of the target. The DP is computed as the percentage of frames where CLE is smaller than a specified threshold. The OS indicates the percentage of frames whose overlap ratio between the estimated bounding box and the ground truth bounding box surpasses a certain threshold. Following the evaluation protocol [10,11], we set the two preset thresholds of the DP and OS to 20 pixels and 0.5 in overall experiments, respectively. In addition, experimental results are reported using the precision plots and success plots under one-pass evaluation (OPE) as in [10,11]. In success plots, the area under the curve (AUC) is adopted to rank the compared trackers in the legend.

Besides OTB2013 and OTB2015, we also implement comparative experiments on VOT2016 [12]. This dataset consists of 60 challenging sequences. The performance is evaluated both in terms of robustness, accuracy and expected average overlap (EAO). The robustness calculates the average number of tracking failures over all sequences. The accuracy computes the average overlapping ratio between the estimated bounding box and the ground truth. EAO ranks the overall performance which takes both accuracy and robustness into account. Readers can refer to [12] for details.

Our methods are implemented in MATLAB 2014a (MathWorks, Natick, MA, USA) for learning and tracking process and C++ for feature extraction. The source codes of compared tracking algorithms are offered by authors, whose parameters are at default values. All the experiments are run on a PC with an AMD A10-5800K 3.8GHz CPU and 8 GB of RAM (Advanced Micro Devices, Sunnyvale, CA, USA).

4.2. Implementation Details

Let $M_o \times N_o$ represent the size of the target bounding box. The global target is divided into 3×3 overlapped patches by taking the patch size and step length as $\left(\frac{M_o}{2}, \frac{N_o}{2}\right)$, that is to say, the parameter $K = 9$. The time period Δt in Equations (4) and (5) is set to five frames and the scale factor γ in Equation (7) is set to 10. The β in Equations (17) and (29) is set to 0.5. For the component tracker CFSP, the histogram of the oriented gradient (HOG) [13] and color names (CN) [20] are applied as the feature representation. The searching window size of $M \times N$ is set to four times the patch size. The learning rate ξ in Equation (32) is set to 0.01 and the threshold $T_{peak} = 0.16$.

For the component tracker CHSP, the surrounding background region R_s is an expanded region of patch with $\frac{1}{2}\left(\frac{M_o}{2} + \frac{N_o}{2}\right)$ as the length and width, while the foreground region R_f is set to 0.8 times the patch size R_o. In Equation (24), $\delta = 10^{-3}$. The learning rate τ in Equation (33) is set to 0.04 and the threshold T_{dis} is set to 0.5/2.5 for gray/color image sequences. For CSPRF, the thresholds T_{cfsp}, T_{chsp} and T_{svm} are set to 0, 0, and 0.5, respectively. The merging factors η_{cfsp} and η_{chsp} are set to 0.6 and 0.3. The SVM classifier is trained by densely drawing samples from a searching window centered at the global target location. The samples with positive label are selected when their overlap ratios with the global target bounding box are above 0.6, and for the samples with negative label, their overlap ratios are below 0.2. For scale estimation, the parameters are the same as the DSST [21] tracker. We keep the above parameters fixed throughout all of the experiments and our proposed CSPRF tracker runs at an average of 5.1 frames per second (FPS).

4.3. Performance Evaluation of the CSPRF Tracker on OTB2013 and OTB2015

Our proposed CSPRF tracker is compared with 10 state-of-the-art trackers including KCF [6], MEEM [4], DSST [21], Staple [14], Staple_CA [8], CSR-DCF [28], SRDCF [7], SAMF [47], LCT+ [17] and RPT [34]. In above trackers, KCF, DSST, and SRDCF are the correlation filters-based trackers. Staple, Staple_CA, CSR-DCF and SAMF introduce color feature as an effective complement to the HOG feature. RPT is the part-based tracker and MEEM is the tracker that uses multiple online SVM classifiers.

4.3.1. Quantitative Evaluation

Figure 5 and Table 1 show overall comparisons between our CSPRF tracker and other 10 trackers on OTB2013 and OTB2015 datasets. It is easily to observe that our CSPRF tracker performs favorably against the compared trackers on both datasets. For the OTB2013 dataset as shown in Figure 5a, the proposed CSPRF tracker achieves the best overall performance both in precision and success plots with a DP score of 87.6% and an AUC score of 65.3%, outperforming the second best tracker LCT+ by 2.9% and 1.8%. For OTB2015 dataset as illustrated in Figure 5b, the CSPRF performs best with a DP score of 83.9% on the precision plot and an AUC score of 61.7% on the success plot, and outperforms the second best Staple_CA by 2.9% and 0.9%, respectively. In contrast to the Staple_CA that only promotes the correlation filter module of Staple, our method improves both the correlation filter and color histogram modules of Staple and, hence, obtains better performance than Staple_CA. Additionally, compared with the Staple tracker, our approach achieves gains of 8.8% and 5.2% in the DP score and 4.6% and 2.6% in the AUC score on both OTB2013 and OTB2015, respectively.

Table 1. Overall performance on the OTB2013 (I) and OTB2015 (II) datasets with the representative mean overlap success (OS) rate at threshold of 0.5, median overlap success (OS) rate, median distance precision (DP) rate, and median center location error (CLE). Best: bold; second best: underline.

		CSPRF	LCT+	DSST	Staple_CA	Staple	SAMF	SRDCF	RPT	KCF	CSR-DCF	MEEM
Meam OS	I	**81.4**	<u>81.2</u>	67.3	76.1	74.2	72.2	78.1	70.2	62.1	75.6	70.8
(%)	II	**75.4**	70.1	61.3	<u>72.8</u>	70.4	67.0	71.2	61.6	55.1	71.2	62.2
Median OS	I	**82.5**	<u>82.3</u>	68.0	77.2	75.1	73.4	78.8	71.9	63.7	76.9	72.9
(%)	II	**76.6**	71.3	62.2	<u>74.5</u>	71.8	68.7	72.3	63.6	56.9	72.3	64.5
Median DP	I	**89.1**	86.1	75.1	85.0	80.2	80.6	82.7	80.5	75.5	83.0	<u>86.7</u>
(%)	II	**85.6**	78.2	69.8	<u>82.7</u>	80.4	77.6	78.3	74.0	71.7	81.5	81.0
Median CLE	I	<u>6.39</u>	7.23	12.2	7.27	8.42	8.72	**4.82**	8.26	11.4	7.98	7.50
(pixel)	II	<u>7.10</u>	9.13	13.1	**7.09**	8.35	9.43	7.75	11.3	14.7	8.50	9.92

Table 1 reports the mean OS (%), median DP (%), median OS (%) and median CLE (pixels) over the OTB2013 and OTB2015 datasets. Our tracker obtains the best results in above three evaluation metrics except that its median CLEs with 6.39 on OTB2013 and 7.10 on OTB2015 are slightly lower than the SRDCF and Staple_CA by 1.57 and 0.01, respectively.

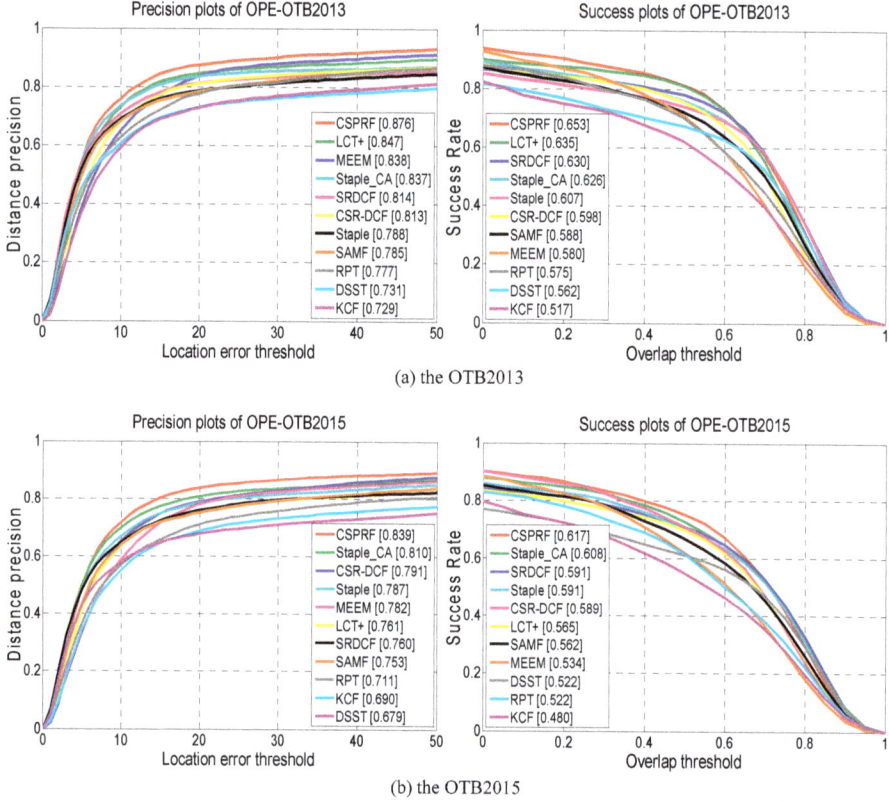

Figure 5. Quantitative evaluation over the OTB2013 and OTB2015 datasets. Precision and success plots using the one-pass evaluation (OPE). The legend of precision plots shows the average distance precision rates (DP) at 20 pixels, and the legend of success plots contains the overlap success scores (OS) with the area under the curve (AUC).

4.3.2. Attribute-Based Evaluation

To facilitate analyzing the strength and weakness of our method in various aspects, we further evaluate the trackers on datasets with 11 attributes. Figure 6 shows the precision and success plots of all compared trackers on OTB2015 with various attributes. Among them, our tracker ranks the best within seven out of 11 attributes including OPE, SV, OCC, DEF, OPR, OV and BC, and achieves a top three performance in terms of IPR and LR. This is attributed to our proposed complete structural patch tracking strategy and the novel updated weight strategy, which can fully emphasize valid cues of the target. Especially, our tracker makes a large margin in terms of BC, DEF, and OCC in the precision plots. This illustrates that our tracker has the distinct advantage in dealing with the background clutter, deformation, and occlusion.

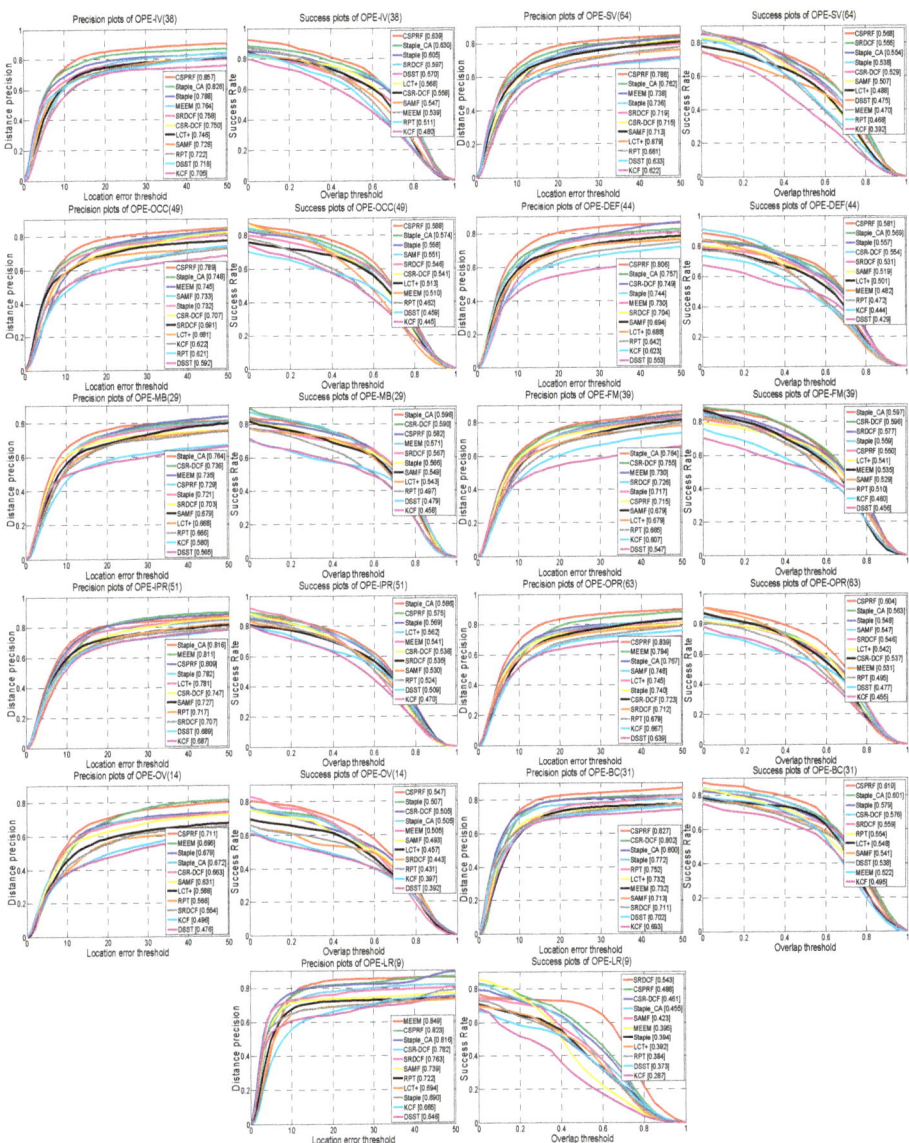

Figure 6. The precision plots and the success plots with 11 attributes on OTB2015. The legend of precision plot contains the average DP score at 20 pixels while the legend of success plot contains the area under the curve (AUC) score for each tracker. The number of sequences for each attribute is shown in brackets.

Table 2 reports the mean DP scores of compared trackers over all 11 attributes on the OTB2013. Our CSPRF tracker obtains the best performance within seven out of 11 attributes including IV, SV, OCC, DEF, OPR, OV, and BC, and achieves the second best performance in IPR with the mean DP score of 82.8%. From Figure 6 and Table 2, it demonstrates that our proposed CSPRF obtains competitive tracking performance against the other state-of-the-art trackers in these challenging attributes.

Table 2. Distance precision scores (%) at a threshold of 20 pixels in terms of individual attributes on the OTB2013. Best: bold, second best: underline.

	CSPRF	LCT+	DSST	Staple_CA	Staple	SAMF	SRDCF	RPT	KCF	CSR-DCF	MEEM
IV(25)	**84.5**	79.2	72.4	<u>80.1</u>	74.2	70.6	71.3	74.1	70.7	71.3	76.9
SV(28)	**84.4**	75.7	71.5	<u>80.5</u>	73.6	73.0	77.1	74.0	65.5	70.0	70.0
OCC(29)	**87.0**	<u>84.6</u>	70.0	80.6	78.3	84.5	81.2	73.5	73.1	79.0	81.6
DEF(19)	**90.1**	<u>87.0</u>	66.3	83.9	78.8	81.9	79.5	72.8	74.6	82.6	84.6
MB(12)	71.1	66.5	54.0	**78.5**	70.8	61.3	<u>72.9</u>	72.6	60.5	72.4	71.3
FM(17)	71.6	66.3	51.8	**76.6**	66.1	65.4	73.0	67.7	57.0	73.2	<u>74.1</u>
IPR(31)	<u>82.8</u>	80.2	75.2	**83.9**	78.8	72.2	75.0	77.7	70.8	74.6	80.9
OPR(39)	**87.6**	<u>84.9</u>	72.0	82.3	77.4	77.8	78.7	77.0	71.5	78.5	<u>84.9</u>
OV(6)	**76.6**	72.8	51.4	69.7	65.0	63.5	70.6	67.8	64.8	66.2	<u>74.4</u>
BC(21)	**84.0**	79.3	69.2	79.0	74.9	71.7	72.7	78.4	72.3	78.8	<u>79.8</u>
LR(4)	80.4	71.7	69.0	<u>97.2</u>	69.5	65.0	76.9	78.1	62.9	65.3	**98.7**

4.3.3. Qualitative Evaluation

Figure 7 illustrates the qualitative comparison of our CSPRF tracker with mentioned 10 trackers on 14 challenging sequences. From these figures, it is clearly observed that our method performs well in all these challenging sequences.

Occlusion. In the Box sequence, the LCT+ quickly drifts to the similar background area from the beginning, and the target is gradually occluded by the Vernier caliper from the 445th frame. When the target reappears in the 490th frame, only our CSPRF, SAMF and MEEM successfully track it while other trackers still stay on the obstruction (Vernier caliper). In the Human3 sequence, LCT+, SAMF, KCF, MEEM, and RPT fail to track the target in the 36th frame. After a short partially occluded duration, all other trackers lose the target as well, only our tracker sticks on it throughout the sequence. In addition, in the Girl2 sequence, only our method can effectively capture it again when the target reappears, while all other compared trackers drift toward the distracter that has the similar appearance as the girl. Here, the success of our tracker is mainly attributed to the confidence updating strategy and the online re-detection mechanism.

Rotation. The target undergoing the in-plane or out-of-plane rotation often causes the variation of target appearance, which will increase the tracking difficulty. In the Skiing sequence, since the target keeps rotating in consecutive frames, most of trackers lose the target in the 19th frame. Only our CSPRF, Staple_CA, and MEEM successfully track the target in the entire tracking period. In the Freeman1 sequence, all the trackers perform well at the beginning, such as frame 30. The target undergoes the out-of-plane rotation at the 140th frame, our tracker and Staple get right estimates in location and scale, and the other trackers all drift to the face of the man, SAMF even loses the target completely. At frame 276, KCF and DSST also lose the target. Another example where the rotation is the main challenge is Sylvester sequence. At frame 1179, only MEEM, LCT+, RPT, and our CSPRF locate the target while other trackers fail to track the target.

Deformation. In Panda sequence, the target suffers from severe deformation. LCT+, SRDCF, KCF, DSST and RPT lose the target at frame 486 and more trackers drift to the background when the panda passes by the tree, whereas our CSPRF, MEEM still track the target (e.g., frames 642, 958). Although the Staple_CA can track the target, it gets inaccurate target location. In Bolt2 sequence, the target undergoes severe deformation as well. Others trackers fail to track the target form the beginning, only our CSPRF, CSR-DCF, Staple and Staple_CA successfully track the target in the whole tracking period. In the Bird2 sequence, many trackers obtain inaccurate target location when the bird turns around at frame 72, and SAMF and DSST fail to track the target at this time. Only our CSPRF, MEEM, and Staple_CA obtain the accurate results in the overall tracking process.

Figure 7. Qualitative evaluation of the proposed algorithm with 10 state-of-the-art methods on 14 challenging video sequences (from left to right and from top to bottom are Box, Human3, Girl2, Skiing, Freeman1, Sylvester, Panda, Bolt2, Bird2, Soccer, Shaking, CarScale, Walking2, and Football, respectively).

Background clutter. The existence of similar-appearing objects to the target in the background makes it challenging to distinguish the target from the background and accurately locate the target. In the Soccer sequence, among all 11 compared trackers, the KCF, LCT+, SRDCF, MEEM, and DSST lose the target at frame 120, and RPT, Staple, and SAMF obtain inaccurate results in terms of location and scale. Only our CSPRF, Staple_CA, and CSR-DCF get the reliable tracking results both in scale and location during the entire tracking period. In the Shaking sequence, Staple_CA, SRDCF, and KCF fail to locate the target and drift to the distracters in the 77th frame. At frame 238, CSR-DCF and SAMF lose the target as well. Only our CSPRF, MEEM, LCT+, and DSST successfully track the target. Although RPT can locate the target, it obtains an incorrect scale estimate. In the Football sequence, most of the compared trackers drift to the distracters at frame 302, only our method, LCT+, MEEM, and SRDCF stick on the target and favorably track the target over all frames.

Scale variation. Due to the KCF and MEEM without handling the scale variation, they do not perform well when the target undergoes large scale variation. The targets in the CarScale and Walking2 sequences undergo the scale variation from beginning to end. In the CarScale sequence, MEEM and KCF obtain inaccurate tracking results in scale in the 174th frame. At frame 205, only our tracker obtains accurate results in scale and location, while many other trackers focus on the head of the car.

In the Walking2 sequence, MEEM, RPT, KCF, and SAMF do not perform well in scale at frame 132. MEEM and LCT+ eventually drift away to the distracter at frame 332. Our tracker with others trackers, including DSST, SRDCF, Staple, Staple_CA, and CSR-DCF, all perform well in scale and location in the whole tracking period.

4.4. Performance Comparison of LGCmF with CSPRF

Since LGCmF exploits the block tracking and response fusion strategies as well, we compare our CSPRF tracker with the LGCmF tracker on OTB2015. Table 3 shows comprehensive performance comparison between these two trackers. Our CSPRF outperforms LGCmF in all evaluation criteria. The reason that our method obtains better results lies in the fact that we adopt a complete block tracking strategy, a novel adaptive hedge algorithm to update the weights and efficient loss metrics in both component trackers.

Table 3. Performance comparison of LGCmF with CSPRF on OTB2015 with the representative mean distance precision (DP) rate at the threshold of 20 pixels, mean overlap success (OS) rate at the threshold of 0.5, median distance precision (DP) rate, median overlap success (OS) rate, median center location error (CLE), and the area under the curve (AUC). Best: bold.

	Mean DP (%)	Mean OS (%)	Median DP (%)	Median OS (%)	Median CLE	AUC
LGCmF	80.6	72.2	82.4	74.1	8.35	59.8
CSPRF	**83.9**	**75.4**	**85.6**	**76.6**	**7.10**	**61.7**

Figure 8 visualizes the tracking results of the LGCmF tracker with our CSPRF tracker on six challenging sequences. CSPRF tracker can perform well when the target objects undergo in-plane rotation (ClifBar), motion blur (BlurCar3), out-of-plane rotation (DragonBaby), background clutter (Dudek), occlusion (Jogging2), and illumination variation (Singer2), whereas the LGCmF fails in all of these sequences.

Figure 8. Visualization of the tracking results of LGCmF and CSPRF trackers on six challenging sequences. (from left to right and from top to bottom are BlurCar3, ClifBar, DragonBaby, Dudek, Jogging2, Singer2).

4.5. Performance Evaluation of Component Trackers CFSP, CHSP

To better understand the improvements of the two component trackers of our CSPRF, in this section we carry out experimental evaluations by comparing with some relevant trackers on OTB2013 and OTB2015.

We compare the tracking performance of CFSP with four relevant trackers, including KCF [6], Staple$_{cf}$ [14], RPT [34], and SAMF [47]. Among them, Staple$_{cf}$ is the part of Staple based on the correlation filter. KCF is the baseline tracker which is used to track each patch in our CFSP. SAMF also employs color names as complementary feature which is the same as our CFSP. In addition, RPT attempts to find the motion trajectory and trackability of random parts.

Figure 9 shows the precision and success plots on the OTB2013 and OTB2015. Overall, our CFSP tracker performs favorably and achieves the best results against the other compared trackers. This demonstrates the effectiveness of the adaptive hedge algorithm and loss terms in CFSP. Specifically, our CFSP significantly improves the Staple$_{cf}$ with gains of 6.0% in the DP score and 2.8% in the AUC score on OTB2013, and with gains of 7.5% in the DP score and 5.0% in the AUC score on OTB2015. Additionally, RPT is also a part of the tracking algorithm based on correlation filters, and our CFSP outperforms the RPT with gains of 1.9% and 5.7% in the DP scores and 2.4% and 5.5% in the AUC scores on OTB2013 and OTB2015, respectively.

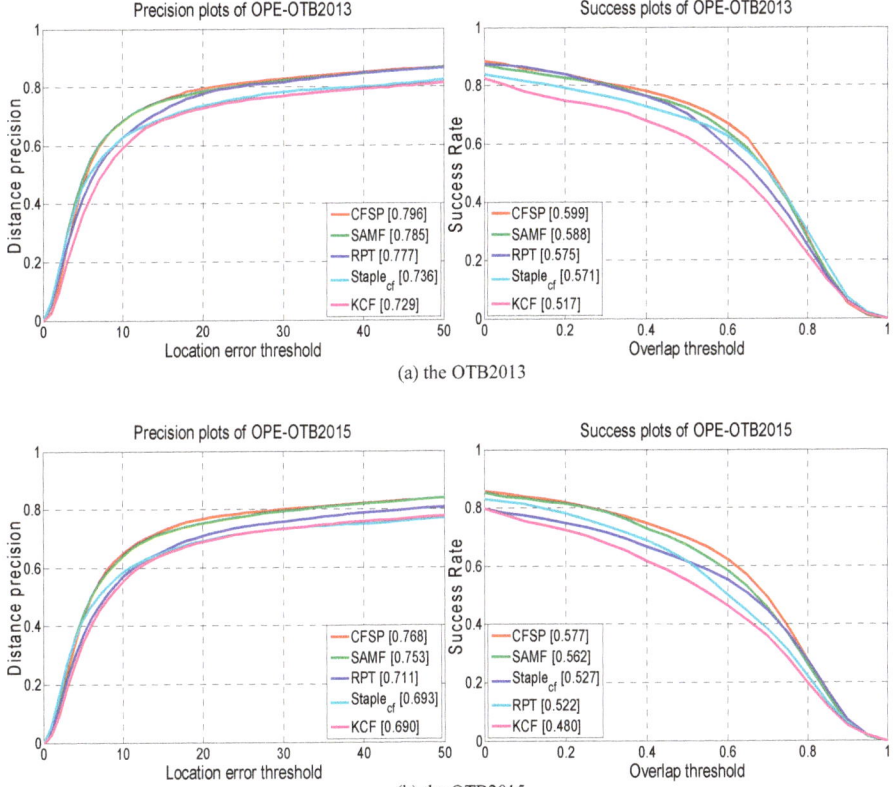

Figure 9. Comparison of the CFSP with four relative trackers on OTB2013 and OTB3015. The legend of the precision plot contains the average DP score at 20 pixels while the legend of success plot contains the area under the curve (AUC) score for each tracker.

We evaluate our component tracker CHSP on OTB2013 and OTB2015 with four relevant trackers including DAT [27], Staple$_{ch}$ [14], PPT [48], and SLC [15]. The Staple$_{ch}$ only contains the part of Staple based on the color histogram. Both PPT and SLC employ part-based color histogram appearance models, while DAT exploits the holistic color histogram appearance model.

Figure 10 visualizes the precision and success plots of our CHSP with four compared trackers. From the figures, we can discover that our CHSP achieves competitive performance against the relevant trackers. Our CHSP is mere inferior to the PPT with losses of 0.5% and 1.8% in the AUC on OTB2013 and OTB2015, respectively. The tracking performance of Staple$_{ch}$ is not satisfactory in the overall evaluation, which ranks at the bottom. Although DAT using the holistic color histogram models owns the similar tracking idea with Staple$_{ch}$, DAT performs better because of adding analysis of the distracters in the tracking process. Specifically, our CHSP outperforms the DAT and Staple$_{ch}$ with gains of 18.4% and 18.8% in the DP scores and 10.4% and 14.4% in the AUC scores on OTB2013 respectively, and with gains of 12.8% and 14.9% in the DP scores and 6.0% and 9.1% in the AUC scores on OTB2015, respectively. SLC employ the same block framework as our CHSP, and its tracking performance has been significantly improved compared to Staple$_{ch}$ and DAT. Our CHSP tracker outperforms the SLC with gains of 5.2% in the DP score and 3.4% in the AUC score on OTB2013, and outperforms the SLC with gains of 3.9% in the DP score and 1.3% in the AUC score on OTB2015. This demonstrates the advantages of the adaptive hedge algorithm and loss terms in CHSP.

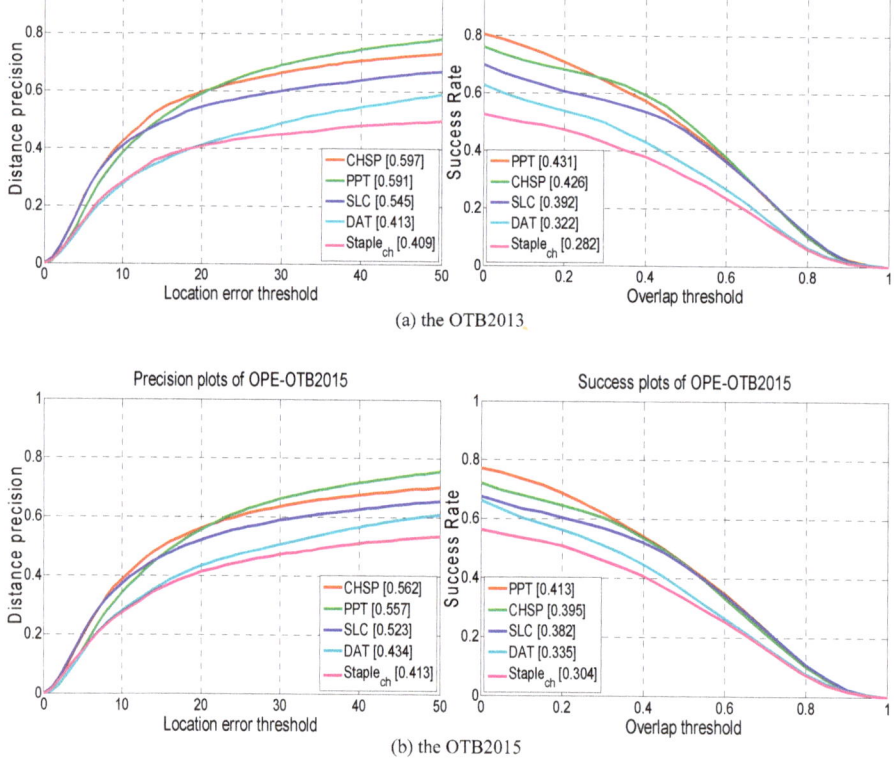

Figure 10. Performance comparison of the CHSP with several relative trackers on OTB2013 and OTB3015. The legend of precision plot contains the average DP score at 20 pixels while the legend of success plot contains the area under the curve (AUC) score for each tracker.

4.6. Performance Evaluation of the CSPRF Tracker on VOT2016

We compare our CSPRF tracker with eight state-of-the-art trackers, including CSR-DCF [28], DAT [27], DSST [21], HCF [40], KCF [6], SRDCF [7], Staple [14], and STRCF [9]. Table 4 lists the tracking results on VOT2016. Our CSPRF performs the second best EAO score of 0.307, only below the CSR-DCF with the best score of 0.332. According to the analysis of [12], the EAO score of our CSPRF is 0.307 which outperforms the definition of the strict state-of-the-art bound 0.251 by 5.6%, and thus it can be regarded as state-of-the-art. And CSPRF achieves some improvement against Staple by a gain of 1.2% in the EAO metric. As for accuracy and robustness, our CSPRF ranks within top three on both two metrics, which demonstrate that our tracker achieves competitive performances against compared trackers.

Table 4. Performance comparison of different trackers on VOT2016 with expected average overlap (EAO), accuracy and robustness. Best: bold, second best: underline.

	CSPRF	CSR-DCF	DAT	DSST	HCF	KCF	SRDCF	Staple	STRCF
EAO	0.307	**0.332**	0.217	0.181	0.220	0.194	0.246	0.295	0.252
Accuracy	0.53	0.52	0.47	0.53	0.45	0.49	0.53	**0.54**	0.51
Robustness	0.97	**0.90**	1.72	2.52	1.42	2.03	1.5	1.35	1.35

5. Conclusions

Based on the success of the Staple tracker, we extend it and propose a novel structural patch complementary tracking algorithm in this paper. We firstly present an adaptive hedge algorithm to overcome the disadvantage of the fixed percentage factor used in the standard hedge algorithm. In the component trackers CHSP and CFSP, we design two reliable loss measurement methods of structural patches, respectively, by which the adaptive hedge algorithm can reliably weigh patches to combine their response maps. The final CSPRF tracker is formulated by selectively merging the response maps of component trackers CHSP and CFSP. In addition, when both of component trackers CHSP and CFSP are unreliable, an online SVM detector is activated to rediscover the target in an extended searching area. Extensive experimental results on OTB2013, OTB2015, and VOT2016 show that the proposed algorithm CSPRF performs favorably against the state-of-the-art trackers in terms of accuracy and robustness. Meanwhile, the CSPRF and the component tracker CHSP have some tracking performance improvements in comparison with the LGCmF and its local layer tracker SLC, respectively. Moreover, the superiorities of two component trackers CHSP and CFSP are justified by comparing with some relevant trackers, in which the CHSP and CFSP have greatly improved in comparison with Staple_{ch} and Staple_{cf}, respectively.

Author Contributions: Z.H. conceived of and performed the experiments, analyzed the data, and wrote the paper. H.Z. reviewed and proofread the manuscript. G.L. supervised the entire study. J.G. provided suggestions for the proposed algorithm. All the authors discussed the results and commented on the manuscript.

Funding: This work is supported by National Natural Science Foundation of China (grant no. 61972307), the Foundation of Preliminary Research Field of China (grant nos. 6140312030217, 61405170206), the 13th Five-Year Equipment Development Project of China (grant no. 41412010202), and the Open Foundation of Shaanxi Key Laboratory of Integrated and Intelligent Navigation under grant no. SKLIIN-20180108.

Conflicts of Interest: The authors declare no conflict of interest.

References

1. Li, X.; Hu, W.; Shen, C.; Zhang, Z.; Dick, A.; Hengel, A.V.D. A survey of appearance models in visual object tracking. *ACM Trans. Intell. Syst. Technol.* **2013**, *4*, 1–48. [CrossRef]
2. Ma, B.; Huang, L.; Shen, J.; Shao, L.; Yang, M.H.; Porikli, F. Visual tracking under motion blur. *IEEE Trans. Image Process.* **2016**, *25*, 5867–5876. [CrossRef] [PubMed]
3. Hare, S.; Saffari, A.; Torr, P.H.S. Struck: Structured output tracking with kernels. In Proceedings of the IEEE International Conference on Computer Vision (ICCV), Barcelona, Spain, 6–13 November 2011; pp. 263–270.
4. Zhang, J.; Ma, S.; Sclaroff, S. MEEM: Robust tracking via multiple experts using entropy minimization. In Proceedings of the European Conference on Computer Vision (ECCV), Zurich, Switzerland, 8–11 September 2014; pp. 188–203.

5. Bolme, D.S.; Beveridge, J.R.; Draper, B.A.; Lui, Y.M. Visual object tracking using adaptive correlation filters. In Proceedings of the IEEE Computer Vision and Pattern Recognition (CVPR), San Francisco, CA, USA, 13–18 June 2010; pp. 2544–2550.
6. Henriques, J.F.; Caseiro, R.; Martins, P.; Batista, J. High-speed tracking with kernelized correlation filters. *IEEE Trans. Pattern Anal. Mach. Intell.* **2015**, *37*, 583–596. [CrossRef] [PubMed]
7. Danelljan, M.; Hager, G.; Khan, F.S.; Felsberg, M. Learning spatially regularized correlation filters for visual tracking. In Proceedings of the IEEE International Conference on Computer Vision (ICCV), Santiago, Chile, 7–13 December 2015; pp. 4310–4318.
8. Mueller, M.; Smith, N.; Ghanem, B. Context-aware correlation filter tracking. In Proceedings of the IEEE Conference on Computer Vision and Pattern Recognition (CVPR), Honolulu, HI, USA, 21–26 July 2017; pp. 1387–1395.
9. Li, F.; Tian, C.; Zuo, W.; Zhang, L.; Yang, M.H. Learning spatial-temporal regularized correlation filters for visual tracking. In Proceedings of the IEEE Conference on Computer Vision and Pattern Recognition (CVRP), Salt Lake City, UT, USA, 18–23 June 2018; pp. 4904–4913.
10. Wu, Y.; Lim, J.; Yang, M.H. Online object tracking: A benchmark. In Proceedings of the IEEE Conference on Computer Vision and Pattern Recognition (CVPR), Portland, OR, USA, 23–28 June 2013; pp. 2411–2418.
11. Wu, Y.; Lim, J.; Yang, M.H. Object tracking benchmark. *IEEE Trans. Pattern Anal. Mach. Intell.* **2015**, *37*, 1834–1848. [CrossRef] [PubMed]
12. Kristan, M.; Leonardis, A.; Matas, J.; Felsberg, M.; Pflugfelder, R.; Cehovin, L.; Vojir, T.; Hager, G.; Lukezic, A.; Fernandez, G.; et al. The visual object tracking VOT2016 challenge results. In Proceedings of the European Conference on Computer Vision Workshops (ECCV), Amsterdam, The Netherlands, 8–16 October 2016; pp. 777–823.
13. Felzenszwalb, P.F.; Girshick, R.B.; McAllester, D.; Ramanan, D. Object detection with discriminatively trained part-based models. *IEEE Trans. Pattern Anal. Mach. Intell.* **2010**, *32*, 1627–1645. [CrossRef] [PubMed]
14. Bertinetto, L.; Valmadre, J.; Golodetz, S.; Miksik, O.; Torr, P.H. Staple: Complementary learners for real-time tracking. In Proceedings of the IEEE Conference on Computer Vision and Pattern Recognition (CVPR), Las Vegas, NV, USA, 27–30 June 2016; pp. 1401–1409.
15. Zhang, H.; Liu, G.; Hao, Z. Robust visual tracking via multi-feature response maps fusion using a collaborative local-global layer visual model. *J. Vis. Commun. Image Represent.* **2018**, *56*, 1–14. [CrossRef]
16. Chaudhuri, K.; Freund, Y.; Hsu, D. A parameter-free hedging algorithm. In Proceedings of the International Conference on Neural Information Processing Systems (NIPS), Vancouver, BC, Canada, 7–10 December 2009; pp. 297–305.
17. Ma, C.; Huang, J.B.; Yang, X.; Yang, M.H. Adaptive correlation filters with long-term and short-term memory for object tracking. *Int. J. Comput. Vis.* **2018**, *126*, 771–796. [CrossRef]
18. Henriques, J.F.; Caseiro, R.; Martins, P.; Batista, J. Exploiting the circulant structure of tracking-by-detection with kernels. In Proceedings of the European Conference on Computer Vision (ECCV), Firenze, Italy, 7–12 October 2012; pp. 702–715.
19. Danelljan, M.; Khan, F.S.; Felsberg, M.; Weijer, J.V.D. Adaptive color attributes for real-time visual tracking. In Proceedings of the IEEE Conference on Computer Vision and Pattern Recognition (CVPR), Columbus, OH, USA, 23–28 June 2014; pp. 1090–1097.
20. Weijer, J.V.D.; Schmid, C.; Verbeek, J.; Larlus, D. Learning color names for real-world applications. *IEEE Trans. Image Process.* **2009**, *18*, 1512–1523. [CrossRef]
21. Danelljan, M.; Hager, G.; Khan, F.S.; Felsberg, M. Accurate scale estimation for robust visual tracking. In Proceedings of the British Machine Vision Conference, Nottingham, UK, 1–5 September 2014; pp. 1–11.
22. Yang, Y.; Zhang, Y.; Li, D.; Wang, Z. Parallel correlation filters for real-time visual tracking. *Sensors* **2019**, *19*, 2362. [CrossRef]
23. Zhang, Y.; Yang, Y.; Zhou, W.; Shi, L.; Li, D. Motion-aware correlation filters for online visual tracking. *Sensors* **2018**, *18*, 3937. [CrossRef]
24. Comaniciu, D.; Ramesh, V.; Meer, P. Kernel-based object tracking. *IEEE Trans. Pattern Anal. Mach. Intell.* **2003**, *25*, 564–577. [CrossRef]
25. Abdelali, H.A.; Essannouni, F.; Essannouni, L.; Aboutajdine, D. Fast and robust object tracking via accept-reject color histogram-based method. *J. Vis. Commun. Image Rep.* **2016**, *34*, 219–229. [CrossRef]

26. Duffner, S.; Garcia, C. PixelTrack: A fast adaptive algorithm for tracking non-rigid objects. In Proceedings of the IEEE International Conference on Computer Vision (ICCV), Sydney, NSW, Australia, 1–8 December 2013; pp. 2480–2487.
27. Possegger, H.; Mauthner, T.; Bischof, H. In defense of color-based model-free tracking. In Proceedings of the IEEE Conference on Computer Vision and Pattern Recognition (CVPR), Boston, MA, USA, 7–12 June 2015; pp. 2113–2120.
28. Lukezic, A.; Vojir, T.; Zajc, L.C.; Matas, J.; Kristan, M. Discriminative correlation filter tracker with channel and spatial reliability. *Int. J. Comput. Vis.* **2018**, *126*, 671–688. [CrossRef]
29. Fan, J.; Song, H.; Zhang, K.; Liu, Q.; Lian, W. Complementary tracking via dual color clustering and spatio-temporal regularized correlation learning. *IEEE Access* **2018**, *6*, 56526–56538. [CrossRef]
30. Nejhum, S.M.S.; Ho, J.; Yang, M.H. Visual tracking with histograms and articulating blocks. In Proceedings of the IEEE Conference on Computer Vision and Pattern Recognition (CVPR), Anchorage, AK, USA, 23–28 June 2008; pp. 1–8.
31. Zhang, T.; Jia, K.; Xu, C.; Ma, Y.; Ahuja, N. Partial occlusion handling for visual tracking via robust part matching. In Proceedings of the IEEE Conference on Computer Vision and Pattern Recognition (CVPR), Columbus, OH, USA, 23–28 June 2014; pp. 1258–1265.
32. Yao, R.; Shi, Q.; Shen, C.; Zhang, Y.; Hengel, A.V.D. Part-based visual tracking with online latent structural learning. In Proceedings of the IEEE Conference on Computer Vision and Pattern Recognition (CVPR), Portland, OR, USA, 23–28 June 2013; pp. 2363–2370.
33. Liu, T.; Wang, G.; Yang, Q. Real-time part-based visual tracking via adaptive correlation filters. In Proceedings of the IEEE Conference on Computer Vision and Pattern Recognition (CVPR), Boston, MA, USA, 7–12 June 2015; pp. 4902–4912.
34. Li, Y.; Zhu, J.; Hoi, S.C.H. Reliable patch trackers: Robust visual tracking by exploiting reliable patches. In Proceedings of the IEEE Conference on Computer Vision and Pattern Recognition (CVPR), Boston, MA, USA, 7–12 June 2015; pp. 353–361.
35. Sun, X.; Cheung, N.M.; Yao, H.; Guo, Y. Non-rigid object tracking via deformable patches using shape-preserved KCF and level sets. In Proceedings of the IEEE International Conference on Computer Vision (ICCV), Venice, Italy, 22–29 October 2017; pp. 5496–5504.
36. Wang, X.; Hou, Z.; Yu, W.; Pu, L.; Jin, Z.; Qin, X. Robust occlusion-aware part-based visual tracking with object scale adaptation. *Pattern Recognit.* **2018**, *81*, 456–470. [CrossRef]
37. Zhang, S.; Lan, X.; Qi, Y.; Yuen, P.C. Robust visual tracking via basis matching. *IEEE Trans. Circuits Syst. Video Technol.* **2017**, *27*, 421–430. [CrossRef]
38. Zhang, L.; Wu, W.; Chen, T.; Strobel, N.; Comaniciu, D. Robust object tracking using semi-supervised appearance dictionary learning. *Pattern Recognit. Lett.* **2015**, *62*, 17–23. [CrossRef]
39. Zhang, S.; Lan, X.; Yao, H.; Zhou, H.; Tao, D.; Li, X. A biologically inspired appearance model for robust visual tracking. *IEEE Trans. Neural Netw. Learn. Syst.* **2017**, *28*, 2357–2370. [CrossRef]
40. Ma, C.; Huang, J.B.; Yang, X.; Yang, M.H. Hierarchical convolutional features for visual tracking. In Proceedings of the IEEE International Conference on Computer Vision (ICCV), Santiago, Chile, 7–13 December 2015; pp. 3074–3082.
41. Qi, Y.; Zhang, S.; Qin, L.; Huang, Q.; Yao, H.; Lim, J.; Yang, M.H. Hedging deep features for visual tracking. *IEEE Trans. Pattern Anal. Mach. Intell.* **2019**, *41*, 1116–1130. [CrossRef]
42. Zhang, S.; Qi, Y.; Jiang, F.; Lan, X.; Yuen, P.C.; Zhou, H. Point-to-set distance metric learning on deep representations for visual tracking. *IEEE Trans. Intell. Transp. Syst.* **2018**, *19*, 187–198. [CrossRef]
43. Danelljan, M.; Bhat, G.; Khan, F.S.; Felsberg, M. Atom: Accurate tracking by overlap maximization. In Proceedings of the IEEE Conference on Computer Vision and Pattern Recognition (CVPR), Long Beach, CA, USA, 16–20 June 2019; pp. 4660–4669.
44. Zhang, S.; Zhou, H.; Yao, H.; Zhang, Y.; Wang, K.; Zhang, J. Adaptive NormalHedge for robust visual tracking. *Signal Process.* **2015**, *110*, 132–142. [CrossRef]
45. Collins, R.T.; Liu, Y.; Leordeanu, M. Online selection of discriminative tracking features. *IEEE Trans. Pattern Anal. Mach. Intell.* **2005**, *27*, 1631–1643. [CrossRef] [PubMed]
46. Crammer, K.; Dekel, O.; Keshet, J.; Shalev-Shwartz, S.; Singer, Y. Online passive-aggressive algorithms. *J. Mach. Learn. Res.* **2006**, *7*, 551–585.

47. Li, Y.; Zhu, J. A scale adaptive kernel correlation filter tracker with feature integration. In Proceedings of the European Conference on Computer Vision Workshops (ECCV), Zurich, Switzerland, 6–12 September 2014; pp. 254–265.
48. Lee, D.Y.; Sim, J.Y.; Kim, C.S. Visual tracking using pertinent patch selection and masking. In Proceedings of the IEEE Conference on Computer Vision and Pattern Recognition (CVPR), Columbus, OH, USA, 23–28 June 2014; pp. 3486–3493.

© 2019 by the authors. Licensee MDPI, Basel, Switzerland. This article is an open access article distributed under the terms and conditions of the Creative Commons Attribution (CC BY) license (http://creativecommons.org/licenses/by/4.0/).

Article

Multiple Event-Based Simulation Scenario Generation Approach for Autonomous Vehicle Smart Sensors and Devices

Jisun Park, Mingyun Wen, Yunsick Sung and Kyungeun Cho *

Department of Multimedia Engineering, Dongguk University-Seoul, Seoul 04620, Korea; jisun@dongguk.edu (J.P.); wmy_dongguk@dongguk.edu (M.W.); sung@mme.dongguk.edu (Y.S.)
* Correspondence: cke@dongguk.edu; Tel.: +82-2-2260-3834

Received: 9 September 2019; Accepted: 12 October 2019; Published: 14 October 2019

Abstract: Nowadays, deep learning methods based on a virtual environment are widely applied to research and technology development for autonomous vehicle's smart sensors and devices. Learning various driving environments in advance is important to handle unexpected situations that can exist in the real world and to continue driving without accident. For training smart sensors and devices of an autonomous vehicle well, a virtual simulator should create scenarios of various possible real-world situations. To create reality-based scenarios, data on the real environment must be collected from a real driving vehicle or a scenario analysis process conducted by experts. However, these two approaches increase the period and the cost of scenario generation as more scenarios are created. This paper proposes a scenario generation method based on deep learning to create scenarios automatically for training autonomous vehicle smart sensors and devices. To generate various scenarios, the proposed method extracts multiple events from a video which is taken on a real road by using deep learning and generates the multiple event in a virtual simulator. First, Faster-region based convolution neural network (Faster-RCNN) extracts bounding boxes of each object in a driving video. Second, the high-level event bounding boxes are calculated. Third, long-term recurrent convolution networks (LRCN) classify each type of extracted event. Finally, all multiple event classification results are combined into one scenario. The generated scenarios can be used in an autonomous driving simulator to teach multiple events that occur during real-world driving. To verify the performance of the proposed scenario generation method, experiments using real driving video data and a virtual simulator were conducted. The results for deep learning model show an accuracy of 95.6%; furthermore, multiple high-level events were extracted, and various scenarios were generated in a virtual simulator for smart sensors and devices of an autonomous vehicle.

Keywords: scenario generation; autonomous vehicle; smart sensor and device; deep learning

1. Introduction

Recently, autonomous vehicles have been a big trend in the development of advanced countries worldwide [1–3]. Especially, studies on the perception system of an autonomous vehicle using smart sensors and devices are being active widely because perception is one of key element of autonomous vehicles. Recently in the autonomous vehicle industry, smart sensors and devices of autonomous vehicles have been trained via virtual self-driving simulators that apply the deep learning technique to reduce development costs and time and secure safety [4–10]. The virtual autonomous driving simulators provide color image (RGB), depth, Lidar, and radar data to train autonomous vehicle's smart devices and sensors [4,5]. To enable an autonomous vehicle to run in real environments, it is critical to train a self-driving car for a variety of driving environments in advance. Furthermore, it is also essential to learn scenarios reflecting a wide range of situations that may occur in the real world.

As an example, when an autonomous vehicle runs on a road in an urban area, the car needs to be trained for scenarios with several people walking on the streets. When an autonomous vehicle runs on an expressway, the car must be trained for scenarios of diverse types of situations that can occur by interaction among cars on the expressway.

Existing studies are based on the scenario generation approaches for autonomous vehicle generated scenarios based on real driving data acquired from the real environment or by using self-driving scenario generation modeling based on expert knowledge [11–16]. However, such approaches require a high ratio of manual processing, which increases the development costs and time for the self-driving simulator. Thus, it is beneficial to investigate the approach for generating the scenarios by using deep learning video analysis for automatically generating a wide range of realistic driving scenarios through the collection and analysis of real driving data without scenario generation modelling.

The deep learning approach for analyzing driving data is limited as it can only analyze the actions of one object [17,18]. As an example, when two individuals are talking and walking, and extraction is to be performed based on a single object, only two walking individuals can be extracted. Such an approach cannot analyze advanced events, including multiple objects and interaction.

This paper proposes an approach to generate the training scenario for autonomous vehicle smart sensors and devices including multiple events while considering multiple objects based on the automatic analysis of a driving video by using two types of deep learning approaches. An event comprises the list of objects included in one specific situation and the actions of each object.

The first step is to extract the areas of objects existing in a driving video input to Faster-region based convolution neural network (Faster-RCNN) [19]. Faster-RCNN is real-time object detection network. Next, the high-level event area is estimated while considering the extracted areas of objects. Then, the events are analyzed using long-term recurrent convolution networks (LRCN) [20] based on the high-level event areas extracted. LRCN classifies the video class by convolutional neural network (CNN) and long short-term memory (LSTM). Finally, the analyzed events are integrated into one scenario. The generated scenario is delivered to the virtual simulator for the learning of an autonomous vehicle, and the relevant scenario is deployed in front of an autonomous vehicle.

This paper contributes to future research as follows. First, a scenario was successfully generated via automatic analysis using deep learning for training and testing of autonomous vehicle's smart sensors and devices. Next, the approach enables the sophisticated analysis of events including interactions among multiple objects as well as the analysis of only a single action by each object. Finally, it is possible to generate higher-level scenarios including multiple events.

Section 2 in this paper describes the existing research on scenario generation for an autonomous vehicle and the video analysis approach based on deep learning. Section 3 discusses the scenario generation approach proposed in this paper, which extracts high-level events using deep learning-based video analysis. Section 4 describes the experiments on the proposed approach and the results, and Section 5 presents the conclusion and directions for further study.

2. Related Works

This section summarizes the existing studies on driving scenario generation approaches and deep learning-based driving video analysis approaches. Then, the necessity for the approach proposed herein is explained.

2.1. Driving Scenario Generation Approach

Several driving simulators has been investigated for development and verification of an autonomous vehicle. The field of driving scenario generation for the operation of autonomous vehicles has recently drawn substantial attention [11–16]. Research on driving scenario generation is largely classified into model-based and data-based scenario generation.

The model-based scenario generation approach defines driving elements, including traffic lane, car, pedestrian, and accident events, in advance as well as scenarios depending on those elements.

In [11] the authors plan movements and generate scenarios by using the action tree of each car based on the accident scenario defined in scripts, and the research in [12] predefines the accident scenario between a car and a pedestrian in the intersection and generates the scenarios. In [13] the authors generate the scenarios based on an analysis of real car accidents and survey data from 'NMVCCS' and in [14] the authors implement the ontology on the driving environment and generate scenarios based on that ontology. For the model-based scenario generation approach above, a more complicated scenario requires higher scenario modeling time and cost. Moreover, it is very difficult to modify or supplement a scenario after it is generated using the approach above.

The data-based scenario generation approach generates a scenario only from real driving data. The research presented in [15] generates the scenarios by using data recorded by experts after analyzing information on a lane type, car, and pedestrian based on a driving video recorded for 30 h on a real road. In [16] the authors acquire real driving data by using laser sensors and cameras and apply the data to the virtual environment simulator. The scenario generation based on real driving data as explained above enables an autonomous vehicle to learn practical scenarios but has disadvantages related to the required time and cost of obtaining real driving data

For model-based and data-based driving scenario generation approaches as described above, it is inevitable that the more diverse types of scenarios that are generated, the greater the required time and cost. Accordingly, this paper attempts to address the disadvantages of existing studies by developing an approach to automatically analyze real driving data and generate diverse types of scenarios, including multiple events using analysis results.

2.2. Deep Learning-Based Driving Video Analysis Approach

The studies analyzing videos by using deep learning have been conducted actively [17,18], and in particular, the dataset for training an autonomous vehicle has been continuously increasing [19,20]. Most studies analyzing deep learning-based videos extracted a specific vector from a series of video frames by using a CNN and integrated the extracted specific vectors around the time axis. However, most studies extracted each object and analyzed only the actions of that object. Furthermore, only one event was analyzed per video. The research presented in [17] extracted RGB image-specific vectors and optical flow vectors per frame by using CNN, entered that extracted specific vectors into CNN to fuse two vectors, and classified it into one event class by using support vector machine (SVM). In [18] the authors extracted specific vectors from RGM images and optical flow images per frame in the video and segmented trajectory data by using CNN. Subsequently, it integrated and estimated three specific vectors and classified the result as one event by SVM.

As described above, the existing studies on deep learning-based driving video analysis analyzed the actions of only one object, rather than advanced events including interaction. They could analyze only a single event per video. This paper proposes an approach to extract and analyze multiple events that are more advanced.

3. Multi-Event-Based Scenario Generation Approach

This paper proposes an approach to generate scenarios for training autonomous vehicle's smart sensors and devices by extracting and analyzing multiple events from driving video using deep learning methods. Figure 1 illustrates the scenario generation process based on the deep learning video analysis approach proposed in this paper. The first step is to extract a high-level event area to detect the objects existing in a video by using Faster-RCNN, which is optimal for detecting objects with the first frame of the input. The objects whose bounding boxes overlap among detected objects are extracted as one event area. Next, the scenario generation step analyzes the images extracted based on the event area in the previous step by using LRCN, which is a type of deep learning-based video classification model, and generates the scenarios for self-driving learning based on the analysis. The generated scenario is finally used as the input data for the self-driving simulator. In the proposed approach, the events are presented as the list of objects and high-level event class included in the relevant events, and scenarios are presented as the list of events. Figure 1 shows the entire process of the proposed approach.

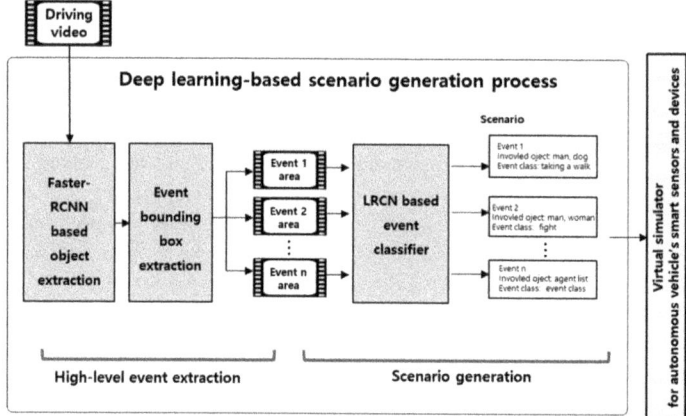

Figure 1. Proposed multi-event-based scenario generation approach.

3.1. High-Level Event Area Extraction Step

Figure 2 shows the process to extract more optimum event areas in the driving data. The process comprises the object detection and event area integration in that sequence. For the object detection task, the first image (frame) is received using the Faster-RCNN approach, and the areas of dynamic objects such as a person, car, and animal, which can be the subject of an event, are extracted. The Faster-RCNN includes Convolutional neural network (ConvNet), Region proposal network (RPN), Region of Interest pooling, regression and classification layer. After extracting an event bounding box based on a single object area, it is difficult to extract high-level events including interactions between objects. The proposed approach enables the extraction of higher-level event areas by integrating neighboring single object bounding boxes into one even bounding box.

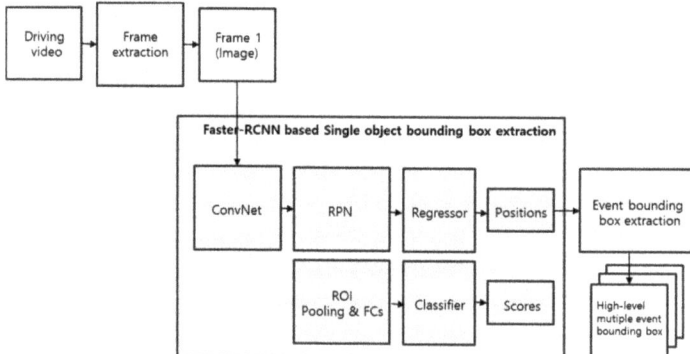

Figure 2. Multiple high-level events extraction process.

Figure 3 illustrates the approach to integrate the object areas detected using Faster-RCNN into the high-level event area. The first step sorts the boxes whose areas are overlapped among bounding boxes of detected objects. Next, the top and left sides of an event bounding box are set to the minimum value among the bounding boxes of overlapped objects, and the right and bottom sides are set to the maximum value among the bounding boxes of overlapped objects. Algorithm 1 is the algorithm to integrate event areas. The overlapped bounding boxes of objects are integrated into one event bounding box through Algorithm 1; subsequently, multiple event areas are extracted based on the integration results.

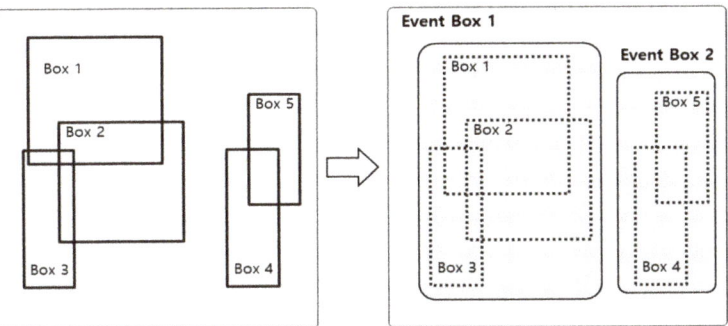

Figure 3. High-level event area extraction process.

Algorithm 1. Event Area Integration Algorithm

E: *An event includes the list of objects included in the event and the class of the event*
O: *Objects including persons, animals, or cars*
Initialize E
GET O
  For each i in O
    IF $E = [\,]$ THEN increment new e
    For each j in E
      IF O_i overlaps E_j THEN
        IF $O_i > E_j$ THEN
          merge O_j into E_j
    ELSE increment new e
    ENDFOR
  ENDFOR

3.2. Scenario Generation Step

The scenario generation process based on multiple event images extracted comprises the LRCN-based event classification task and the scenario generation task depending on the classification results. As shown in Figure 4, the deep learning model structure classifying events based on LRCN comprises the combination of CNN extracting the features of the extracted images and LSTM learning the sequential data. The specific feature vectors per frame are extracted via CNN after receiving individual frames of each event image based on the extracted event areas. Next, the result values acquired after entering the specific vectors per frame to LSTM in consecutive order, which are classified into the event label via the Fully Connected Layer. As the event areas include only a part of the full image, the specific vectors in the first frame of the original video on the full area as well as the feature value of event area frame are entered into the last Fully Connected Layer to include the features of full images, including weather and road type.

Multiple event images are classified by repeating the process above and stored as one scenario. A scenario is the list of events, and each event includes the types of objects contained in the relevant event and the high-level event class of the relevant event. A list of scenario elements is presented in Table 1.

After a scenario is generated in the structure described above, the relevant data is transferred to the virtual simulator, as illustrated in Figure 5. The input scenarios execute the events in front of an autonomous vehicle depending on the object list and action contained in each event. The virtual simulator operates the input scenario and then the autonomous vehicle learn the scenario by training their virtual sensing device data such as RGB-D, Lidar, and Radar data.

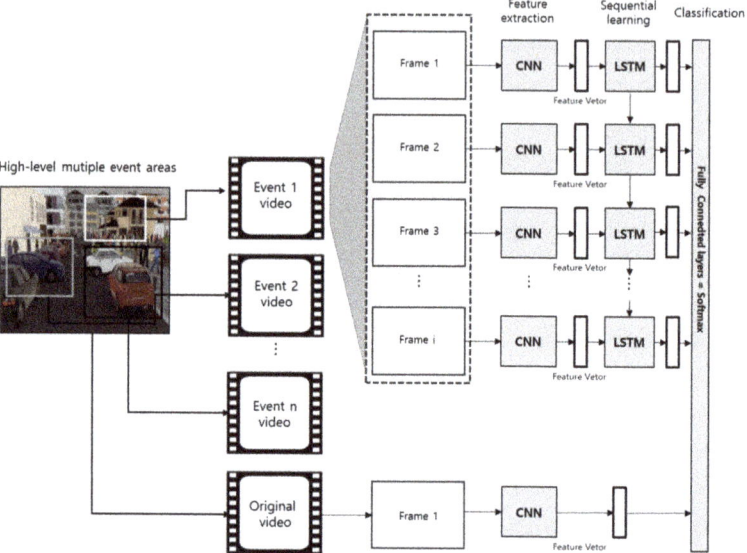

Figure 4. Long-term recurrent convolution networks (LRCN)-based event classification model structure.

Table 1. List of scenario elements.

Elements	Symbols	Description
Scenario	s = (e list)	One scenario includes multiple events
Event	e = (o list, event class)	An event includes the list of objects included in the event and the class of the event
Object	o = object	Objects including persons, animals, or cars
High-level event class	c = event class	Types of events occurring in the driving video

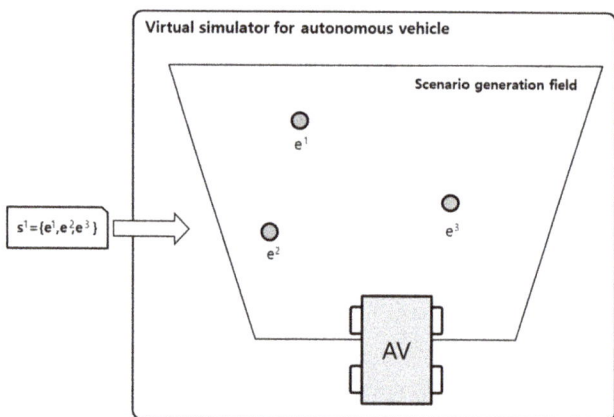

Figure 5. Execution of multiple events in a virtual simulator through the scenario input.

4. Experiments and Analysis

This section describes the experiments and analysis of the scenario generation approach based on the deep learning image analysis proposed herein to verify its performance. To this end, the experimental environment is described and learning data is presented. The results of the algorithm extracting multiple event areas are compared to those of the existing Faster-RCNN. Next, the image analysis algorithm performance proposed herein is compared to that of the existing RCNN algorithm and analyzed. Finally, the final extracted scenario was executed in the simulator, which was constructed for the experiment, and the results are analyzed.

4.1. Experiment Environment and Training Data

The proposed method's development environment was implemented on a computer with Intel i5, Nvidia GTX 1070 GPU, and DDR 5 H/W. The scenario generation model utilizing the deep learning-based video analysis was implemented in Keras (Backend-Tensorflow), which is a deep learning library. The scenario generated using the proposed approach was finally applied to the virtual simulator, which was made by us, based on Unity for autonomous vehicle's smart sensors and devices to train. Artificial intelligence objects such as people, animals, and cars exist in the virtual simulator and act based on artificial intelligence according to the input scenario. Based on the input scenario, human, animal, and vehicle agents are operated in front of an autonomous car. The autonomous vehicle's virtual sensing device is trained by using RGB, depth, Lidar, and Radar data. Figure 6 shows the virtual simulator environment screenshot.

Figure 6. Virtual simulator environment for an autonomous vehicle to train.

Studies analyzing driving videos via deep learning have been actively conducted using public driving datasets [21,22]. However, the public driving data have only single action labels. Accordingly, the experiment in this paper collected videos, including events that occurred on roads or streets, and labelled their ground truth. In total, 725 videos were collected and classified into 23 classes. Table 2 summarizes the event class types. The event classes have high-level event classes, including single actions of cars, animals, and people and the interactions among them.

As shown in Table 3, nine object types were identified from the analysis on the objects included in each event.

Table 2. List of event types.

High-Level Event Class	No. of Video Clips
human_push_car	36
human_motocyling	27
human_hugging	85
vehicle_changeLane	16
human_wave_hand	20
human_pet_animal	12
vehicle_turn	24
human_checkVictim	22
vehicle_stop	50
human_walk	53
vehicle_pass_by	70
human_crossroad	39
human_run	13
human_wait	81
human_getoff	15
human_check Car	5
human_use_phone	7
human_fight	10
human_phone_call	27
human_talk	12
human_smoking	53
human_trash_collecting	24
human_sit	24
23	725

Table 3. List of object types.

Object Types (Total Nine Types)
Person, car, bike, motorbike, bus, truck, bird, cat, dog

4.2. High-Level Event Area Extraction Results

This subsection analyzes the Faster-RCNN-based event image extraction results. Although only the areas of each object are extracted, as shown in Figure 7, when extracting event areas only by using the existing Faster-RCNN, it is verified that the high-level event areas including objects that are correlated one another are extracted when the event area integration algorithm is applied as well, as shown in Figure 8.

Figure 7. Faster-region based convolution neural network (RCNN) results.

Figure 8. Faster-RCNN based event area integration algorithm results.

4.3. LRCN-Based Event Classification Result

To analyze the extracted event images, the image analysis model was implemented based on the LRCN model combining CNN and LSTM. Next, an autonomous vehicle learned using the collected data and the accuracy of event classification was evaluated. Cross-validation, one of the methods to measure the effectiveness of classification performance in the field of computer vision recognition, was adopted to verify the learning model in this study. Cross-validation is a representative method to measure the accuracy by comparing the estimates with actual values when verification data is entered into the model after learning. Table 4 presents the confusion matrix with estimates and actual values.

Table 4. Confusion of estimates and actual values.

Confusion Matrix		Actual Values	
		Positive	Negative
Estimates	Positive	True Positive (TP)	False Positive (FP)
	Negative	False Negative (FN)	True Negative (TN)

Accuracy indicates how close the measured values are to the true values. Equation (1) estimates the accuracy based on the confusion matrix in Table 4.

$$\text{Accuracy} = \frac{TP + TN}{TP + FN + FP + FN}. \quad (1)$$

We applied Inception-v3 [23] which is a pre-trained model of CNN to LRCN. The training data was divided into 600 for training and 125 for testing. The input data size is 240 × 240 and the batch sizes are 34 for 200 epochs and two for 600 epochs. Figure 9 shows the confusion matrix of the result.

The Table 5 presents the comparison results of LRCN and the proposed method. The proposed approach's classification accuracy exceeds 96.5%.

Table 5. Comparison results of classification models.

	LRCN [19]	LRCN + Full Area	LRCN (Inception-v3) + Full Area (Proposed Approach)
Classification Accuracy	78.2	80.5	95.6

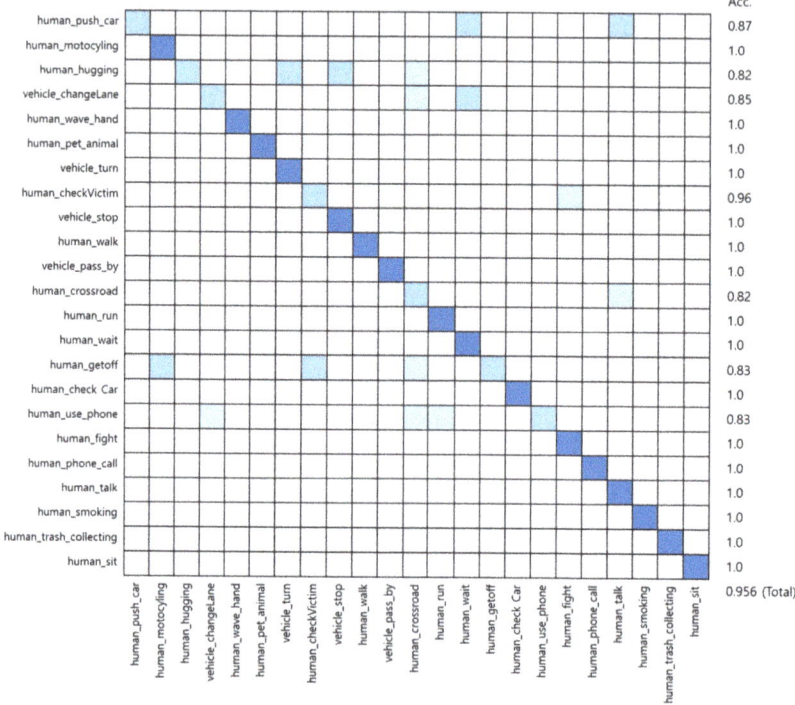

Figure 9. Confusion matrix of the LRCN result.

4.4. Scenario Generation and Implementation Results

Using the trained proposed model, we generated four scenarios as below. Tables 6–9 show the input original video data which was taken in real world and the scenario generated by analyzing the driving video using deep learning and the results of implementing that scenario in the simulator. The implementation of this scenario verified that the objects detected from real driving data were analyzed per event unit and saved to the scenario file, and relevant multiple events were generated through artificial intelligence objects in the virtual simulator based on the scenario.

Table 6. Scenario Generation Result #1.

Table 6. *Cont.*

Output data (= scenario)
s1 = {e1(human_motocyling), e2(vehicle_changeLane)}
Final result in simulator

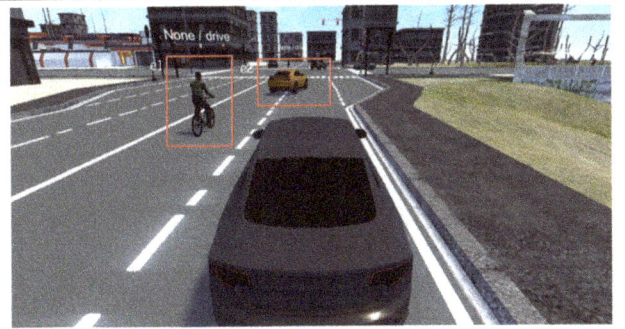

Table 7. Scenario Generation Result #2.

Input Data

Output data (= scenario)
s2 = {e1(vehicle_turn), e2(vehicle_stop)}
Final result in simulator

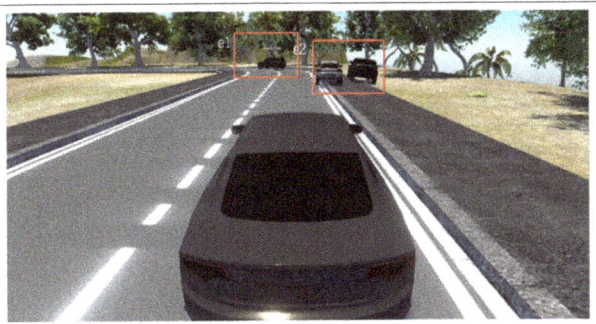

Table 8. Scenario Generation Result #3.

Input Data

Output data (= scenario)
s3 = {e1(vehicle_stop), e2(human_motocyling e3(vehicle_stop)}
Final result in simulator
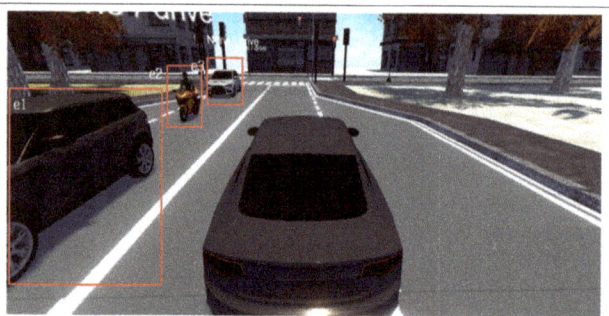

Table 9. Scenario Generation Result #4.

Input data

Output data (= scenario)
s4 = {e1(human_pet_animal), e2(vehicle_stop)}
Final result in simulator

Table 9. *Cont.*

5. Conclusions

This paper proposed an approach to automatically analyze real driving data by using the deep learning image analysis method without complicated scenario generation modeling and then generated a scenario for smart sensor and devices of an autonomous vehicle such as camera, Lidar, or Radar to train in virtual simulator, including multiple events based on the automatic analysis results. The approach proposed in this paper includes multiple events by extracting them from one driving image and enables high-level event analysis, including interactions among objects, not rather than analyzing only a single action of an object.

The experiment achieved an accuracy of 95.6% by training the model using the dataset constructed in this paper and evaluating the event analysis model classified into 23 classes. Accordingly, it was verified that multiple high-level events could be acquired from a single video as compared to the existing deep-learning algorithm. Furthermore, it was observed that multiple events extracted were saved as one scenario and executed in a similar manner as the input driving data in the virtual self-driving simulator.

Further studies must investigate extraction of dynamic events while tracking dynamically moving objects by analyzing all consecutive frames of a video when extracting the event areas. Moreover, further studies will attempt to determine the approach to enable analysis on a wide range of elements, including the movement direction and speed of an object, individual actions, weather, and road conditions included in the events as well as the event types.

Author Contributions: J.P. and M.W. wrote the paper and the source codes. Y.S. supervised experiment and editing. K.C. provided full guidance. All authors read and approved the final manuscript.

Funding: This research was supported by a grant from Defense Acquisition Program Administration and Agency for Defense Development, under contract #UE171095RD and the MSIT (Ministry of Science, ICT), Korea, under the High-Potential Individuals Global Training Program) (2019-0-01585) supervised by the IITP (Institute for Information & Communications Technology Planning & Evaluation).

Conflicts of Interest: The authors declare no conflict of interest.

References

1. Rasouli, A.; Tsotsos, J.K. Autonomous vehicles that interact with pedestrians: A survey of theory and practice. *IEEE Trans. Intell. Transp. Syst.* **2019**. [CrossRef]
2. Shi, W.; Alawieh, M.B.; Li, X.; Yu, H. Algorithm and hardware implementation for visual perception system in autonomous vehicle: A survey. *Integration* **2017**, *59*, 148–156. [CrossRef]
3. Okuda, R.; Kajiwara, Y.; Terashima, K. A survey of technical trend of ADAS and autonomous driving. In Proceedings of the Technical Papers of 2014 International Symposium on VLSI Design, Automation and Test, Hsinchu, Taiwan, 28–30 April 2014.
4. Dosovitskiy, A.; Ros, G.; Codevilla, F.; Lopez, A.; Koltun, V. CARLA: An open urban driving simulator. *arXiv* **2017**, arXiv:1711.03938.

5. Shah, S.; Dey, D.; Lovett, C.; Kapoor, A. Airsim: High-fidelity visual and physical simulation for autonomous vehicles. In *Field and Service Robotics*; Springer: Berlin, Germany, 2018; pp. 621–635.
6. Pan, X.; You, Y.; Wang, Z.; Lu, C. Virtual to real reinforcement learning for autonomous driving. *arXiv* **2017**, arXiv:1704.03952.
7. Hong, Z.W.; Yu-Ming, C.; Su, S.Y.; Shann, T.Y.; Chang, Y.H.; Yang, H.K.; Ho, B.H.; Tu, C.-C.; Chang, Y.-C.; Hsiao, T.-C.; et al. Virtual-to-real: Learning to control in visual semantic segmentation. *arXiv* **2018**, arXiv:1802.00285.
8. Li, P.; Liang, X.; Jia, D.; Xing, E.P. Semantic-aware grad-gan for virtual-to-real urban scene adaption. *arXiv* **2018**, arXiv:1801.01726.
9. Shalev-Shwartz, S.; Shammah, S.; Shashua, A. Safe, multi-agent, reinforcement learning for autonomous driving. *arXiv* **2016**, arXiv:1610.03295.
10. Song, W.; Zou, S.; Tian, Y.; Fong, S.; Cho, K. Classifying 3D objects in LiDAR point clouds with a back-propagation neural network. *Hum. Centric Comput. Inf. Sci.* **2018**, *8*, 29. [CrossRef]
11. Gajananan, K.; Nantes, A.; Miska, M.; Nakasone, A.; Prendinger, H. An experimental space for conducting controlled driving behavior studies based on a multiuser networked 3D virtual environment and the scenario markup language. *IEEE Trans. Hum. Mach. Syst.* **2013**, *43*, 345–358. [CrossRef]
12. Xu, Y.; Yan, X.; Liu, D.; Xiang, W. Driving Scenario Design for Driving Simulation Experiments Based on Sensor Trigger Mechanism. *Inf. Technol. J.* **2012**, *11*, 420–425. [CrossRef]
13. Chrysler, S.T.; Ahmad, O.; Schwarz, C.W. Creating pedestrian crash scenarios in a driving simulator environment. *Traffic Inj. Prev.* **2015**, *16*, S12–S17. [CrossRef] [PubMed]
14. McDonald, C.C.; Tanenbaum, J.B.; Lee, Y.C.; Fisher, D.L.; Mayhew, D.R.; Winston, F.K. Using crash data to develop simulator scenarios for assessing novice driver performance. *Transp. Res. Rec.* **2012**, *2321*, 73–78. [CrossRef] [PubMed]
15. Van der Made, R.; Tideman, M.; Lages, U.; Katz, R.; Spencer, M. Automated generation of virtual driving scenarios from test drive data. In Proceedings of the 24th International Technical Conference on the Enhanced Safety of Vehicles (ESV), Gothenburg, Sweden, 8–11 June 2015.
16. Bagschik, G.; Menzel, T.; Maurer, M. Ontology based scene creation for the development of automated vehicles. In Proceedings of the 2018 IEEE Intelligent Vehicles Symposium (IV), Changshu, China, 26–30 June 2018.
17. Kataoka, H.; Satoh, Y.; Aoki, Y.; Oikawa, S.; Matsui, Y. Temporal and fine-grained pedestrian action recognition on driving recorder database. *Sensors* **2018**, *18*, 627. [CrossRef] [PubMed]
18. Kataoka, H.; Suzuki, T.; Oikawa, S.; Matsui, Y.; Satoh, Y. Drive video analysis for the detection of traffic near-miss incidents. In Proceedings of the 2018 IEEE International Conference on Robotics and Automation (ICRA), Brisbane, Australia, 21–25 May 2018.
19. Ren, S.; He, K.; Girshick, R.; Sun, J. Faster r-cnn: Towards real-time object detection with region proposal networks. In Proceedings of the neural information processing systems, Montréal, QC, Canada, 7–12 December 2015.
20. Donahue, J.; Anne Hendricks, L.; Guadarrama, S.; Rohrbach, M.; Venugopalan, S.; Saenko, K.; Darrell, T. Long-term recurrent convolutional networks for visual recognition and description. In Proceedings of the IEEE conference on computer vision and pattern recognition, Boston, MA, USA, 7–12 June 2015.
21. Geiger, A.; Lenz, P.; Urtasun, R. Are we ready for autonomous driving? the kitti vision benchmark suite. In Proceedings of the 2012 IEEE Conference on Computer Vision and Pattern Recognition, Washington, DC, USA, 16–21 June 2012.
22. Cordts, M.; Omran, M.; Ramos, S.; Rehfeld, T.; Enzweiler, M.; Benenson, R.; Schiele, B. The cityscapes dataset for semantic urban scene understanding. In Proceedings of the IEEE conference on computer vision and pattern recognition, Las Vegas, NV, USA, 26 June–1 July 2016.
23. Szegedy, C.; Vanhoucke, V.; Ioffe, S.; Shlens, J.; Wojna, Z. Rethinking the inception architecture for computer vision. In Proceedings of the IEEE conference on computer vision and pattern recognition, Las Vegas, NV, USA, 26 June–1 July 2016.

© 2019 by the authors. Licensee MDPI, Basel, Switzerland. This article is an open access article distributed under the terms and conditions of the Creative Commons Attribution (CC BY) license (http://creativecommons.org/licenses/by/4.0/).

Article

Distributed Reliable and Efficient Transmission Task Assignment for WSNs

Xiaojuan Zhu [1], Kuan-Ching Li [2,*], Jinwei Zhang [1] and Shunxiang Zhang [1]

[1] School of Computer Science and Engineering, Anhui University of Science and Technology, Huainan 554, China; xjzhu@aust.edu.cn (X.Z.); jwzhang@aust.edu.cn (J.Z.); sxzhang@aust.edu.cn (S.Z.)
[2] Dept. of Computer Science and Information Engineering, Providence University, Taichung 43301, Taiwan
* Correspondence: kuancli@pu.edu.tw; Tel.: +886-4-2632-8001

Received: 4 September 2019; Accepted: 10 November 2019; Published: 18 November 2019

Abstract: Task assignment is a crucial problem in wireless sensor networks (WSNs) that may affect the completion quality of sensing tasks. From the perspective of global optimization, a transmission-oriented reliable and energy-efficient task allocation (TRETA) is proposed, which is based on a comprehensive multi-level view of the network and an evaluation model for transmission in WSNs. To deliver better fault tolerance, TRETA dynamically adjusts in event-driven mode. Aiming to solve the reliable and efficient distributed task allocation problem in WSNs, two distributed task assignments for WSNs based on TRETA are proposed. In the former, the sink assigns reliability to all cluster heads according to the reliability requirements, so the cluster head performs local task allocation according to the assigned phase target reliability constraints. Simulation results show the reduction of the communication cost and latency of task allocation compared to centralized task assignments. Like the latter, the global view is obtained by fetching local views from multiple sink nodes, as well as multiple sinks having a consistent comprehensive view for global optimization. The way to respond to local task allocation requirements without the need to communicate with remote nodes overcomes the disadvantages of centralized task allocation in large-scale sensor networks with significant communication overheads and considerable delay, and has better scalability.

Keywords: wireless sensor networks; task assignment; distributed; reliable; energy-efficient

1. Introduction

Task assignment is an essential issue in wireless sensor networks (WSNs). In such multi-sensor systems, task allocation is based on the deadline and the priority of tasks, and different tasks are reasonably assigned to the sensor nodes for best perception performance. However, task allocation schemes that only aim at reducing energy consumption are not sufficient for several applications with high-reliability requirements. That is, whether tasks with high-reliability requirements in WSNs such as measurement, monitoring, and configuration can be completed in a timely manner depends on how the transmission tasks in WSNs can be reliably and efficiently assigned. As is known, energy-savings and high reliability are two conflicting goals, and it is thus challenging to assign high reliability and low energy tasks to appropriate nodes in WSNs.

In recent years, task priority assignment in WSNs has attracted the attention of researchers and satisfactory research results have been achieved. Most existing methods for practical applications are centralized algorithms, as the centralized task assignment method based on a single sink node still has the problem of poor reliability and scalability, so offline execution is selected due to computational complexity. Besides, the communication overhead and delay of task assignment will be significant in large-scale network scenarios or frequent network and dynamic changes. Therefore, it is very urgent to investigate novel distributed, reliable, and efficient task assignment methods for WSNs.

Aiming to solve the reliable and efficient distributed task assignment problem in WSNs, task-aware reliable and efficient task allocation for WSNs is investigated in this paper. The contributions of this paper are summarized as follows:

1) A global network view consisting of a multi-level view (including physical topology, routing topology, and task view) is constructed as a conceptual basis for global optimization;
2) A reliability evaluation model is established from the perspective of the task, and the constraint factors and objective functions of the task allocation are analyzed;
3) We propose the transmission-oriented reliable and energy-efficient task allocation (TRETA)-Cluster, in which the sink assigns reliability to all cluster heads according to the reliability requirements, and the cluster head performs local task assignment according to the assigned phase target reliability constraints. Simulation results show that TRETA-Cluster reduces the communication cost and latency of task assignment compared to centralized task allocation;
4) We propose the TRETA-Multi-Sink, where the global view is obtained by fetching local views from multiple sinks. Multiple sinks have a consistent comprehensive view of global optimization. The ways to respond to local task assignment requirements without the need to communicate with remote nodes overcomes the disadvantages of centralized task assignment in large-scale WSNs with significant communication overhead and delay, and have better scalability.

The remainder of this paper is structured as follows. Section 2 introduces the related work, the task allocation problem in WSNs is described in Section 3, and a global network view consisting of a multi-level view is proposed in Section 4. Next, we present a reliable and efficient task allocation algorithm for WSNs in Section 5, including centralized task allocation and two distributed task assignments. The performance is evaluated by theoretical analysis and simulation in Section 6, and finally, concluding remarks and future work are discussed in Section 7.

2. Related Work

Many research results on distributed task assignments for WSNs have been achieved, although most methods aim solely at maximizing the network lifetime [1,2]. Yu Wanli et al. propose an optimal online task assignment algorithm DOOTA (Distributed Optimal On-line Task Allocation) that is suitable for multi-partition scheduling by taking into consideration the energy cost of communication, calculation, sensing, and sleep activities [1]. To solve the problem of task allocation on the Internet of Things-based applications, considering special functions and design features, Khalil Enan et al. proposed a new task group and a virtual object-based framework and then adopted a meta-heuristic solving method [2] by modeling the problem as a single-objective optimization problem with the primary goal of minimizing energy consumption.

In addition to considering the extension of the network lifetime, part of the work also takes into account the quality of service (QoS) requirements [3,4]. Using fuzzy inference systems, Ghebleh Reza et al. proposed a method for discovering multi-criteria resource discovery in the context of distributed systems [4]. The main disadvantage of this is that the computational overhead increases rapidly as more requests and resources are invoked. Several researchers have applied game theory to WSNs in order to solve task allocation problems [5,6]. To solve the selfish behavior of some nodes, some have adopted the cooperative enforcement games strategy [7]. Despite that, the QoS requirements they satisfy are mainly focused on minimizing completion time, load balancing, data sampling rate, and accuracy, among others. Among the requirements to be considered, there is no research work on reliability requirements. Alternatively, some researchers have addressed the problem of distributed task allocation in WSNs from cloud-based architectures [8,9]. Though, due to the distance between the cloud platform and the user, the group perception has higher waiting times.

Based on the self-organizing characteristics of WSNs, Ye et al. divided the implementation techniques of distributed task allocation into two types [10]: one is based on reinforcement learning [11], and another is a collaboration based on nodes—the local interaction between nodes in a collaborative manner to achieve a self-organizing task assignment. For example, based on the auction method [9,12] and the distributed task allocation based on the negotiation method [13], the two methods can achieve optimal results since they are obtained through negotiation between the two parties, which differs from other methods in that they use only a specific algorithm or a specific set of algorithms to obtain results. It happens that substantial communication overhead cannot be avoided during the negotiation process, so the auction-based and negotiation-based approaches are not suitable for resource-constrained WSNs.

The task assignment based on auction and negotiation has a significant communication load in a typical distributed task allocation due to the frequent interaction between nodes and unsuitable resource-limited WSNs. Moreover, the task allocation based on reinforcement learning also faces the problem of longer convergence time. In research that utilizes intelligent algorithms to solve the optimization goals of task assignment, such as particle swarm optimization, it is easy to fall into the optimal local problem. Wang et al. proposed a globally optimized task allocation algorithm based on an ant colony algorithm that requires a global pheromone matrix to obtain the optimal solution [14]. However, the authors missed mentioning how to construct a three-dimensional (3D) path pheromone storage space and the cost for the construction of this 3D space.

The research status presented above prompted the authors to investigate the distributed reliable task allocation problem in WSNs in order to develop an optimal global solution for reliable and efficient task allocation problems at a minimal cost. As is known, the transmission task is the most crucial issue in WSNs and thus the goal is to minimize the energy consumption of task allocation in WSNs based on the deadline and the reliability of tasks.

Two task allocation strategies under the constraints of reliability and task deadline are proposed in this paper, named TRETA-Cluster and TRETA-Multi-Sink. The term "task" represents the "transmission task" in the subsequent parts of this paper.

3. Description of Reliable and Efficient Task Allocation in WSNs

3.1. Energy Consumption Model

The total energy consumption of a task m_i is the sum of the energy consumed on all nodes that undertook such a task. That is, the energy consumption of a node includes computational and communication energy consumption associated with that task. For calculation purposes, the energy model depicted in [15] is used in this paper, so e_{ij} represents the energy consumed by the task m_i in node N_j. When the node N_j is not selected by the task m_i, $e_{ij} = 0$. Otherwise, e_{ij} is composed of the calculated energy consumption and communication energy consumption of node N_j. The calculated energy consumption of N_j, namely e_{ij}^{comp}, is given by Equation (1):

$$e_{ij}^{comp} = e_j \times t_{ij}^{comp} \qquad (1)$$

where e_j represents the average processing energy consumption of the nodes in the network, and t_{ij}^{comp} is the calculation time of task m_i.

The communication energy consumption of a node, namely e_{ij}^{comm}, includes the energy consumption for transmitting and receiving data packets. According to the commonly used communication energy model [6], the energy consumption of transmitting and receiving data of length l_{ij} bit at the distance d is calculated by Equations (2) and (3):

$$e_{ij}^t = \left(e_{elec} + \varepsilon_{amp} \times d^2\right) \times l_{ij} \qquad (2)$$

$$e_{ij}^r = e_{elec} \times l_{ij} \qquad (3)$$

where e_{elec} is the energy consumption of operating the radio model for each bit, and ε_{amp} is the coefficient of the transmit amplifier. The communication energy consumption of the node is calculated by Equation (4):

$$e_{ij}^{comm} = e_{ij}^{t} + e_{ij}^{r} \quad (4)$$

Thus, the energy consumption of task m_i in node N_j is then calculated using Equation (5).

$$e_{ij} = e_{ij}^{comp} + e_{ij}^{comm} \quad (5)$$

3.2. Constraints of Task Reliability and Deadline

3.2.1. Task Deadline Constraint

Let P_i, S_i and C_i represent the task period of m_i, the task start time and the task end time (deadline) respectively. The assignment of task m_i thus needs to be satisfied with the deadline constraint:

$$C_i \geq P_i + S_i \quad (6)$$

3.2.2. Task Reliability Constraint

As is known, a WSN consists of a large number of randomly distributed sensor nodes. Performing different tasks in the same network will also show different task reliability due to factors such as network traffic and transmission path; thus, a scheme that matches the task assignment of WSNs with task reliability requirements becomes necessary. Task assignment needs to satisfy Equation (7):

$$R_{WSN} \geq R_s \quad (7)$$

where R_{WSN} represents the transmission reliability of the current WSN, whose value can be obtained from the evaluation model of reliable transmission given in Equations (8) or (10). Also, R_s represents a threshold for the application of WSN transmission reliability.

Reliability is an important indicator to measure the QoS of WSNs. It has been well studied in the past and is mainly regulated by the successful delivery rate of the link packets. Users are more concerned with the quality of the transmission of a task than the quality of the link in many applications [16]. Since a task is contained by many packets, the user is concerned whether the end-to-end event is successfully perceived rather than the successful delivery rate of the individual node's data packets. Therefore, it is necessary to measure the reliability of WSNs from the perspective of transmission and how the tasks in the network are utilized based on the analysis of the granularity.

For two typical topologies in WSNs, mesh (planar) and cluster-based (hierarchical), the cognitive data transmission process of these conventional topologies is analyzed. A task reliability evaluation model is then established based on the transmission path.

(A) Clustered Topology

Clustered topology is widely used in a variety of applications due to its higher energy efficiency and scalability. The heads of different clusters compose the backbone layer of WSNs. When evaluating a transmission task, it is supposed there are L task-related clusters in the sensing area. The reliability of clustered WSNs at time t is:

$$ClusterR_{s_i,sink}(t,i) = \prod_{c=1}^{L} R_A^c(t,i) \times R_T^c(t,i) \quad (8)$$

where $R_A^c(t,i)$ refers to the reliability of the i-th transmission performed by cluster c. In most clustered WSNs, a cluster is designed with one cluster head, and the transmission between cluster head and members is single-hop [17]. Since the packet delivery rate is within one hop in a cluster, the error between the average and the actual value is smaller. Therefore, $R_A^c(t,i)$ can be modeled by a k-out-of-n system under the multisource environment.

Let $R_T^c(t,i)$ represent the reliability of the head of cluster c, which successfully sends the collected data of i-th transmission to sink at time t. This stage is considered successful as long as there is at least one path whose packet delivery rate is higher than the transmission threshold. Thus, the head of cluster c which successfully sends the collected data of the i-th transmission to sink at time t can be modeled by a parallel system.

(B) Mesh Topology

In a mesh WSN, there are multiple source nodes randomly located in the perceived area. The sink is the destination in the uplink transmission, as multiple source nodes collect data packets independently yet transmit them to the sink via intermediate nodes. The transmission reliability $MeshR_{s_i,sink}(t,i)$ in mesh WSN is shown in Equation (9), where K is the number of source nodes, and $R_{s_i,sink}(t,i)$ represents the reliability of the transmission task from the source s_i to the sink in the network at time t.

$$MeshR_{s_i,sink}(t,i) = \sum_{s_i=1}^{K} R_{s_i,sink}(t,i) \qquad (9)$$

Let l represent the number of paths in the minimal path sets from the source s_i to sink. At time t, $R_{s_i,sink}^i(t,i)$ is the probability that there exists at least one path whose packet delivery rate is greater than the threshold of i-th transmission in the disjoint minimal path sets (from s_i to sink) is shown in Equation (10).

$$R_{s_i,sink}^i(t,i) = 1 - \prod_{j=1}^{l}[1 - R_{path(s_i,sink)}^j(t,i)] \qquad (10)$$

3.3. Reliable and Efficient Task Allocation Problem Model

WSNs can be represented by graph $G(V,E)$. $M(t)$ represents the set of tasks to be allocated at time t, and $m_i \in M(t)$ represents the i-th task in $M(t)$. In a task allocation for WSNs, the two processes of task mapping and task scheduling are included. First, each task is mapped to the sensor node in the graph $G(V,E)$ that is represented by the function $\varphi(m_i) : M \to V$. When a task is assigned to multiple nodes, communication task scheduling is performed between the nodes, and the process of path allocation is represented by the function $\phi(m_i) : L \to E$. Therefore, the reliable and efficient WSNs task allocation problem can be abstracted into the following constraint optimization problems:

Input: $TG(V,E), RT(S,Ps), M(t)$
Output: $\varphi(m_i) : M \to V$, $\phi(m_i) : L \to E$

Satisfying Equation (11):

$$\begin{aligned} \min \ &Eng\varphi(m_i) = \sum_{\tau \in V} e_{ij}, \quad m_i \in M(t) \\ s.t \ &R_{WSN} \geq R_s, \forall \tau \in V \\ &C_i \geq P_i + S_i \end{aligned} \qquad (11)$$

where $TG(V,E)$ represents the physical topology graph of the network, V represents the set of sensor nodes in the network, E represents the set of links between the nodes in the network, $RT(S,Ps)$ represents the routing topology of the network, S represents the source node set and Ps represents the path set of all source nodes to the sink in the network. $Eng(\varphi(m_i))$ is the energy consumption of task m_i that is equal to the sum of the energy consumption on all nodes in the network assigned to task m_i. From this, Equation (11) indicates that the goal of the task allocation to WSNs is to minimize energy consumption under the constraints of reliability and deadline. Given that this is a nonlinear mixed-integer programming problem, it can be solved using a heuristic algorithm.

4. Global Network View

The physical topology of a WSN, the routing topology, and the set of tasks to be allocated at time t are known as input data assigned by the task in Equation (11). Additionally, the sensor nodes in a WSN learn local topologies through topology discovery, which are periodically sent to sink nodes to form the global physical topology of the network.

4.1. Route Topology

We proposed the route topology inference (RTI) in [18]. The algorithm framework is shown in Figure 1. Since this method is not limited by the routing protocol adopted by the current network, and only uses the packet tracking hybrid active detection method to construct a transmission path from the source node to the sink, an online global routing topology view for a WSN can be provided and shown.

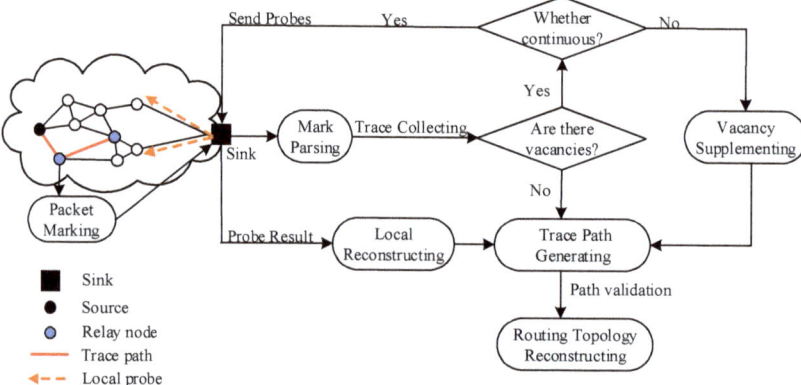

Figure 1. The overall framework of route topology interference (RTI).

The core idea of packet tracing is that, on the transmission path from the source node to sink, the forwarding nodes selectively mark part packets according to the marking rule upon receiving the packets. At the end of the sampling period, the sink can establish trace lists from the source node to the sink by parsing the marked packets. As the sink determines the one-to-one correspondence trace in the trace list according to the packet with the same hop to the source node, it retrieves the trace list from the source node. The trace list returned by mark parsing may be incomplete due to packet loss or insufficiently marked packets. Also, the vacant traces in the trace list can be supplemented by auxiliary inference or active detection [18].

RTI increases the memory load, though it has a significant advantage in the correctness and convergence of reconstruction. Moreover, the relay node does not need to mark all packets, as it only marks the packets based on conditions, and for one path, each packet is marked only by one relay node.

The reconstructed WSN routing topology is shown as a graph, where the nodes on each path and links between the nodes are added to the graph, and the source node identifying a path is added to the link also. If a source node has multiple paths to the sink, the remaining nodes or edges are added to the routing topology based on the tracking path after the first path is added, so the marked routing topology view from the source node to the sink is generated.

4.2. Multi-Level Global View

Once the routing topology is acquired, the task logical topology of different tasks in the current network is further abstracted according to the task mark in the data packet. As shown in Figure 2, the architecture that provides a conceptual basis for global optimization of network management and the multi-level global view architecture consisting of the physical topology, the routing topology, and the task logical topology is obtained. The routing topology layer can generate different routing topologies for applications according to different node sets in the physical topology layer (i.e., different shadows in Figure 2). Also, links of different thicknesses in the task logic topology reflect the current traffic of the link.

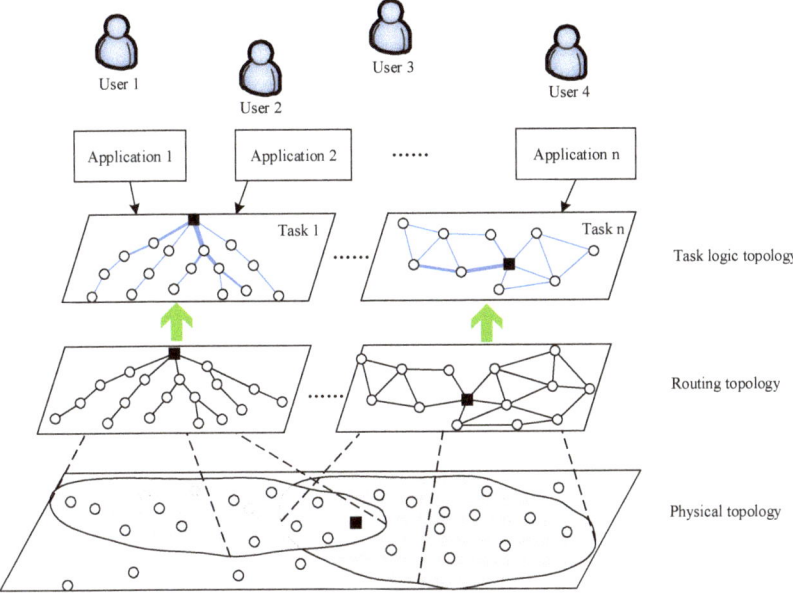

Figure 2. The multi-level global view of a wireless sensor network (WSN).

5. Reliable and Efficient Task Allocation Algorithm for WSNs

5.1. Centralized Task Assignment

As presented before, the aims at this research are to minimize the energy consumption of task allocation under the constraints of reliability and deadlines in the task allocation model problem presented in Section 3, where $TG(V, E)$, $RT(S, Ps)$ and $M(t)$ are the three inputs as presented in Equation (11).

TRETA is performed by the sink based on the given task reliability and deadline constraints through a global view, and its framework consists of five modules, enumerated as: data collection, task reliability evaluation, topology management, event management, and task assignment. The data collection module collects state information of each node, such as node ID (Identity Document), node residual energy information, node neighbor, and packet loss rate information, among others; the node collection module stores network state information such as nodes, paths and task information (transmitted from the application layer) in the corresponding node table, topology information table, state information table, and task information table in the database for extraction by other modules; and the topology information table in the data collection module stores the global physical topology of the network.

To establish a global view of the current network state, a topology management module in the framework of TRETA is designed. In addition to the global physical topology, the topology management module obtains the routing topology of the global network and then constructs a two-level global view.

The event module manages events and triggers other module updates when a drive event occurs. To achieve reliable and efficient task allocation in WSNs, the driver events concerned include topology change (nodes join or leave), as the reliability of the task is lower than the threshold and new tasks to be allocated.

Based on the modules presented above, a reliable and efficient task allocation algorithm, namely TRETA, is proposed, as depicted in Figure 3.

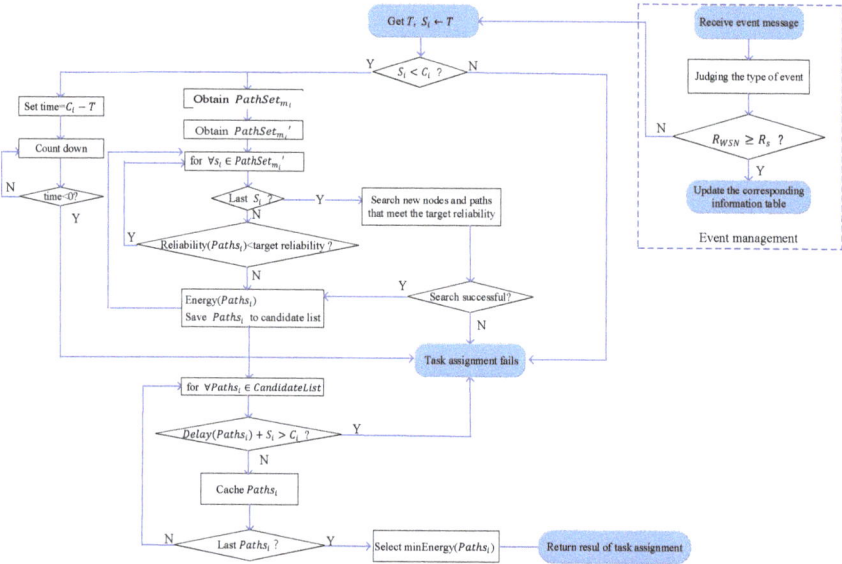

Figure 3. Transmission-oriented reliable and energy-efficient task allocation (TRETA) algorithm.

During the execution of the network and once the defined driving events occur, the event management module triggers the update of the data collection module, and the sink determines whether the reliability of the updated network is less than the target reliability. If yes, the task allocation module is notified to re-assign; otherwise, only the global view corresponding to the event is updated.

5.2. Distributed Task Assignment

Since TRETA performs task assignments from the perspective of global optimization, it has the disadvantage of significant communication overhead and delays in large-scale WSN. Aiming to obtain reliable and efficient distributed task allocation in WSNs, two distributed task assignments for WSNs based on TRETA are proposed.

5.2.1. Distributed Reliable and Efficient Task Allocation in a Hierarchical Topology

A hierarchical topological diagram of a clustered WSN, where the sink is the center of the entire network, is depicted in Figure 4a. The cluster head saves the collected local physical routing topology as well the state information of the member nodes in the local cluster, passing them next to the sink. Then, the sink node can obtain the physical topology of the entire network by merging all the topologies received, so the topology management module receives a two-level global view.

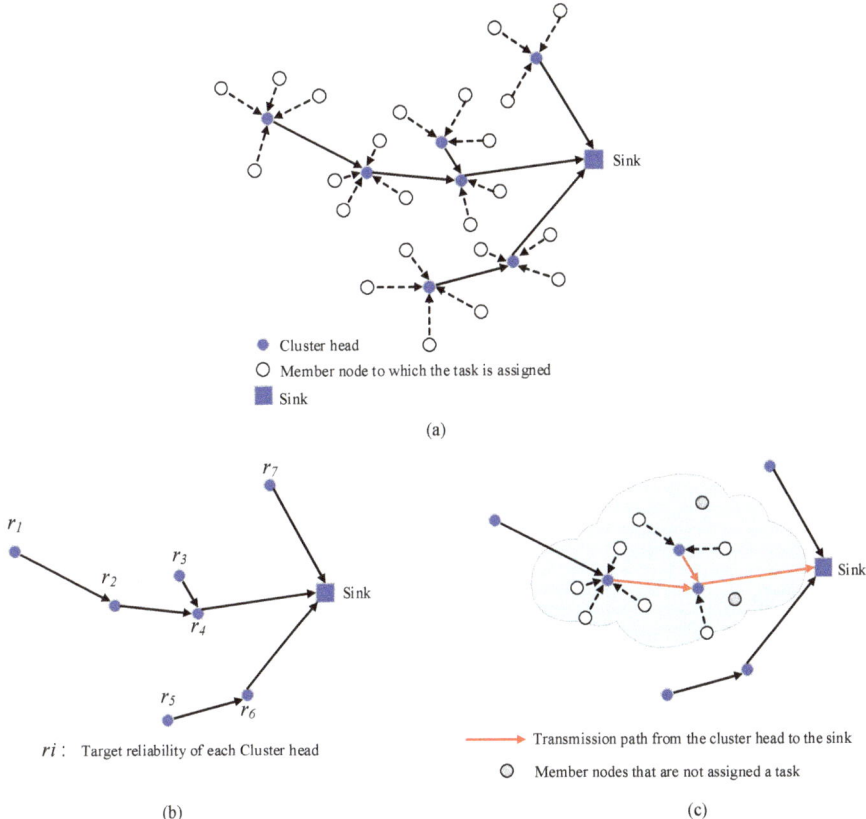

Figure 4. Reliable and efficient distributed task allocation based on clustering: (**a**) Cluster-based WSNs; (**b**) Sink assigns reliability to the cluster head; (**c**) Cluster head performs local task allocation.

Based on the above observations, we present a distributed reliable and efficient task assignment algorithm based on TRETA for clustered WSNs, named TRETA-Cluster, which divides the task allocation into two phases: backbone network and intra-cluster allocation.

1) The sink performs the reliability distribution at the backbone network composed of the cluster heads according to the target reliability, where reliability r_i is obtained at each cluster head, as shown in Figure 4b. Combined with the target reliability and the task deadline constraint, the sink calls the TRETA algorithm to select the cluster head with the smallest total energy consumption that satisfies the task deadline and the target reliability constraint, and the transmission path of the cluster head to the sink.
2) The cluster head calls the TRETA to select the nodes in the cluster and the intra-cluster paths according to the obtained intra-cluster reliability index and local view to realize the task allocation of the cluster.

If the task allocation is successful, the allocation result is returned; otherwise, the cluster head selects the node that satisfies the target reliability according to the local physical topology. In case the node is found, the allocation result is returned. The task allocation fails and returns if otherwise, as shown in Figure 4c.

As a topology change occurs, the reliability may become lower than the threshold or the cluster head can be interrupted. In this case, the task allocation is re-executed by the sink or the cluster head according to the rule, as seen in Algorithm 1, TRETA-Cluster, Lines 17–20.

The reliability allocation [19,20] is to assign the target reliability of a task to the appropriate subsystems, components, and nodes of the system to determine the reliability of each component. Referring to the reliability allocation algorithm in [21], we assign the target reliability of the task to the two stages and use them as the target reliability of each stage (Algorithm 1, TRETA-Cluster, Lines 5–6; RA is the reliability distribution function).

Algorithm 1. TRETA-Cluster
Input: $TG(V, E)$, $RT(S, Ps)$, $M(t)$, R_s, C_i
Output: $\varphi(m_i) : M \rightarrow V$, $\phi(m_i) : L \rightarrow E$
1 for $\forall m_i, m_i \in M(t)$
2   get T, $S_i = T$
3   if $S_i < C_i$
4   for $\forall CH, CH \in RT(S, Ps)$ //CH represents the cluster head
5           $R_{CH} = RA(RT(S, Ps))$
6           $R_{CM} = RA(RT(S, Ps))$
7   for $\forall CH \in RT_b(S, Ps)$
8   call TRETA $(TG_b(S, Ps), RT_b(S, Ps), R_{CH})$
9   for $\forall$ Cluster $\in RT(S, Ps)$
10   call TRETA $(TG_c(S, Ps), RT_c(S, Ps), R_{CM})$
11   if call TRETA $(TG_c(S, Ps), RT_c(S, Ps), R_{CM}) \neq \emptyset$
12   return allocation result
13   get T
14   if $T > C_i$, return line 1
15   if CH fault
16   sink update $TG(V, E)$ and $RT(S, Ps)$
17   if $R_{WSN} < R_s$, then
18   call TRETA $(TG_{b'}(V, E), RT_{b'}(S, Ps), R_{CH'})$
19   elseif the failed cluster is re-clustered
20   sink update $TG(V, E)$ and $RT(S, Ps)$
22   if member of CH change
23   sink update $TG(V, E)$ and $RT(S, Ps)$
24   if $R_{WSN} \geq R_s$, then return
25   else
26   call TRETA $(TG_{c'}(V, E), RT_{c'}(S, Ps), R_{CM})$
27   if call TRETA $(TG_{c'}(V, E), RT_{c'}(S, Ps), R_{CM}) \neq \emptyset$
28   return allocation result
29   else return line 1
30 if $R_{WSN} < R_s$, then
31   return line 1
32 return

It can be noted that the local task allocation of the cluster head may select the node participating in the task, and since the path from the cluster head to the sink has been obtained in the first stage, the selection from the node to the path is completed in two phases in a task assignment.

Whenever the nodes in the cluster change (join or leave), the cluster head transmits the topology change to the sink. The sink determines whether the node change affects the reliability of the task. If not, it returns. Otherwise, the cluster head node tries to re-select the node in the cluster after the change. Additionally, if the intra-cluster target reliability constraint can be met, the cluster head re-performs the local task assignment. Otherwise, the sink performs the task allocation again.

When the sink finds that the task transmission is below the reliability threshold, the task allocation is performed again. However, when the cluster head fails, the cluster will disconnect from other parts of the network and the task assignment policy issued by the sink cannot reach the cluster, bringing challenges to the reliable task allocation of WSNs. The proposed processing scheme follows: the sink updates the two-level global view and determines whether the fault of the cluster has an impact on the target reliability of the current task. If so, the task is reassigned according to the updated two-level global view. Otherwise, the sink only updates the two-level global view that reduces the frequency of the update. After the faulty cluster re-selects the cluster head, the cluster head collects the cluster topology and the sink node updates the global view and reperforms the task assignment.

5.2.2. Distributed Reliable and Efficient Task Allocation in Planar Topology

For a large-scale WSN, perceptual information needs to go through multi-hop communication to the sink. Considering the overhead of establishment, maintenance routing by node, and long-distance multi-hop communication delay in planar topology to ensure the reliability and scalability of WSNs, some researchers have proposed the multi-sink deployment scheme [22–25] or software-defined network (SDN) controller [26].

The deployment strategies of multiple sinks in a planar topology include static and dynamic, random, and scheduled deployment methods. As an illustration, the static and mobile deployment of multiple sinks is shown in Figure 5.

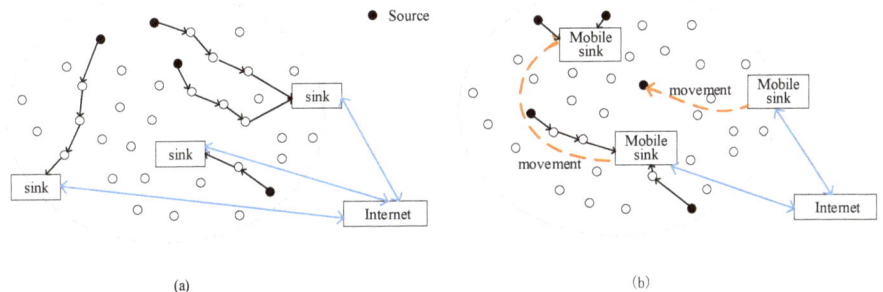

Figure 5. Multi-sink deployed WSNs. (**a**) Static multi-sink WSNs; (**b**) mobile multi-sink WSNs.

In a WSN with multiple sinks deployed, this section proposes a distributed task assignment strategy based on multiple sinks. Multiple sinks in this strategy have a global view of the entire network, enabling local requirements to be globally optimized and without the need to communicate with remote nodes, reducing the communication cost and latency of task assignments.

(A) Consistency of Multi-Sink Global View

Due to the geographical distribution and asynchronous operation features, when the sink publishes the global view, the update time of multiple sinks is not synchronized due to network delay or other reasons, resulting in inconsistency in the forwarding or processing of the task data. To share the global view among multiple sinks, it is necessary to consider the consistency problem of the global views between multiple sinks. In a resource-constrained WSN, how to efficiently share global view among multiple sinks to achieve fast and efficient global task allocation is still one of the significant questions.

In Eric Brewer's consistency, availability and partition tolerance (CAP) theory, it has been proved that consistency, availability, and partition tolerance in a distributed system cannot be considered together [27]. Partition occurs easily in WSNs. If strong consistency is guaranteed, the availability of the network cannot be guaranteed at the same time, and the communication cost of realizing the strong consistency of multiple sink global views in large-scale WSNs is high, the final consistency is chosen to reduce the communication cost of synchronization between nodes.

In an existing final consistency technology, Dynamo, the storage platform of the key-value pattern of Amazon has been paid more attention by many researchers [28,29]. Dynamo proposes an NWR model that guarantees eventual consistency, where N represents the number of copies of data being saved, R represents the number of copies required for each read success, and W represents the number of copies as are necessary for each write success. By setting R and W, when $R + W > N$, it produces a system similar to Quorum.

Quorum is widely used in distributed storage systems [30–33]. It is a set and a subset of all copies C, where two parameters W and R are pre-defined for N copies, and $N = |C|$, $W \geq 1$, $R \leq N$. The quorum set of writing is as follows:

$$S_W = \{Q | Q \subseteq C \wedge |Q| = W\} \tag{12}$$

and the quorum set of reading as:

$$S_R = \{Q | Q \subseteq C \wedge |Q| = R\} \tag{13}$$

If $R + W > N$, there is an intersection between S_W and S_R, so any read operation can return the latest write and guarantee strong consistency. Conversely, if $R + W \leq N$, then the two elements overlap with probability $P_{overlap}$ [29], as in Equation (14):

$$\begin{aligned} P_{overlap} &= 1 - \frac{\binom{N}{W}\binom{N-W}{R}}{\binom{N}{W}\binom{N}{R}} \\ &= 1 - \frac{\binom{N-W}{R}}{\binom{N}{R}} \end{aligned} \tag{14}$$

For a write request, it is first sent to the replica as a coordinator, and then the coordinator propagates the write to all other replicas. After getting at least $W - 1$ responses, the coordinator returns. Read requests are handled in the same way as write requests, although the coordinator should wait for at least $R - 1$ responses. If the read replica set overlaps with the write replica set, the read request can return the most recent value. Otherwise, it will return stale data.

Similarly, this paper proposes a method to ensure the final consistency of multiple global views. In this method, each sink maintains a log in addition to the global view that includes the version of the current global view (new), the version of the previous global view (old), and the update operation from the old version to a new version.

In a multi-sink WSN, if the network status changes, the changed local sink sends an update request to other sinks. If at least $W - 1$ sinks return a response, the update request is considered successful, and the sink submits an update and broadcasts the update to other sinks. Next, the previously responded $W - 1$ sinks update the current global view version, the previous version of the global view, and the update operation from the old version to the new version in the log. As soon as a sink node initiates a task assignment, the global view needs to be read. In addition to reading the locally saved global view, the sink node issues a read request to the other sinks, then waits for $R - 1$ responses, and finally returns the read result.

Since the task allocation has a deadline constraint, to reduce the delay of reading the global view, the algorithm sets $R = 2$ to read the global view of two sinks. Therefore, the complexity of guaranteeing consistency is pushed to the write operation; that is, for the network state update process which is not sensitive to delay, as long as $R + W > N$ is satisfied, the final consistency can be guaranteed. If concurrent partial updates occur in the network at the same time, the concurrency control protocol is used for processing.

(B) Task Assignment Based on Multi-Sink WSNs

In a planar WSN topology where multiple sinks important, these sinks form a multicast group, and the multiple sinks are equal. To simplify the design, it is assumed that multiple sinks can remain fully connected and the sinks do not fail, as each sink collects the local view, including the physical topology and routing topology. The local views are periodically exchanged so that multiple sinks in the multicast group gradually obtain a global view of the network.

The concept of TRETA-Multi-Sink follows: after each sink has acquired the global view and there is a task to be allocated, if the current time of the system is less than the deadline of the task to be assigned, the sink closest to the task can be selected to call the TRETA algorithm for local optimization allocation. Since the sink has a global view and despite a local allocation, the sink can perform a globally optimized task assignment. In the algorithm, multiple sinks select the final consistency when exchanging global views to reduce the communication cost of synchronization between sinks. If the current time of the system is higher than the deadline of the task to be assigned, the current task allocation is interrupted.

Specifically, for the task m_i to be assigned, if the current time of the system is less than the deadline of the task, the nearest sink of task m_i is found through the physical topology information and is represented by $sink_j$, which reads the local and adjacent global view version. To ensure the final consistency of the global view between multiple sinks, the reading number of copies is R according to the NWR model (we let $R = 2$); if the versions are different, it will read the global view with the latest version. Otherwise, it will read the local - global view of $sink_j$. Then, $sink_j$ calls the TRETA algorithm for local optimization allocation. During this process, if the current time of the system exceeds the task deadline, the task allocation is interrupted.

When the network update occurs, if the scope of the update involves only one sink (represented by $sink_k$), then $sink_k$ multicasts the update request to other sinks. According to the NWR model, the number of copies that need to be written to the update is $W - 1$. W in this paper should satisfy $W > N - 2$.

If the number of sinks in response to the update request is higher than $W - 1$, $sink_k$ submits the view update, and other sinks in the multicast group update the local-global view accordingly. Otherwise, the update of $sink_k$ is deleted to avoid inconsistency with other different sink views.

When the sink finds that the task transmission is lower than the reliability threshold, the task allocation is re-executed. The specific process of Algorithm 2, TRETA-Multi-Sink is as follows:

Algorithm 2. TRETA -Multi-sink

Input: $TG_j(V,E)$, $RT_j(S,Ps)$, $M(t)$, R_s, C_i
Output: $\varphi(m_i) : M \to V$, $\phi(m_i) : L \to E$
1 $TG(V,E) = \text{Merge}(TG_j(V,E))$, $j = 1,2,\ldots n$
2 $RT(V,E) = \text{Merge}(RT_j(V,E))$, $j = 1,2,\ldots n$
3 for $\forall m_i, m_i \in M(t)$
4   get T, $S_i = T$
5   if $S_i < C_i$
6          $sink_j = \{sink|\text{Mindistinct}(m_i, sinks), sink \in TG_j(V,E)\}$
7   read $sink_j.viewID$ and $sink_{jn}.viewID$
8   if $sink_j.viewID \neq sink_{jn}.viewID$
9   read view of latest$(sink_j.viewID, sink_{jn}.viewID)$
10   else read $sink_j.RT_j(S,Ps)$
11   call TRETA $(TG_j(V,E), RT_j(S,Ps), M(t), R_s)$
12   get T
13   if $T > C_i$, return line 1
14   if $TG(V,E)$ change happens and $Num(\text{changed sin k}) = 1$
15             $sink_k$ record change in cache
16             $sink_k$ multicast update-request to other sinks
17   if $Num\left(sink_k^{res}\right) > W - 1$
18                 $sink_k$ submit an update
19   update$\left(sink_k.TG_j(V,E), sink_k.RT_j(S,Ps), sink_k.log\right)$
20   update$\left(sink_k^{res}.TG_j(V,E), sink_k^{res}.RT_j(S,Ps), sink_k^{res}.log\right)$
21   else delete $sink_k.record$
22   if $R_{WSN} < R_s$, then
23   return line 3
24 return

6. Performance Analysis

6.1. Theoretical Analysis

In this section, the performance of distributed reliable and efficient task allocation is analyzed primarily from the algorithm complexity, communication load, and delay of the task assignment.

6.1.1. Algorithm Complexity

TRETA-Cluster is divided into two phases. In the first phase, the sink performs task allocation for the backbone network composed of the cluster heads and the sink, with the worst-case calculation complexity $O((\alpha N)^3)$, where α represents the ratio of the size of the backbone network to the size of the entire network system. During the second phase, the cluster head is responsible for the task allocation within the cluster. The calculation complexity of this process is $O((\gamma N)^3)$, where γ represents the ratio of the cluster member nodes size to the entire network size. Based on the above process, the calculation complexity of distributed task allocation based on clustering topology is $O(N^3)$.

In TRETA-Multi-Sink, the calculation complexity of multiple sinks in local task allocation is the same as TRETA, although the processing of global view consistency is added to the network update in TRETA-Multi-Sink, which includes the write operation of multiple sinks when global view update occurs and read operation when task assignment occurs. Assuming N is the network size, β is the ratio of the part of the network to be updated, and W is the number of global views that need to be written, the calculation complexity of the write operation is $O(\beta WN)$. Let the number of global views that need to be read is R ($R = 2$); then, the calculation complexity of the read operation is $O(2N)$. Based on the above process, the calculation complexity of the distributed task allocation algorithm is $O(N^3)$.

The two distributed task allocation algorithms above reduce the size of the network. As noted, the distributed task allocation has the polynomial computation complexity in the background of global optimization.

6.1.2. Communication Load of the Distributed Task Assigned

In the distributed task allocation strategy for the clustered topology, the sink multicasts the allocation policy to the selected cluster heads that are responsible for the task allocation within the cluster. Therefore, the communication load depends on the amount of information in the sink's allocation policy. In this paper, the communication load of TRETA-Cluster is measured by the number of packets in the allocation policy by multicast, as shown in Equation (15):

$$CM = Num_{cp} \tag{15}$$

For the multi-sink topology, the communication load includes: (1) multi-casting the global view between the sinks, namely Num_{gv}, where Num_{gv} represents the communication load for transmitting one global view, and a represents the number of times of global views transmitted; (2) ensuring global view consistency between multiple sinks, i.e., the write load WM at update time and the communication load RW at read time, as shown in Equation (16):

$$DM = aNum_{gv} + WM + RW \tag{16}$$

where only the version and update operation of the global view is transferred when the update is written, and performed only when the network is updated, as shown in Equation (17):

$$WM = \begin{cases} VUM \\ 0 \end{cases} \tag{17}$$

where VUM indicates the write load of the version and update operation of the global view when it is updated, and 0 means no update.

When the sink reads the global view, it needs to read two copies of the global view; that is, the global view of the local and that of the nearest sink in the network, as required by the final consistency. If the local version of the global view is newer, there is no need to communicate with other sinks, and thus the communication load of the read is 0. Otherwise, one global view needs to be transmitted. Therefore, the communication load RW when reading is as shown in Equation (18):

$$RM = \begin{cases} 0 \\ Num_{gv} \end{cases} \tag{18}$$

6.1.3. Delay of Distributed Task Assignment

In order to implement the task assignment, two known premises are required: the physical topology of the entire network and the routing topology of the network established in advance. They are considered in the initialization process for the task assignment, so the delay is not considered at initialization.

The task assignment delay for a clustered topology is calculated by Equation (19):

$$RT_{d\_c} = RT'_c + MC + CHD \tag{19}$$

where RT_{d_c} represents the delay of task assignment for the clustered topology, RT'_c represents the delay of the sink allocating a task to the subgraph composed of the cluster heads, MC represents the delay of the multicast performed by the sink, and CHD represents the delay in the allocation of the task by the selected cluster head.

Assuming that each sink has obtained a global view in a multi-sink deployment environment, this stage can be implemented by periodically swapping local views with each other through multiple sinks, requiring it as initialization prior to the task assignment. The delay for distributed task allocation based on multiple sinks is calculated by Equation (20):

$$RT_{d\_d} = \overline{RT_c''} + C_{rt}. \qquad (20)$$

where RT_{d_d} represents the delay of TRETA-Multi-Sink, $\overline{RT_c''}$ represents the average of the delays of the tasks assigned by multiple sinks locally, and C_{rt} indicates the time to ensure the final consistency of the view, as shown in Equation (21):

$$C_{rt} = \begin{cases} W_{rt} \\ R_{rt} \end{cases} \qquad (21)$$

where W_{rt} indicates the time of writing of the global view whenever it is updated and R_{rt} indicates the time of reading of the global view when the distributed task is assigned.

6.1.4. Energy Resilience of the Transmission in WSNs

The nodes can go down for various reasons, e.g., the time of life or for the specific protocol used. In order to evaluate the energy resilience of the transmission in WSNs, we quoted an index η [34] as shown in Equation (22), where N_p indicates the number of packets received by the sink after a fixed time, S indicates the number of initial active nodes, and D indicates the number of dead nodes after a fixed time.

$$\eta = \frac{N_p}{S-D} \qquad (22)$$

The index gives an indication of how efficient the network is in allowing information to be delivered considering both the number of packets that are running in the network and the number of nodes that are going out over time.

6.2. Simulation

6.2.1. Simulation Design

In order to carry out a reliable and efficient task assignment, two distributed task assignment strategies are proposed, namely TRETA-Cluster and TRETA-Multi-Sink. In this section, a simulation is carried out using TOSSIM (TinyOS Simulator). In order to evaluate energy consumption, this paper expands TOSSIM and adds a power consumption model. The performance of the strategies is analyzed and compared in three aspects: different network size, task arrival rate, and network update rate.

In order to compare with the distributed task allocation, simulation of the centralized task allocation TRETA is also performed in the same network environment. The performance metrics analyzed include the energy allocated, the delay of successful allocation during the deadline, and the success rate of the task assignment. In addition, the simulation is carried out under the cluster topology and the multi-sink-based topology for the distributed task assignment.

1) Cluster topology

    The sensing area is 100 m × 100 m, and the number of sensor nodes is 50 to 300. There is only one sink in the cluster topology, with the sink node located in the center of the sensing area and remaining nodes randomly deployed.

2) Multi-sink-based topology

    The deployment strategies of multiple sinks in a multi-sink-based topology include static and dynamic, random, and scheduled deployment methods. In the proposed simulation experiment, the static uniform deployment method is selected, with the number of sinks set to five, and the sensing area of the square is evenly divided into four sub-areas. One of the five sinks is located in the center of the sensing area, while the remaining four sinks are located in the center of the respective remaining four sub-areas. The five sinks remain fully connected, and the case where the sinks fail is not considered.

The main simulation parameters are shown in Table 1. Three of the abbreviations (e_{elec}, ε_{amp}, and e_j) represent the energy consumption of operating the radio model for each bit, the coefficient of the transmit amplifier, and the average calculated the energy consumption of the nodes in the network, respectively. Kenneth et al. mentioned that wireless communication is usually the most energy-consuming process in traditional WSN applications [16]. Specifically, the energy required for a single bit transmission is 1000 times the energy consumed by calculating a single bit in classic 32-bit architecture, and thus the value of e_j is set to 0.05 nJ/b.

Table 1. Main simulation parameters.

Parameter	Value
Bandwidth	250 kbps
e_{elec}	50 nJ/b
ε_{amp}	100 pJ/b/m^2
Maximum transmission range of the node	50 m
The initial energy of the node	2 kJ
e_j	0.05 nJ/b

The parameters of the task to be assigned are shown in Table 2, and three aspects are included: task type, transmission parameters, and task environment. Each simulation is executed for 120 minutes and repeated 500 times. For each performance index, we used the Monte Carlo method to obtain simulation results. The specific process is as follows: first, in the TOSSIM simulation environment, Python is used to generate a random number of each task parameter in the value interval based on a Poisson distribution. Secondly, each random variable is directly sampled, and simulation experiments and calculations are performed according to the task assignment strategy, and the optimal strategy of task assignment is obtained. Finally, statistical analysis is performed on the test results to obtain the average value of each evaluation index of the task assignment strategy.

Table 2. Task parameters to be assigned.

Parameter	Value
Transmission direction	Upstream
Task interval	10 s
Task duration	[100 s, 300 s]
Target reliability	90%
Packet size	500 bytes
Number of packets per task	[100, 1000]
Packet loss rate	0–10%
Topology change rate	0–10%

6.2.2. Simulation Results and Analysis

The performance evaluation indicators in this research include the energy consumption of task assignment, the delay of successful allocation within the deadline, and the success rate of the task assignment. The success rate of task assignment represents the ratio of the number of tasks successfully assigned to the total number of tasks to be assigned in a simulation cycle, while the other two indicators are calculated according to Equations (5), (19), and (20). For comparison purposes, the simulation of centralized task allocation in the same network environment is conducted and represented by TRETA.

(A) Energy Consumption for Task Assignment

It is shown in Figure 6, the comparison of energy consumption among the three allocation strategies is proposed: the centralized task allocation TRETA and the distributed task assignment strategies TRETA-Cluster and TRETA-Multi-Sink. As seen in this figure, the energy consumption increases with the increase of node size. Among them, the energy consumption of TRETA is highest, since the global state of the network is converged to a single sink through long-distance multi-hop in large-scale WSNs. On the other hand, the energy consumption of TRETA-Multi-Sink is relatively small, as it is based on the multi-sink being able to collect the local state and merge it into the global view by multicasting. TRETA-Cluster is performed in two stages. When the task cannot meet the target reliability, if the reselected node in the cluster can reach the stage target reliability, the sink does not need to re-allocate the task, so the energy consumption of TRETA-Cluster is less than both TRETA and TRETA-Multi-Sink.

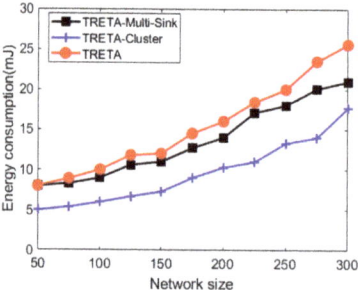

Figure 6. Distributed task assignment energy consumption.

(B) Delay of Successful Assignment within the Deadline

The delay of task assignment in the size of the network from 50 to 300 nodes is analyzed. As can be seen in Figure 7, TRETA has a substantial delay. Due to long-distance multi-hop aggregation to a single sink, achieving the global state of the cluster topology causes considerable delays in large-scale WSNs.

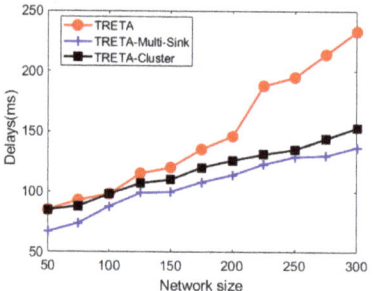

Figure 7. Task allocation delay.

(C) The Success Rate of Task Assignment

The success rate of task assignments under different task arrival rates is analyzed. Such a rate represents the number of tasks waiting to be allocated per second, and the success rate of the task allocation represents the ratio between the number of successfully assigned tasks in one simulation cycle and the total number of tasks to be assigned. As seen in Figure 8, the success rate of the task allocation decreases as the task arrival rate increases, and the rise in the task arrival rate leads to the increase in link conflict in the network. Since the TRETA-Multi-Sink can be distributed to multiple sinks for local processing according to the task area, its success rate in the task assignment is better than TRETA and TRETA-Cluster. The higher the task arrival rate, the more prominent the advantage.

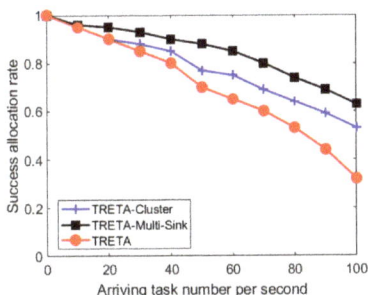

Figure 8. The success rate of task assignment under different task arrival rates.

(D) Energy Consumption and Delay of Task Assignments in the Dynamic Update of the Network

Figure 9 shows the energy consumption of the three proposed strategies for different network update ratios. It is noted that as the network update ratio increases, the energy consumption is significantly increased. Since the network update in TRETA is brought to the only sink through the long-distance multi-hop, the communication load of the network is increased. Moreover, the energy consumption of TRETA is higher than the other two strategies in most cases. With the network update ratio greater than 25%, the energy consumption of TRETA-Multi-Sink increases rapidly, surpassing TRETA and TRETA-Cluster, becoming the highest among the three strategies due to the exchange of global views between multiple sinks and the operation of ensuring global view consistency under a high network update rate.

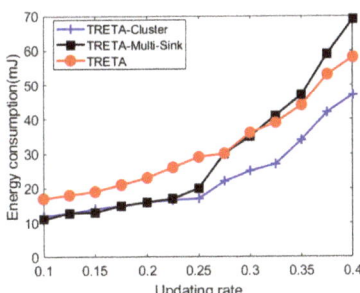

Figure 9. Energy consumption of task assignment under different network update rates.

Figure 10 shows the delays of the three proposed strategies for different network update ratios. As the network update ratio increases, the delays in the three strategies increase. Under the current network simulation environment, analysis shows that the delays of TRETA-Cluster and TRETA-Multi-Sink are smaller than TRETA when the network update ratio is lower than 25%, which is due to the fact that TRETA-Cluster and TRETA-Multi-Sink have a small number of updates in the network, they are processed locally without affecting the reliability of the task target, and a complete network update is not required. Nevertheless, when the network update ratio exceeds 25%, the delay of TRETA-Multi-Sink increases rapidly and becomes higher than TRETA, which is caused by the operation of ensuring global view consistency at such a high update rate.

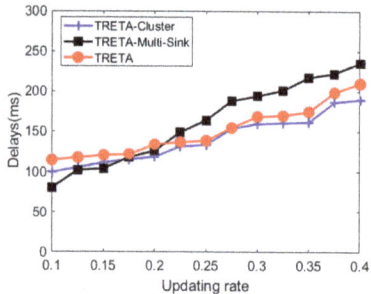

Figure 10. Delay in task assignment under different network update rates.

(E) Energy Resilience of the Transmission for Task Assignments

In order to evaluate the energy resilience of the transmission in WSNs, we quoted an index η in [34]. Since TRETA-Cluster is a distributed modification of TRETA for clustering topologies, we compared TRETA-Cluster and LEACH (Low Energy Adaptive Clustering Hierarchy) with index η. Simulation experiments were carried out under clustering topology. Let the number of initial active nodes be 500. For the same transmission task, we sample and count the number of packets received by sink nodes and the number of dead nodes at different time points. Then, we use Equation (22) to calculate the index η. Figure 11 shows the variation of the index η of the two algorithms over time. It can be seen from Figure 11 that the index η of both methods increases with time. Compared with LEACH, the growth rate of index η of TRETA-Cluster is more prominent, since TRETA-Cluster adds more redundancy packets than LEACH to meet the target reliability.

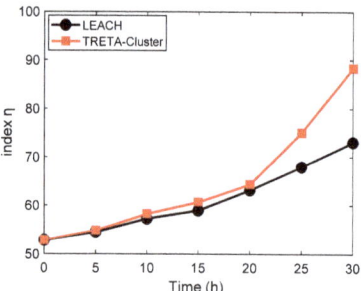

Figure 11. Energy resilience index of TRETA-Cluster and LEACH.

7. Concluding Remarks and Future Work

To solve the problems of poor scalability and reliability, significant communication overheads and delays in centralized task assignment for WSNs, we propose two distributed reliable and efficient transmission task allocations for WSNs, namely TRETA-Cluster and TRETA-Multi-Sink. The performance of these strategies is analyzed and evaluated, then compared with the centralized task assignment TRETA.

In TRETA-Cluster, the cluster head can perform local task assignment according to the stage target reliability constraint that reduces the communication cost and delay compared with TRETA. To summarize, the advantages include: (1) the cluster head can accurately select the number of nodes and the specific node ID in the cluster to complete the task, thereby saving energy; (2) as the topology changes occur in the cluster, the cluster head can select the replacement node according to the stage target reliability which has adaptive features. The advantages of TRETA -Multi-Sink are two-fold: (1) the global view is generated by merging the local views of multiple sinks that reduce the delay in obtaining the global view compared to the centralized acquisition method, and (2) multiple sinks respond to the local task assignment requirements in a globally optimized manner and do not need to communicate with remote nodes. It shows better scalability as well as overcoming the shortcomings of centralized task assignment in large-scale WSNs with significant communication overheads and delays.

From the analysis of both theoretical and simulation results, we show that the proposed distributed task allocation strategies are promising and superior to the centralized task allocation under the same network environment in terms of energy consumption, delay, and success rate.

From the observations of the limitations in distributed strategies when the network update ratio is higher than 25%, as TRETA-Multi-Sink no longer has the lead (which is caused by ensuring the consistency of the global view), we target the re-design and improvements of TRETA-Multi-Sink as a future direction to improve its efficiency. Another direction of investigation is to apply and adapt the proposed strategies in Cluster of Things (CoT) and Edge environments, where the communication conditions and environments highly yet dynamically vary all the time. Finally, this paper only researches the situation where the global view does not have concurrent updates. The efficiency of the distributed task allocation strategy when the network is concurrently updated will also be included as a future work direction.

Author Contributions: Methodology and investigation, X.Z.; Validation and data curation, J.Z. and S.Z.; investigation, Writing—original draft preparation, X.Z.; Writing—supervision, review, and editing, K.-C.L.

Funding: This work is jointly supported by the National Key Research and Development Program (Grant No. 2018YFC0604404), National Natural Science Foundation of China (Grant No. 51504010) and the Fundamental Research Funds for the Central Universities (Grant No. PA2019GDPK0079).

Conflicts of Interest: The authors declare no conflict of interest.

Nomenclature

Symbol	Description
s_i	i-th source node
m_i	The task to be assigned, which is the element in the set $M(t)$
e_{ij}	Energy consumed by the task m_i on the node N_j
e_{ij}^{comp}	Calculated energy consumption of N_j
e_j	Average processing energy consumption of the nodes in the network
t_{ij}^{comp}	Calculation time of the task m_i
e_{ij}^{comm}	Communication energy consumption of the node N_j
e_{ij}^{t}	Energy consumption of transmitting data of length l_{ij} bit at the distance d
e_{elec}	Energy consumption of operating the radio model for each bit
ε_{amp}	Coefficient of the transmit amplifier
e_{ij}^{r}	Energy consumption of receiving data of length l_{ij} bit at the distance d
P_i	Task period of m_i
S_i	Start time of task m_i
C_i	Deadline for task m_i
R_{WSN}	Mission reliability of the network
R_s	Target reliability
$ClusterR_{s_i,sink}(t,i)$	i-th transmission Reliability of clustered WSNs from s_i to the sink at time t
$R_A^c(t,i)$	Reliability of the i-th transmission performed by cluster c.
$R_T^c(t,i)$	Reliability of the head of cluster c which successfully sends the collected data of i-th transmission to sink at time t.
$MeshR_{s_i,sink}(t,i)$	i-th transmission reliability in mesh WSNs from s_i to the sink at time t
$R_{s_i,sink}^i(t,i)$	Probability that there exists at least one path whose packet delivery rate is greater than the threshold of i-transmission in the disjoint minimal path sets (from s_i to sink)
$R_{path(s_i,sink)}^j(t,i)$	i-th transmission reliability in mesh WSNs of j-th path from s_i to the sink at time t
$TG(V,E)$	Physical topology of the network

$RT(S, Ps)$	Routing topology of the network
$M(t)$	Task set to be assigned at time t
$\varphi(m_i)$	Mapped function of m_i
$Eng(\varphi(m_i))$	Energy consumption of the task m_i
$RT_b(S, Ps)$	Physical topology of the backbone network
$TG_b(S, Ps)$	Routing topology of the backbone network
$TG_c(S, Ps)$	The physical topology of the intra-cluster
$RT_c(S, Ps)$	Routing topology of the intra-cluster
$TG_{b'}(V, E)$	Physical topology after backbone network update
$RT_{b'}(S, Ps)$	Routing topology after backbone network update
$TG_{c'}(V, E)$	Physical topology of the intra-cluster after update
$RT_{c'}(S, Ps)$	Routing topology of the intra-cluster after update
$TG_j(V, E)$	Local physical topology of the j-th sink in a multi-sink topology
$RT_j(S, Ps)$	Local routing topology of the j-th sink in a multi-sink topology
$sink_j$	The closest sink to the task m_i
$sink_{jn}$	Sink adjacent to $sink_j$
$sink_k$	Sink involving topology changes
$viewID$	The version ID of the global view
$PathSet_{m_i}$	Path set of the source node to the sink in the coverage area of the task m_i
$PathSet_{m_i}'$	Path set after deleting the paths with link conflict in $PathSet_{m_i}$
$Paths_i$	Transmission path of each source node s_i to sink in $PathSet_{m_i}'$
R_{s_i}	Reliability of $Paths_i$
E_{s_i}	Energy consumption of $Paths_i$
$path_{R_s}$	Path to meet the target reliability R_s
R_{CH}	Target reliability of cluster head to sink after reliability allocation
$R_{CH'}$	Updated R_{CH}
R_{CM}	Target reliability of cluster head after reliability allocation
$sink_k^{res}$	Sink respond to update request of $sink_k$
$sink_k^{res}.RT_j(S, Ps)$	Routing topology of the j-th sink in $sink_k^{res}$
$sink_k^{res}.TG_j(V, E)$	Physical topology of the j-th sink in $sink_k^{res}$
$sink_k.record$	Update operation record of $sink_k$
$sink_k^{res}.log$	Log of sink which respond to update
S_W, S_R	Quorum set of writing and reading
$P_{overlap}$	Overlapping probability of S_W and S_R
CM	Communication load
Num_{cp}	Number of packets in the allocation policy by multicast,
Num_{gv}	Communication load for transmitting one global view
WM, RW	Writing load, reading load
VUM	Write load of version and update operation of the global view when it is updated
RT_{d_c}	Delay of task assignment for the clustered topology
RT_c'	Delay of the sink allocates a task to the subgraph composed of the cluster heads
MC	Delay of the multicast performed by the sink
CHD	Delay in the allocation of the task by the selected cluster head
RT_{d_d}	Delay of TRETA-Multi-Sink
$\overline{RT_c''}$	Average of the delays of the tasks assigned by multiple sinks locally
C_{rt}	Time to ensure the final consistency of the view
W_{rt}	Time of writing the global view whenever it is updated
R_{rt}	Time of reading the global view when the distributed task is assigned
η	Index for energy resilience of the transmission
N_p	Number of packet received by the sink after fixed time
S	Number of initial active nodes
D	Number of nodes death after a fixed time

References

1. Yu, W.; Huang, Y.; Garcia-Ortiz, A. Distributed optimal on-line task allocation algorithm for wireless sensor networks. *IEEE Sens. J.* **2018**, *18*, 446–458. [CrossRef]
2. Khalil, E.A.; Ozdemir, S.; Tosun, S. Evolutionary task allocation in internet of things-based application domains. *Future Gener. Comput. Syst.* **2018**, *86*, 121–133. [CrossRef]
3. Li, W.; Delicato, F.C.; Pires, P.F.; Lee, Y.C.; Zomaya, A.Y.; Miceli, C.; Pirmez, L. Efficient allocation of resources in multiple heterogeneous wireless sensor networks. *J. Parallel Distrib. Comput.* **2014**, *74*, 1775–1788. [CrossRef]
4. Ghebleh, R.; Ghaffari, A. A multi-criteria method for resource discovery in distributed systems using deductive fuzzy system. *Int. J. Fuzzy Syst.* **2017**, *19*, 1829–1839. [CrossRef]
5. Alskaif, T.; Zapata, M.G.; Bellalta, B. Game theory for energy efficiency in wireless sensor networks: Latest trends. *J. Netw. Comput. Appl.* **2015**, *54*, 33–61. [CrossRef]
6. Pilloni, V.; Navaratnam, P.; Vural, S.; Atzori, L.; Tafazolli, R. Tan: A distributed algorithm for dynamic task assignment in WSNs. *IEEE Sens. J.* **2014**, *14*, 1266–1279. [CrossRef]
7. Tang, C.; Li, X.; Wang, Z.; Han, J. Cooperation and distributed optimization for the unreliable wireless game with indirect reciprocity. *Sci. China Inf. Sci.* **2018**, *60*, 110205. [CrossRef]
8. Duan, X.; Zhao, C.; He, S.; Cheng, P.; Zhang, J. Distributed algorithms to compute Walrasian equilibrium in mobile crowdsensing. *IEEE Trans. Ind. Electron.* **2017**, *64*, 4048–4057. [CrossRef]
9. Pilloni, V.; Atzori, L.; Mallus, M. Dynamic involvement of real-world objects in the IoT: A consensus-based cooperation approach. *Sensors* **2017**, *17*, 484. [CrossRef]
10. Ye, D.Y.; Zhang, M.J.; Vasilakos, A.V. A survey of self-organization mechanisms in multiagent systems. *IEEE Trans. Syst. Man Cybern. Syst.* **2017**, *47*, 441–461. [CrossRef]
11. Khan, M.I.; Xia, K.; Ali, A.; Aslam, N. Energy-aware task scheduling by a true online reinforcement learning in wireless sensor networks. *Int. J. Sens. Netw.* **2017**, *25*, 244–258. [CrossRef]
12. Nguyen, C.L.; Hoang, D.T.; Wang, P. Data collection and wireless communication in internet of things (IoT) using economic analysis and pricing models: A survey. *IEEE Commun. Surv. Tutor.* **2016**, *18*, 2546–2590.
13. de la Hoz, E.; Gimenez-Guzman, J.; Marsa-Maestre, I.; Orden, D. Automated negotiation for resource assignment in wireless surveillance sensor networks. *Sensors* **2015**, *15*, 29547–29568. [CrossRef] [PubMed]
14. Wang, L.; Wang, Z.; Hu, S.; Liu, L. Ant colony optimization for task allocation in multi-agent systems. *China Commun.* **2013**, *10*, 125–132. [CrossRef]
15. Liu, Q.; Yin, X.; Yang, X.; Ma, Y. Reliability evaluation for wireless sensor networks with chain structures using the universal generating function. *Qual. Reliab. Eng. Int.* **2017**, *33*, 2685–2698. [CrossRef]
16. Cai, J.; Song, X.; Wang, J.; Gu, M. Reliability analysis for a data flow in event-driven wireless sensor networks. *Wirel. Pers. Commun.* **2014**, *78*, 151–169. [CrossRef]
17. He, Y.; Gu, C.; Han, X.; Cui, J.; Chen, Z. Mission reliability modeling for multi-station manufacturing system based on Quality State Task Network. *Proc. Inst. Mech. Eng. Part O J. Risk Reliab.* **2017**, *231*, 701–715. [CrossRef]
18. Zhu, X.J.; Lu, Y.; Zhang, J.; Wei, Z. Routing topology inference for wireless sensor networks based on packet tracing and local probing. *IEICE Trans. Commun.* **2019**, *102*, 122–136. [CrossRef]
19. Xin, Y.W.; Xiao, Y.W.; Narayanaswamy, B. Reliability allocation model and algorithm for phased mission systems with uncertain component parameters based on importance measure. *Reliab. Eng. Syst. Saf.* **2018**, *180*, 266–276.
20. Garg, H.; Sharma, S.P. Multi-objective reliability-redundancy allocation problem using particle swarm optimization. *Comput. Ind. Eng.* **2013**, *64*, 247–255. [CrossRef]
21. Li, R.; Wang, J.; Liao, H.; Huang, N. A new method for reliability allocation of avionics connected via an airborne network. *J. Netw. Comput. Appl.* **2015**, *48*, 14–21. [CrossRef]
22. Deng, R.; He, S.; Chen, J. An online algorithm for data collection by multiple sinks in wireless sensor networks. *IEEE Trans. Control Netw. Syst.* **2018**, *5*, 93–104. [CrossRef]
23. Basagni, S.; Carosi, A.; Petrioli, C.; Phillips, C.A. Coordinated and controlled mobility of multiple sinks for maximizing the lifetime of wireless sensor networks. *Wirel. Netw.* **2011**, *17*, 759–778. [CrossRef]
24. Lee, E.; Park, S.; Lee, J.; Oh, S.; Kim, S.H. Novel service protocol for supporting remote and mobile users in wireless sensor networks with multiple static sinks. *Wirel. Netw.* **2011**, *17*, 861–875. [CrossRef]

25. Lanny, S.; Kenneth, N.B.; Cormac, J.S. Planning the deployment of multiple sinks and relays in wireless sensor networks. *J. Heuristics* **2015**, *21*, 197–232.
26. Banerjee, A.; Hussain, D.M.A. SD-EAR: Energy aware routing in software defined wireless sensor networks. *Appl. Sci.* **2018**, *8*, 1013. [CrossRef]
27. Gilbert, S.; Lynch, N. Brewer's conjecture and the feasibility of consistent, available, partition-tolerant web services. *Acm Sigact News* **2002**, *33*, 51–59. [CrossRef]
28. Eventually Consistent-Revisited. Available online: http://www.allthingsdistributed.com/2008/12/eventually_consistent.html (accessed on 4 September 2019).
29. DeCandia, G.; Hastorun, D.; Jampani, M.; Kakulapati, G.; Lakshman, A.; Pilchin, A.; Sivasubramanian, S.; Vosshall, P.; Vogels, W. Dynamo: amazon's highly available key-value store. In *ACM SIGOPS Operating Systems Review*; ACM: New York, NY, USA, 2007; Volume 41, pp. 205–220.
30. Vashisht, P.; Sharma, A.; Kumar, R. Strategies for replica consistency in data grid—A comprehensive survey. *Concurr. Comput. Pract. Exp.* **2018**, *29*, e3907. [CrossRef]
31. Wang, X.; Sun, H.; Deng, T. On the tradeoff of availability and consistency for quorum systems in data center networks. *Comput. Netw.* **2015**, *76*, 191–206. [CrossRef]
32. Raychoudhury, V.; Cao, J.; Wu, W. K-directory community: Reliable service discovery in MANET. *Pervasive Mob. Comput.* **2011**, *7*, 140–158. [CrossRef]
33. Guerraoui, R.; Vukolic, M. Refined quorum systems. *Distrib. Comput.* **2010**, *23*, 1–42. [CrossRef]
34. Leccese, F.; Cagnetti, M.; Giarnetti, S.; Petritoli, E.; Luisetto, I.; Tuti, S.; Leccisi, M.; Pecora, A.; Maiolo, L.; Đurović-Pejčev, R.; et al. Comparison between routing protocols for wide archeological site. In Proceedings of the IEEE International Conference on Metrology for Archaeology and Cultural Heritage, Cassino, Italy, 22–24 October 2018; ISBN 978-1-5386-52.

© 2019 by the authors. Licensee MDPI, Basel, Switzerland. This article is an open access article distributed under the terms and conditions of the Creative Commons Attribution (CC BY) license (http://creativecommons.org/licenses/by/4.0/).

Article

An Audification and Visualization System (AVS) of an Autonomous Vehicle for Blind and Deaf People Based on Deep Learning

Surak Son, YiNa Jeong and Byungkwan Lee *

Department of Computer Engineering, Catholic Kwandong University, Gangneung 25601, Korea; sonsur@naver.com (S.S.); lupinus07@nate.com (Y.J.)
* Correspondence: bklee@cku.ac.kr; Tel.: +82-33-649-7573

Received: 14 October 2019; Accepted: 12 November 2019; Published: 18 November 2019

Abstract: When blind and deaf people are passengers in fully autonomous vehicles, an intuitive and accurate visualization screen should be provided for the deaf, and an audification system with speech-to-text (STT) and text-to-speech (TTS) functions should be provided for the blind. However, these systems cannot know the fault self-diagnosis information and the instrument cluster information that indicates the current state of the vehicle when driving. This paper proposes an audification and visualization system (AVS) of an autonomous vehicle for blind and deaf people based on deep learning to solve this problem. The AVS consists of three modules. The data collection and management module (DCMM) stores and manages the data collected from the vehicle. The audification conversion module (ACM) has a speech-to-text submodule (STS) that recognizes a user's speech and converts it to text data, and a text-to-wave submodule (TWS) that converts text data to voice. The data visualization module (DVM) visualizes the collected sensor data, fault self-diagnosis data, etc., and places the visualized data according to the size of the vehicle's display. The experiment shows that the time taken to adjust visualization graphic components in on-board diagnostics (OBD) was approximately 2.5 times faster than the time taken in a cloud server. In addition, the overall computational time of the AVS system was approximately 2 ms faster than the existing instrument cluster. Therefore, because the AVS proposed in this paper can enable blind and deaf people to select only what they want to hear and see, it reduces the overload of transmission and greatly increases the safety of the vehicle. If the AVS is introduced in a real vehicle, it can prevent accidents for disabled and other passengers in advance.

Keywords: autonomous vehicle; audification; sensor; visualization; speech to text; text to speech

1. Introduction

Autonomous cars represent a key area of the fourth industrial revolution. Various carmakers around the world are actively conducting research with the aim of producing fully autonomous vehicles, and advances in information and communications technology (ICT) are greatly speeding up the development of autonomous vehicle technology. A fully autonomous vehicle means that people are not involved in driving at all, the car drives on its own and immediately deals with a variety of risk factors. Autonomous vehicles can be classified into five stages. Level 0 is the stage where the driver performs all actions to drive the vehicle, and there is no autonomous driving at all. Level 1 is the passive stage, where the vehicle automatically handles acceleration, steering, etc. In addition, vehicles in this stage have a lane-keeping assist (LKA), which automatically returns the vehicle to the original lane when it gets out of the lane without turning on a turn signal, and cruise control (CC), which maintains a specified speed. Level 2 is the stage where the vehicle automatically decelerates and even operates the brakes, which is slightly more advanced than Level 1.

Level 3 is a semi-autonomous driving stage, which includes all the functions of Level 2 and analyzes the road situation using advanced sensors or radar so that the car can drive a certain distance on its own without driver intervention. Level 4 is the stage where a self-driving vehicle can safely reach the designated destination without the driver's intervention. However, this is not a perfect stage because it cannot completely guarantee safety. At Level 5, the driver does not exist, there are only passengers, and the vehicle performs all movements on its own. The vehicle uses artificial intelligence and various sensors to cope with all possible road situations [1].

Many companies, including Google®, Tesla, and Mercedes-Benz, are testing fully autonomous driving vehicles on the road. Google's self-driving car, however, collided with a large bus, while Tesla's autonomous driving vehicle crashed into a bicycle and caused a fire [2,3]. Fully autonomous driving vehicles are not yet available in the test phase. In particular, since occupants of fully autonomous vehicles do not drive directly on their own, it is not easy to recognize the situation before and after a vehicle accident, and also, they are likely to be negligent in checking the vehicle. Therefore, fully autonomous vehicles should frequently inform the occupants of their analysis results through artificial intelligence-based self-diagnosis. If all passengers of a fully autonomous vehicle are deaf or blind, there is no way to inform them of the results of the self-diagnosis analysis, which increases the risk of an accident. In 2016, Google succeeded in piloting a self-driving vehicle with a blind person, but even then, he was in the vehicle with a sighted person [4].

To address these problems, this paper proposes an audification and visualization system (AVS) for blind and deaf people. The AVS consists of a data collection and management module (DCMM), an audification conversion module (ACM), and a data visualization module (DVM). The DCMM stores and manages the data collected from the vehicle. The ACM has a speech-to-text submodule (STS) that recognizes a user's speech and converts it to text data, and a text-to-wave submodule (TWS) that converts text data to voice. The DVM visualizes the collected sensor data, fault self-diagnosis data, etc., and places visualized data according to the size of the vehicle's display.

The composition of this paper is as follows: Section 2 describes the existing studies related to the AVS in this paper. Section 3 details the structure and operation of the AVS. Section 4 compares it with the existing methods to analyze performance. Section 5 discusses the conclusion of the proposed AVS and future research directions.

2. Related Works

2.1. Hidden Markov Model

Li et al. proposed a new algorithm combining the hidden Markov model (HMM) and Bayesian filtering (BF) techniques to recognize a driver's intention to change lanes. The grammar recognizer in the algorithm was inspired by speech recognition, and the output value of the algorithm is preliminary classified behavior. The behavior classification value, the final output of BF, is generated using the current and previous output of the HMM. This algorithm was validated using a naturalistic dataset. The proposed HMM–BF framework can meet 93.5% and 90.3% recognition accuracy for right and left lane changes, respectively, which is a significant improvement over the HMM-only algorithm [5].

Liang et al. proposed a new filter model-based hidden Markov model (FM-HMM) for intrusion detection system (IDS) to decrease the overhead and time for detection without impairing accuracy. This work was the first to model the state pattern of each vehicle in vehicle ad hoc networks as an HMM and quickly filter the messages in the vehicle instead of detecting these messages. The FM-HMM consists of three modules: The schedule module generates the parameters of the HMM for adjacent vehicles by using the Baum–Welch algorithm [6]; the filter module predicts the future state of an adjacent vehicle by using several HMMs; and the update module updates the parameters of the HMM using the timeliness method [7].

Saini et al. introduced two additional kernels based on convex hull and the Ramer–Douglas–Peucker (RDP) algorithm and proposed a trajectory classification approach, which supervises a combination of global and segmental HMM-based classifiers. To begin with, the HMM is used for global classification of categories to provide state-by-state distribution of trajectory segments. The trajectory and global recognition that completed classification improved the classification results. Finally, global and segmental HMM are combined using a generic algorithm. They experimented with two public datasets, commonly known as T15 and MIT, and achieved accuracy of 94.80% and 96.75%, respectively, on these datasets [8].

Siddique et al. presented a self-adaptive sampling (SAS) method for mobile sensing data collection. SAS regulates the sampling rate using the flow state of vehicles estimated in individual lanes, classifying the estimated flow state into four categories (free flow, stopped, acceleration, and deceleration) using the HMM and identifies stopping and movement in the lane using support vector machine (SVM). The identification of vehicle flow conditions is used to change the sampling rate. The SAS method can reduce the total amount of data by 67–77% while retaining the most important data points [9].

Liu et al. proposed a controller integration approach that adopts behavior classification to improve the ability of leading vehicles to cope with outside obstacles. This approach, based on the HMM, detects if there is any hazard behavior in the neighboring vehicles. The detected behaviors are transmitted to the model predictive controller in a driving vehicle. A behavior-guarded cost function of the controller is designed to increase the stability against danger while driving. The effect of the state deviation of the lead vehicle in the convoy is studied based on leader-to-formation stability characteristics. Furthermore, a nonlinear bound is also given to specify the performance of the proposed controller [10].

Mingote et al. proposed a novel differentiated neural network with an alignment mechanism for text-dependent speaker verification. They did not extract the embedding of speech from the global average pooling of the temporal dimension. Because the proposed neural network uses phonetic information for verification, it maintains the temporal structure of each phrase by replacing a redirection mechanism with the alignment model of a phonetic phrase. They applied convolutional neural networks (CNNs) to the front end and learn the neural network that produces super-vectors of each word, whose pronunciation and syntax are distinguished at the same time. This choice has the advantage that super-vectors encode phrases and speaker information, which showed good performance in text-dependent speaker verification tasks [11].

Wang et al. proposed a method for determining vehicle driving status from time-ordered trajectory data using the HMM. This method is preprocessed to discard track sequences with insufficient length to ensure the usefulness of linear smoothing and least squares fitting. A directional area segmentation algorithm was proposed to extract the directional angle of the vehicle from the preprocessed orbital sequences, and it obtains and patterns the various driving states of the vehicle in real time. Finally, multiple observations based on the Baum–Welch algorithm can obtain the optimal HMM model parameters for each track pattern at a particular traffic site and then determine the real-time vehicle driving state by matching with the trained HMM model above [12].

Kato et al. proposed a car tracker based on a hidden HMM/Markov random field (MRF)-based segmentation method that is capable of classifying each small region of an image into three categories (vehicles, shadows of vehicles, and background) from a traffic-monitoring video. The temporal continuity of the different categories for one small location is modeled as a single HMM along the time axis, independent of the neighboring regions. In order to incorporate spatially dependent information among neighboring regions into the tracking process, at the state-estimation stage, the output from the HMMs is regarded as an MRF and the maximum a posteriori criterion is employed in conjunction with the MRF for optimization. At each time step, the state estimation for the image is equivalent to the optimal configuration of the MRF generated through a stochastic relaxation process [13].

Wang et al. proposed a novel framework called chain of road traffic incident (CRTI) for predicting accidents. CRTI observes the moving features of a driving vehicle, which are the external performance of a road transport system that reflects the "health states" (safety states) of a given time. A two-stage modeling procedure for CRTI is then proposed using a scenario-based strategy. A support vector machine is used to classify leaving versus remaining in lane scenes, and Gaussian mixture-based hidden Markov models are developed to recognize accident versus non-accident pattern CRTIs given the classified scene [14].

Jazayeri et al. presented a comprehensive approach to localize target vehicles with video under various environmental conditions. Extracted geometric features from the video are continuously projected onto a 1-D profile and constantly tracked. This method compensates for the complexity of the vehicle shape, color, and type recognition using time information and models the driver's field of vision probabilistically according to the features of the background and the movement of the vehicle. In the proposed method, the HMM is used to probabilistically track vehicles apart from the background [15].

2.2. Vehicle for Disabilities

Choromański et al. presented an original concept of an urban transport system based on a hybrid vehicle that can move by a human-driven electric vehicle or a special pod car vehicle (right of way a, b, or c). The system was developed at the Warsaw University of Technology and is referred to as the hybrid vehicle and transit system for urban application (HVTASUA). The system was designed not only for ordinary drivers, but also for elderly people and those who lack driving skills, such as people with physical disabilities. Based on this system, an original design for vehicles and standardization of human-machine interface (HMI) are proposed. This HVTSUA is integrated with Ford's already developed Eco-Car system. Integrating these two elements and equipping them with new technology became the basis for a system with new quality [16].

The aim of Bennett et al. was to investigate possible barriers to the use of autonomous vehicles (AVs) that are perceived by people with intellectual disabilities. A structural topic modelling (STM) approach was employed to analyze 177 responses of mentally disabled people to an open-ended question about AV travel intentions. Results from the STM, together with data on the sample participants' level of internal locus of control, generalized anxiety, age, gender, prior knowledge of AVs, and level of individual disability were then incorporated into a structural equation model constructed to relate attitudinal topics identified by the STM to the participants' willingness to travel in AVs. Three categories of attitudes toward AVs arose from the STM, relating to freedom, fear, and curiosity. Two of the three themes, freedom and fear significantly predicted the participants' willingness to use driverless vehicles. The freedom theme was significantly explained by generalized anxiety, intensity of disability, and prior knowledge of AVs. The concept of fear depended significantly on generalized anxiety and prior knowledge, and also on the locus of control and (female) gender. The theme of curiosity was influenced by locus of control and prior knowledge [17]. They employed a mixed research methodology to assess attitudes toward AVs in a UK sample of individuals with physical disabilities affecting their mobility. Participants were asked in an open-ended way to express their ideas about AVs, and their responses were analyzed using STM. Outputs for the STM analysis were then adopted in a structural equation model (SEM) constructed to predict the willingness of the participants to travel in driverless vehicles. The results were compared with those obtained from a control group of people without physical disabilities. The attitudes of people with disabilities toward AVs were significantly different from those of respondents without disabilities. Attitudes toward AVs among people with disabilities were significantly influenced by their level of interest in new technologies, generalized anxiety, intensity of disability, prior knowledge of AVs, locus of control, and action orientation. A latent class analysis confirmed the relevance of these variables as determinants of the views of people with disabilities on AVs [18].

Xu et al. proposed an intelligent guided vehicle prototype for blind people. The system integrates ultrasound and photoelectric detection using ARM as a controller and processor, and automatically navigates to the destination. This system consists of four modules: ultrasonic detection, photoelectric detection, voice prompt, and automatic control. The ultrasonic detection module detects reflected signals on the road using the distance between the road block and the vehicle. The photoelectric detection module recognizes the road and tracks the vehicle. The voice prompt module and automatic control module allow blind people to enter voice commands and control the vehicle [19].

3. Proposed Method

3.1. Overview

An autonomous driving vehicle performs a number of driving functions using various sensors. Although the sensor data measured in the vehicle varies, there is a limit to the type of data displayed on the instrument cluster and the way it is indicated. For example, the vehicle's instrument cluster displays some information about the vehicle, such as speed, RPM, etc., but much other information, such as air pressure, door opening and closing, etc., is still not accurately displayed. If all of this information were to be displayed on the instrument cluster, it would be difficult for occupants to look at so much information at a glance. It goes without saying that if only blind or deaf people are riding in self-driving cars, it would be more difficult to identify the information. The status information of the vehicle should be visualized more accurately for the deaf and be audified for the blind.

To solve these problems, this paper proposes an audification and visualization system (AVS) of an autonomous vehicle for blind and deaf people based on deep learning, which uses self-diagnosis results published previously [20,21] and the sensor data collected from vehicles, a graphical library to visualize the data desired by deaf people and audify the data desired by blind people. Figure 1 shows the structure of the AVS, which consists of three modules.

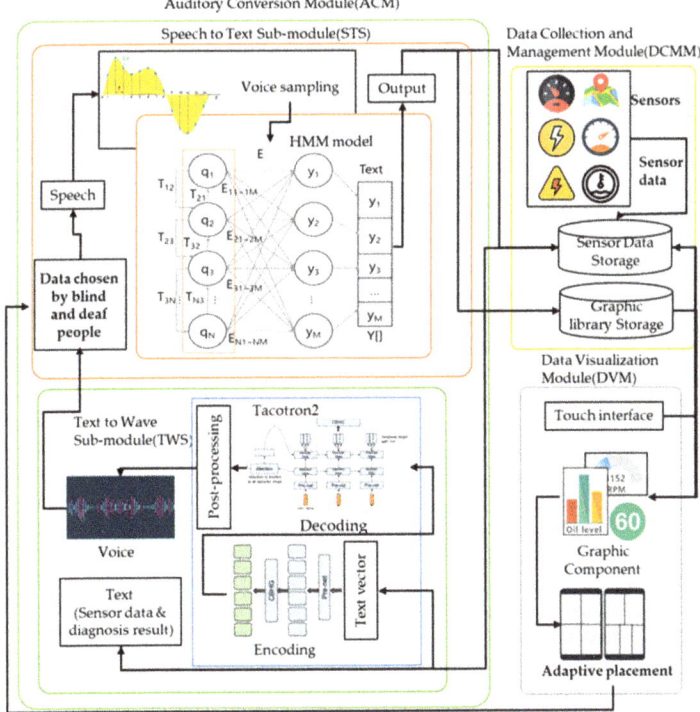

Figure 1. Structure of the system. HMM: hidden Markov model.

The data collection and management module (DCMM) stores and manages all the sensor data collected from the vehicle, self-diagnosis data, and graphics libraries in in-vehicle storage. The auditory conversion module (ACM) receives speech and provides voice output about the vehicle's condition. The ACM consists of a speech-to-text submodule (STS) that recognizes speech and converts it to text data and a text-to-wave submodule (TWS) that converts data to voice. The STS learns HMM point-to-point using the in-vehicle NVIDIA px2 driver and receives a sensor name from the user by using the learned HMM. The TWS converts text data to voice using Tacotron2 and outputs it. The data visualization module (DVM) visualizes the data received from the DCMM and places the visualized data, the graphic components, on an in-vehicle display.

The AVS operates as follows. First, the AVS receives information visualized through the touch interface or audified through a speech recognizer. If someone uses speech recognition, the ACM converts the speech to text using the HMM of the STS and sends the converted text to the CMM. The CMM transmits the vehicle's sensor data and self-diagnosis data [20,21] to the ACM, and the ACM audifies the information received from the DCMM.

If someone uses the touch interface, the DCMM selects the sensor data in the sensor data storage and the graphics functions in the graphic library storage and transmits them to the DVM. The DVM visualizes the data entered by the DCMM and the vehicle's self-diagnostic data and positions the visualized data on the vehicle's display adaptively to inform the person of the vehicle's condition.

3.2. Design of a Data Collection and Management Module

The DCMM stores the sensor data and Python seaborn packages collected from the vehicle. It consists of a transceiver that receives in-vehicle sensor messages, sensor data storage where the received messages are stored, and graphic library storage where the graphic libraries are stored. The transceiver consists of a CAN transceiver that receives CAN messages, a MOST transceiver that receives MOST messages, a FlexRay transceiver that receives FlexRay messages, and a LIN transceiver that receives LIN messages. Each received message is stored in the sensor data storage, as shown in Figure 2.

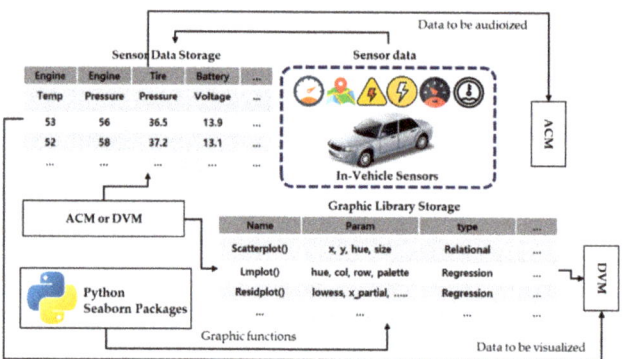

Figure 2. Structure of a data collection and management module.

When the ACM or DVM requests the audified or visualized data to the DCMM, the DCMM retrieves them from sensor data storage and transmits them to the ACM or DVM again and also provides the functions necessary for the DVM to visualize the data in graphic library storage.

The DCMM then stores the condition of the vehicle using the previously published self-diagnosis system and transmits it to the ACM and DVM, so that the self-diagnosis results of the autonomous vehicle are visualized or audified.

3.3. Design of an Auditory Conversion Module

The auditory conversion module (ACM) receives the user's speech and provides voice output of the vehicle's condition. Figure 3 shows the flow of the ACM. The ACM consists of a speech-to-text submodule (STS) that recognizes speech and converts it into text data and a text-to-wave submodule (TWS) that converts text data to voice. The ACM receives the data that has to be audified by using the STS. The ACM transmits the received data to the DCMM, and the DCMM again transmits the data that has to be audified to the ACM. The ACM then generates sentences with the data, which have to be audified using the TWS and delivers them after changing to voice. For example, if a person says, "Tell me the brake status when the sound is strange," the STS recognizes it and requests the information about the brake sensors and the vehicle's self-diagnosis information to the DCMM. The DCMM transmits the brake information and the vehicle's self-diagnosis information to the TWS, and the TWS informs the user after turning them into voice.

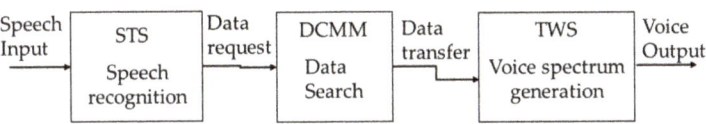

Figure 3. Structure and data flow of ACM.

3.3.1. Speech-to-Text Submodule (STS)

The STS uses the HMM to recognize speech. Unlike a recurrent neural networks (RNN), which reflects all previous states, the HMM recognizes the user's voice based on the Markov chain, which relies on the immediately previous state. Because the STS recognizes in a short word the data that have to be visualized in a short word, the HMM is more efficient than the RNN. The STS learns the HMM independent of the cloud and receives speech from the vehicle's microphone and outputs the text data by using a point-to-point learning method and NVIDIA PX2 driver. An in-vehicle solid-state drive (SSD) transmits training data to the NVIDIA PX2. The NVIDIA PX2 learns the HMM using it. The learned HMM converts speech to text data. The ACM transmits the converted text data to the DCMM. Figure 4 shows the learning order of the HMM.

$$P(q_i|q_1, \ldots, q_{i-1}) = P(q_i|q_{i-1}). \tag{1}$$

Equation (1) represents a Markov chain. The Markov chain can compute the results of perfect input and output. However, the HMM, which has to compute the results with only a person's input, uses the Markov chain assuming that the results are hidden. In general, the HMM is represented in the form of a directed graph, as shown in Figure 5. In the graph, q_i represents a hidden state and y_j refers to the observed value from q_i. The HMM consists of <Q, Y, π, T, E>, and Table 1 shows the components of the HMM.

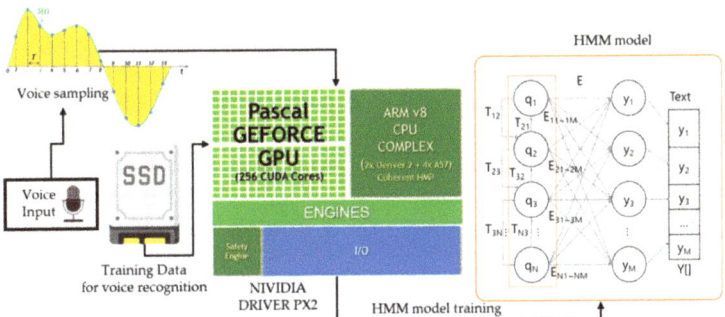

Figure 4. Point-to-point learning order of the hidden Markov model (HMM).

Table 1. Components of HMM.

Set Name	Set Contents	Meaning
Q	{$q_1, q_2, \ldots, q_N$}	Set of hidden states
Y	{$y_1, y_2, \ldots, y_M$}	Set of observed values in a hidden state
π	{ $\pi_1, \pi_2, \ldots, \pi_N \vert R^N$}	Set of initial probabilities $p(q_i)$ with the probability of initial state q_i
T	{$T_{12}, T_{21}, \ldots, T_{NM}, T_{MN} \vert R^{N \times N}$}	Set of transition probabilities $p(q_j \vert q_i)$ indicating the probability of moving from q_i to q_j
E	{$E_{11}, E_{12}, \ldots, E_{NM} \vert R^{N \times M}$}	Set of assignment probabilities $p(y_j \vert q_i)$ indicating the probability that y_j will occur in q_i
θ	{π, T, E}	HMM parameter

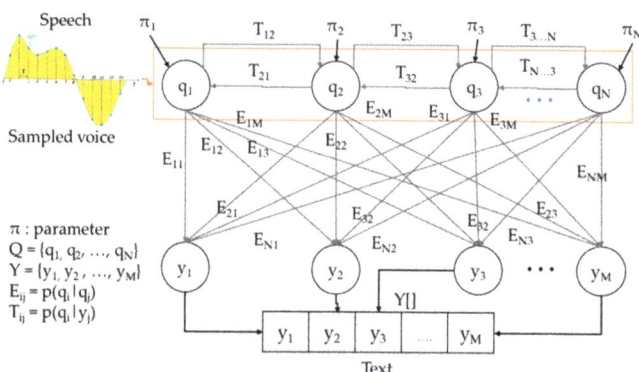

Figure 5. Flowchart of HMM.

When θ is given in the HMM, the STS computes a probability about a given observed value by using a dynamic programming-based forward algorithm and a list of the states with the highest probability by using the Viterbi algorithm. Viterbi Algorithm is a dynamic programming technique for finding the most likely sequence of hidden states. Here, the hidden states refer to observed value of the HMM.

When the list of states is computed, the STS learns the HMM's θ by using the Baum–Welch algorithm and the training data. In other words, the STS computes the probability of observed values and the list of states by specifying the initial θ and learns the HMM based on them. Because the STS outputs part names by receiving the user's speech and the observed STS value, Y becomes the part name of a vehicle that can be audified, and these are shown in Table 2.

Table 2. Observed values that can be output from the speech-to-text submodule (STS).

Observed Value	Data Name	Observed Value	Data Name
y_1	Speed	y_7	Driving distance
y_2	RPM	y_8	Timing belt
y_3	Tire	y_9	Spark plug
y_4	Steering wheel	y_{10}	Air conditioner
y_5	Engine oil	y_{11}	Brake pad
y_6	Coolant		

To begin with, the STS sets a random initial θ and computes the observed value $P(Y|\theta)$. Equation (2) computes the first observed value, y_1, and Equation (3) computes y_1–y_m:

$$P(y_1|\theta) = \sum_{i1} p(q_{i1}) p(y_1|q_{i1}), \qquad (2)$$

$$P(Y|\theta) = \sum_{i\,1}\sum_{i\,2}\cdots\sum_{i\,m} p(q_{i\,1})p(y_1|q_{i\,1})p(q_{i\,2}|q_{i\,1})p(y_2|q_{i\,2})\cdots p(q_{i\,m}|q_{i\,m-1})p(y_m|q_{im}). \quad (3)$$

However, because this method has the time complexity of $O(n^m)$, it is impossible to compute the observed values inside the vehicle. Therefore, this paper uses the forward algorithm in the HMM to reduce the computation time of the STS. The key idea of the forward algorithm is to store duplicate computation results in a cache and fetch them when necessary. The forward algorithm defines the new variable $\alpha_t(p_j)$ in Equation (4). Using the defined $\alpha_t(p_j)$, Equation (3) is simplified to Equation (5):

$$\alpha_t(q_j) = p(y_1, y_2, \ldots, y_t, s_t = q_j|\theta), \quad (4)$$

$$\alpha_t(q_j) = \sum_{i=1}^{N} a_{t-1}(q_i) p(q_j|q_i) p(y_t|q_j). \quad (5)$$

When the observed values are computed, the STS traces back the hidden states with the observed values and makes the back-traced states into one array by using the Viterbi algorithm. Algorithm 1 represents the pseudo-code in which the STS computes observed values using the forward algorithm and an array of hidden states by using the Viterbi algorithm.

Algorithm 1. Computation of observed values and array of states.

Input: initial probabilities π, transition probabilities T,
       emission probabilities E, number of states N,
       observation Y = $y_1, y_2, \ldots, y_m$;
Forward(π, T, E, Y){
  for(j=1; jN; j++){
    $\alpha_1(q_j) = p(q_j)p(y_1|q_j)$
  }
  for(t=2; t<=T, t++){
    for(j=1; j<=N; j++){
      $\alpha_t(q_j) = \sum_{i=1}^{N} a_{t-1} p(q_i) p(q_j|q_i) p(y_t|q_j)$
    }
  }
  $p(Y|\pi, T, E) = \sum_{i=1}^{N} a_T(q_j)$
  return $p(Y|\pi, T, E)$
}
Viterbi(π, T, E, Y){
  for(j=1; j<=N; j++){
    $v_1(q_j) = p(q_j)p(y_1|q_j)$
  }
  for(t = 2; t<=T, t++){
    for(j=1; j<=N; j++){
      $v_t(q_j) = \max\limits_{q \in Q} v_{t-1}p(q)p(q_j|q)p(y_t|q_j)$
      $S[t] = \arg\max\limits_{q \in Q} v_{t-1}p(q)p(q_j|q)p(y_t|q_j)$
    }
  }
  return S[]
}

The STS learns the HMM by using the computed observed values, an array of states, and the Baum–Welch algorithm. The HMM's learning is to make the best parameter, θ^*. The Baum–Welch algorithm computes a parameter θ using correct Y and Q in Equation (6):

$$P(Y, Q|\theta) = \prod_{k=1}^{N}\left(p(q_1^k)p(y_1^k|q_1^k)\prod_{t=2}^{M}(p(q_t^k|q_{t-1}^k)p(y_t^k|q_t^k))\right). \quad (6)$$

However, because the STS does not know the exact Q for Y, the STS converts Equation (6) to Equation (7) by taking the log on both sides of Equation (6) and computes $Q(\theta, \theta')$ by substituting Equation (7) with Equation (8):

$$\log p(Q, Y|\theta) = \sum_{k=1}^{N}\{p(q_1^k)p(y_1^k|q_1^k) + \sum_{t=2}^{M}\log(p(q_t^k|q_{t-1}^k)) + \sum_{t=2}^{M}\log(p(y_t^k|q_t^k))\}p(Q, Y|\theta'), \quad (7)$$

$$Q(\theta, \theta') = \sum_{q_1, q_2, \ldots, q_N} \log\{p(Q, Y|\theta)\}p(Q, Y|\theta'). \quad (8)$$

In Equation (7), θ means a current parameter and θ' is the immediately previous one. The STS substitutes Equation (7) with Equation (8) and generates $L(\theta, \theta')$ that can compute $Q(\theta, \theta')$ using the Lagrange multiplier method [22]. Equation (9) indicates $L(\theta, \theta')$:

$$L(\theta, \theta') = Q(\theta, \theta') - \omega_\pi\left(\sum_{i=1}^{N}p(q_i) - 1\right) - \sum_{i=1}^{N}\left(\sum_{j=1}^{N}p(q_j|q_i) - 1\right) - \sum_{i=1}^{N}\omega_{E_i}\omega_{T_i}\left(\sum_{j=1}^{M}p(y_j|q_i) - 1\right). \quad (9)$$

In Equation (9), $p(q_i)$ means π_i of θ, $p(q_j|q_i)$, T_{ij} of θ, $p(y_j|q_i)$, E_{ij} of θ, and ω, Lagrange multiplier. Because the STS has to find θ that maximizes $L(\theta, \theta')$, it computes the optimal parameter θ^* by differentiating $L(\theta, \theta')$ with π_i, T_{ij}, and E_{ij}. Equation (10) computes the optimal π_i value, Equation (11) the optimal T_{ij} value, and Equation (12) the optimal E_{ij} value:

$$\pi_i = \frac{\sum_{k=1}^{K}p(Y, q_1^k = q_i|\theta')}{\sum_{j=1}^{N}\sum_{k=1}^{K}p(Y, q_1^k = q_i|\theta')}, \quad (10)$$

$$T_{ij} = \frac{\sum_{k=1}^{K}\sum_{t=2}^{T}p(s_{t-1}^k = q_j, s_t^k = q_j|Y^k, \theta')}{\sum_{k=1}^{K}\sum_{t=2}^{T}p(s_t^k = q_j|Y^k, \theta')}, \quad (11)$$

$$E_{ij} = \frac{\sum_{k=1}^{K}\sum_{t=1}^{T}p(s_t^k = q_i|Y^k, \theta')I(y_t^k = q_j)}{\sum_{k=1}^{K}\sum_{t=1}^{T}p(s_t^k = q_i|Y^k, \theta')}. \quad (12)$$

If Equations (10)–(12) are used, it is possible to compute the optimal parameter θ^* from the previous parameter θ. When the STS computes the optimal parameter θ^* using the initial parameter, the STS receives speech and transmits a list of text data for audification to the DCMM.

3.3.2. Text-to-Wave Submodule

The text-to-wave submodule (TWS) audifies the sensor data transmitted from the DCMM. It executes text-to-speech (TTS) using Google's Tacotron2 [23]. Because the Tacotron2 operates on the basis of an RNN encoder–decoder model, in this paper Tacotron2 is learned by using the LJ Speech Dataset [24] and estimates whether the sensor data of the vehicle are accurately output to voice. Figure 6 shows the flow of the TTS.

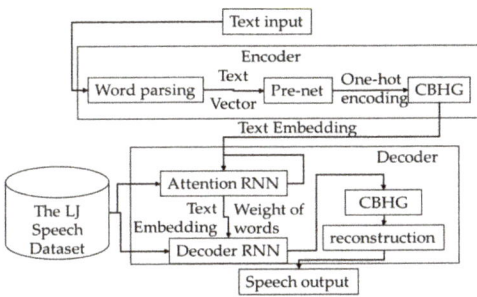

Figure 6. Structure of TWS.

To begin with, the TWS receives sentences and breaks the words in the encoder. When a vector of the words is entered in the encoder, the pre-net of the Tacotron2 one-hot encodes the vector. Here, one-hot encoding means the process of converting a text vector into an array of 0 and 1 that the encoder can recognize easily.

When the pre-net one-hot encodes the text vector, the TWS transmits it to the Tacotron2's convolution bank + highway net + bidirectional gated recurrent unit (CBHG) network, a neural network model. The convolution bank extracts the features of text from one-hot-encoded text vectors, the highway net deepens the neural network model, and the bidirectional gated recurrent unit (GRU) generates text embedding by considering the previous and subsequent information of vectors processed in pre-net. The input sequence of the Convolution Bank is K sets of 1-D convolutional filters. Where 1 set of 1-D convolutional filters consists of $C_1, C_2, C_3 \ldots$ and C_K filter. The input sequence is max-pooled. The output of the Convolution Bank is delivered to the Highway Net which extracts high-level features from it. Finally, the Bidirectional GRU extracts sequential features using the order of the input sequence. Figure 7 shows the structure of the CBHG.

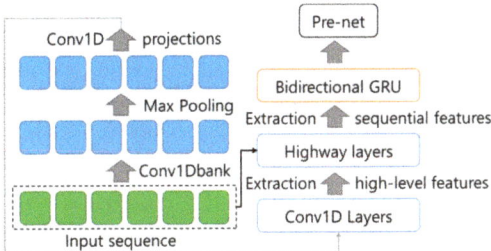

Figure 7. The structure of the convolution bank + highway net + bidirectional gated recurrent unit CBHG.

The generated text embedding is sent to an attention model. The attention model determines which is more complex and which is more important in the transmitted text embedding by using the attention RNN. This attention model enables the Tacotron2 to do point-to-point learning and converts to voice the text vectors that are not learned. Figure 8 shows the composition of the attention and decoder RNNs. Equation (13) indicates the attention value used in the attention RNN:

$$\text{Attention }(Q, K, V) = \text{Attention value}, \tag{13}$$

where Q means the query of the hidden states that decoder cells have at the time of t, K means the keys of the hidden states that encoder cells have at all times, and V means the values of the hidden states that encoder cells have at all times.

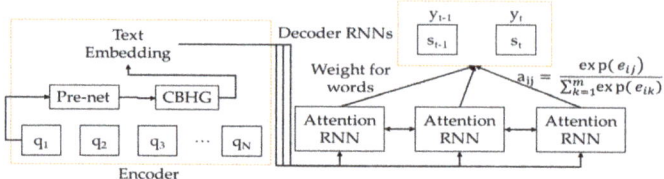

Figure 8. Importance of words computed in attention RNN.

The attention RNN sends the attention values, the weight of each word, to the decoder RNN. Attention value a is obtained from Equation (14):

$$a_{ij} = \frac{\exp(e_{ij})}{\sum_{k=1}^{m} \exp(e_{ik})}. \quad (14)$$

Here, a_{ij} means the weight of the input states. An input state with high weight has a greater influence on the output state. The a_{ij} is normalized to 1. The m in Equation (14) means the number of words entered in the decoder, and a_{ij} indicates how similar the ith vector of text embedding and the jth vector of the encoder are in the previous step. The e is computed in Equation (15). In Equation (15), α represents a constant to optimize the similarity between s(i–1) and q_j like the learning rate of a neural network model, s(i–1) represents the text embedding vector of the previous step when the decoder predicts the ith word, and q_j represents the jth column vector of the encoder.

$$e = \alpha(s_{i-1}, q_j). \quad (15)$$

The text embedding that the weight computed in the attention RNN is added to is sent to three decoder RNNs. The decoder RNNs convert the text embedding to a spectrum form by using a sequence-to-sequence method. The converted text embedding is output to voice via the CBHG module.

3.4. Design of a Data Visualization Module

The data visualization module (DVM) receives all the sensor data collected from the vehicle and visualizes only the data that deaf people desire by using an adaptive component placement algorithm that adjusts and places the graphical user interface (GUI) components properly on the vehicle's display. Figure 9 shows the structure of the DVM.

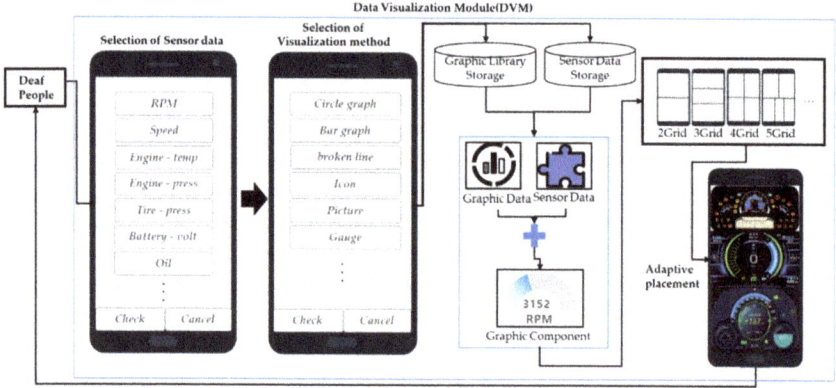

Figure 9. Structure of data visualization module.

The visualization process of the DVM is as follows. First, people can select the sensor data to visualize. The types of sensor data are represented by buttons. Second, they decide how they want to visualize the sensor data of their own choosing. The sensor data that will be visualized has to be selected before selecting the visualization method. For example, if someone wants to display RPM as a bar graph, they choose RPM from the sensor data and click the bar graph from the visualization method. The DVM generates a GUI component combining the entered sensor data and visualization method. Algorithm 2 indicates the process by which the DVM receives the necessary information for visualization.

Algorithm 2. Data visualization algorithm.

input: int number_kind, list data_value[], String data_name[], String
      graph_name;
init: Button check_sensors[number_kind];
      List graph[];
      List param_data[];
      int k=0;
for(int i = 0; i<kind; i++){
  check_sensors[i]=data_value[data_name[i]];
  check_sensors.enable;
}

if(ClickEvent(data_name) && ClickEvent(graph_name)){
  check_sensors[data_name].disable;
  param[k].input(key : data_value[data_name],value :
  graph[graph_name],);
}

if(ClickEvent(send)){
  send(param[]);
}
if(ClickEvent(cancel)){
  check_sensors[data_name].enable;
  param[k].delete(key : data_value[data_name],value :
  graph[graph_name],);
}

Here, *number_kind* means the number of data types output to the display, *data_value[]* means sensor values, *data_name[]* means sensor names, and *Graph_name* is a visualization method. The selected data types are stored in *data_value[data_name]* and the selected visualization method in *value: graph[graph_name]*. *ClickEvent(send)* means that the selection and visualization of sensor data is over, and *ClickEvent(cancel)* means that all selected contents are deleted. When someone finishes all selections, the DVS generates a GUI component by combining the selected sensor name, sensor data, and visualization method.

The DVS then receives and visualizes the vehicle's failure self-diagnosis from the DCMM. The visualization method of failure self-diagnosis is not selected manually, it is visualized using a gauge graph. The DVS sets the gauge graph to a range from 0 to 1 to visualize the failure self-diagnosis. Because the type of display and the resolution size used by each vehicle vary, it is difficult to adapt to various environments unless the sizes of visualization components vary. Thus, the DVM generates adaptive components that adjust the size of the GUI components and displays them on the vehicle's display so that visualization components can be used without problems in various display environments. Algorithm 2 shows the process of dividing the display and placing GUI components.

Algorithm 3. Adaptive component placement.

SetComponent(Object[] component[], int compNumber, int compX[], int compY[]){

int i = 0;

Rect grid[] = GridPartition(compNumber, vertical, 1.5);

for(i = 0; i<= compNumber; i++){
   if(compNumber == 1)
     displayComponent(component[i], grid[i]);
   else{
    if((compX[i]<compY[i]) && (grid[i].x<grid[i].y)){
     displayComponent(component[i], grid[i]);
    }
    elseif((compX[i]>compY[i]) && (grid[i].y<grid[i].x)){
     displayComponent(component[i], grid[i]);
    }
    elsief((compX[i]<compY[i]) && (grid[i].y<grid[i].x)
    && (component[i].type == digit)){
     component[i] = ReplaceXandY(component[i]);
     displayComponent(component[i], grid[i]);
    }
    else if((compX[i]<compY[i]) && (grid[i].y<grid[i].x)
    && (component[i].type != digit)){
     grid[i] = ReplaceGrid(horizontal, 1.5);
     displayComponent(component[i], grid[i]);
}}}}

Here, *compNumber* means the number of components, *compX* means the size of the component's x-axis, and *compY* the size of the y-axis. *GridPartition (compNumber, Vertical, 1.5)* means that when a grid is divided, it is divided vertically in proportion to the number of components. In $i = 0; i < compNumber; i++$, i is the number of components in the grid, and grid segmentation is done as large as the number of components.

If the order of a component is ith, it is placed in the ith grid. For example, if the third component is placed on the display, it means that the component is placed on the third grid. *(compX[i] < compY[i]) && (grid[i].x < grid[i].y)* means that if the x-axis size of the component and grid is less than the y-axis size, the component is represented on the display as is. *(compX[i] > compY[i] && (grid[i].y < grid[i].x)* means that if the x-axis size of the component and grid is greater than the y-axis size, the display is still represented on the display as is. *(compX[i] < compY[i]) && (grid[i].y < grid[i].x)* means that if the x-axis size of the component is less than the y-axis size and the x-axis size of the grid is greater than the y-axis size, the component's x-axis and y-axis are swapped together to be represented on the display. *(compX[i] < compY[i]) && (grid[i].y < grid[i].x)* means that if the x-axis size of the component is greater than the y-axis size and the x-axis size of the grid is less than the y-axis size, the component's x-axis and y-axis are swapped together to be represented on the display.

4. Performance Analysis

In this paper, four experiments were conducted to validate the AVS. The first experiment was conducted in the vehicle and in the cloud with the HMM model to validate learning efficiency in the NVIDIA px2 driver environment. The second experiment was conducted to compare the learning time of HMM, RNN, and long short-term memory (LSTM). The third experiment was conducted to compare the computational time of Tacotron2, Deep Voice, and Deep Mind. The fourth experiment was conducted to compare the time taken for the vehicle's instrument counter and DVM to visualize the vehicle's information in real time. The experiment was conducted on a PC with Intel i5-7400 CPU, GTX1050 GPU, 8GB RAM, and Windows 10 Education OS.

4.1. Performance Analysis of HMM Learning

Figure 10 shows the time it took for the STS to learn the HMM in the cloud and in a vehicle. In this experiment, the time to learn the HMM was measured with 1 to 10 test datasets. In the vehicle, the number of test datasets did not have a significant impact on the learning time, but the learning time in the cloud was increased as the number of test sets increased. The trend line of learning time from the vehicle is 0.78182 and that of from the cloud is 2.84242. In the vehicle, the more learning data the HMM have, the more effective the HMM learning is. In addition, since the average learning time in a vehicle is about 2.5 times faster than in the cloud, this paper proposes using the HMM in the vehicle rather than in the cloud.

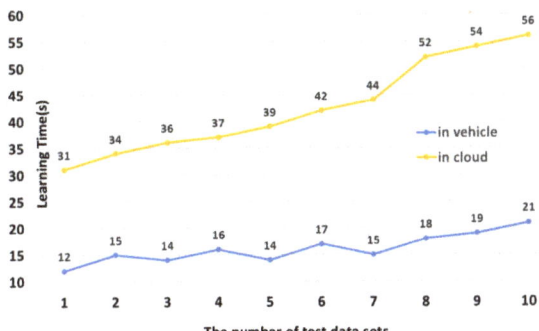

Figure 10. Performance analysis of HMM learning.

4.2. Performance Analysis of STS

Figure 11 shows the learning times of HMM, RNN, and LSTM. The learning time for each model was measured, increasing the sentences in the test dataset from 500 to 13,000. The sentences used in the experiment are included in the LJ training data sets. The experimental results show that the HMM was able to recognize speech faster than RNN and LSTM by 25% and by 42.86%, respectively. When the number of the test sentences was small, there was little difference in the learning time between the RNN and the HMM, but as the number of the test sentences grew larger, the time of the HMM learning become faster. Therefore, the HMM is the most appropriate to ensure the real-time of AVS speech recognition and voice transmission.

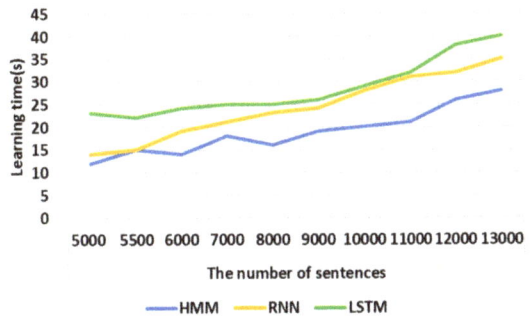

Figure 11. Performance Analysis of HMM. LSTM, long short-term memory.

Figure 12 shows the accuracy of HMM when the sentences with noise were entered into the HMM. The sentences used in the experiment were included in the LJ training data sets and were not used in the HMM's training. The accuracy is computed by comparing the difference between when a sentence is entered into the HMM and when it is output. The experiment was conducted using 1 to 10 sentences, and the sentences with the noise of 0 dB, 40 dB, 60 dB, 80 dB, and 120 dB were entered into the HMM.

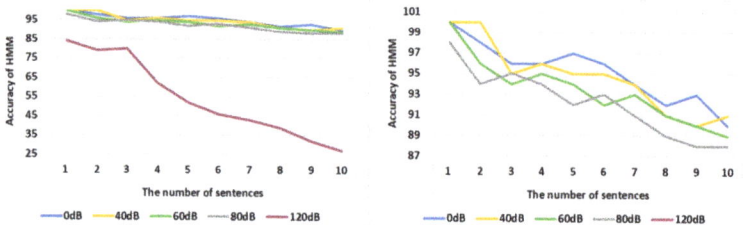

Figure 12. The accuracy of the HMM according to noise.

In the left experiment of Figure 12, because the accuracy was significantly reduced in case of 120 dB, the 120 dB was excluded in the right experiment. The average accuracy of the HMM was 95% when sentences without noise were input into the HMM. That of the HMM was 95% when sentences with 40 dB noise were input into the HMM. The average accuracy of the HMM was 93% when sentences with 60 dB noise were input into the HMM. The average accuracy of the HMM was 92% when sentences with 80 dB noise were input into the HMM. The average accuracy of the HMM was 54% when sentences with 120 dB noise were input into the HMM. The average noise is between 60 and 70 dB when the vehicle is travelling at a speed of 60 to 80 km/h. The accuracy of the HMM can be above 90% even when 80 dB noise is mixed in a sentence, so the HMM can recognize the speech well even if noise occurs in the vehicle.

4.3. Performance Analysis of TWS

Figure 13 shows the computational time it took for the TTS engines Tacotron2, Deep Voice, and Deep Mind to convert text into voice when sentences were increased from 1 to 10. The sentences used in this experiment are those not used for learning among LJ Training data sets. Tacotron2 converted text into voice about 20 ms faster than Deep Voice and about 50 ms faster than Deep Mind. The TWS converted the vehicle's data into voice using Tacotron2, so real time could be guaranteed. The computational time of the Deep Mind is about 1.5 times slower than the Deep Voice and Tacotron2. As test data sets get more, the difference of computational time between the Tacotron2 and Deep Voice gets bigger. The trend line of the Tacotron2 is 0.69091 and that of the Deep Voice line is 1.95758. Therefore, the AVS should use Tacotron2 to ensure real-time voice transmission.

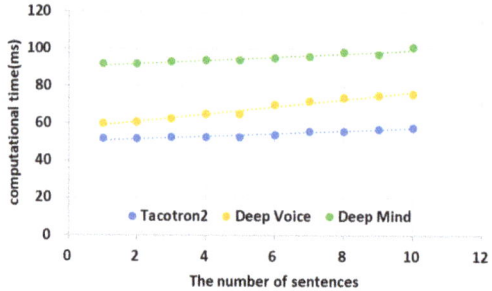

Figure 13. Performance analysis of HMM.

4.4. Performance Analysis of DVM

Figure 14 compares the time it took for the sensor data to be displayed on the vehicle's instrument counter to analyze the performance of the DVM. Performance analysis used 1 to 10 datasets for visualization. The average time for the DVM to visualize sensor data was about 665 ms and for the instrument counter was about 667 ms. The DVM was about 2 ms faster than an existing instrument counter; it can visualize information without compromising the real time of an existing instrument counter. However, the instrument counter cannot display information more accurately than the DVM because it should display only the information that is visible in driving, and as test data sets gets more, it takes more time for the counter to visualize sensor data than DVM. When the number of sensor data sets to be visualized was 10, the DVM was about 20 ms faster than the instrument counter in visualizing sensor data.

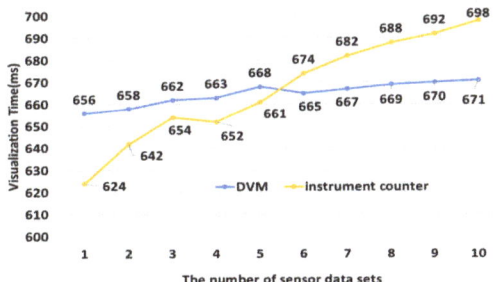

Figure 14. Performance analysis of DVM.

As a result, the AVS not only visualizes data intuitively without compromising the real time of an existing instrument, but also enables the vehicle data to be heard in real time for the blind. Therefore, the AVS will better prevent accidents than existing systems and help blind people check the condition of the vehicle.

5. Conclusions

When blind and deaf people use fully autonomous vehicles as passengers, an intuitive and accurate visualization screen for the deaf should be provided, and an audifying system with speech-to-text (STT) and text-to-speech (TTS) functions should be provided for the blind. This paper proposes an audification and visualization system (AVS) for blind and deaf people that can store the graphic data collected from vehicle cameras, etc., and the in-vehicle sensor data and fault self-diagnosis data in the vehicle and select the desired data by using a touch interface and a speech recognizer. The AVS consists of three modules. The DCMM stores and manages the data collected from the vehicle. The ACM has a speech-to-text submodule (STS) that recognizes a user's speech and converts it to text data, and a text-to-wave submodule (TWS) that converts the text data to speech. The DVM receives sensor data, self-diagnosis information, and a graphics library from the DCMM for visualization through the touch interface. The DVM provides an adaptive position of visualized data to fit the display on the vehicle.

We conducted four experiments to validate the AVS. The first experiment was conducted in the vehicle and in the cloud with the HMM model to validate learning efficiency. The second experiment was conducted to compare the learning times of HMM, RNN, and LSTM. The third experiment was conducted to compare the computational time of Tacotron, Deep Voice, and Deep Mind. The fourth experiment was conducted to compare the time it took for the vehicle's instrument counter and DVM to visualize the vehicle's information in real time. According to the experimental results, the HMM was about 2.5 times faster than the cloud when it was learned in the NVIDIA px2 driver, the HMM used in STS was learned about 25% faster than the RNN and about 42% faster than the LSTM, and

Tacotron2 used in TWS converted text to voice 20 ms faster than Deep Voice and 50 ms faster than Deep Mind. Finally, the DVM visualized the data 2 ms faster than existing instrument clusters. Therefore, the AVS can express more information without compromising the real time of existing systems.

However, the AVS was not tested on actual vehicles and mean opinion score (MOS) measurements, which are voice quality criteria, were not done when data was converted to voice. If the AVS is applied to real vehicles, it will not only make autonomous vehicles more readily available to blind and deaf people, but will also increase the stability of existing autonomous vehicles. In addition, vehicles without displays were not considered because the AVS would generate a GUI based on the displays inside the vehicle. The future AVS should objectively evaluate the quality of data converted into voice by MOS measurements and be applied to actual vehicles, and how the system will be applied to devices other than smartphones or in-vehicle displays should be studied.

Author Contributions: Conceptualization, B.L. and Y.J.; methodology, S.S.; software, S.S.; validation, B.L. and Y.J.; writing—original draft preparation, S.S. and Y.J.; writing—review and editing, B.L.; visualization, B.L.; supervision, B.L.; project administration, B.L.; funding acquisition, B.L.

Funding: This work was supported by a National Research Foundation of Korea (NRF) grant funded by the Korean government (MSIT) (No. NRF-2018R1A2B6007710).

Acknowledgments: In this section you can acknowledge any support given which is not covered by the author contribution or funding sections. This may include administrative and technical support, or donations in kind (e.g., materials used for experiments).

Conflicts of Interest: The authors declare no conflict of interest.

References

1. SAE International(www.sae.org). Levels of Driving Automation. Available online: https://www.sae.org/news/press-room/2018/12/sae-international-releases-updated-visual-chart-for-its-%E2%80%9Clevels-of-driving-automation%E2%80%9D-standard-for-self-driving-vehicles (accessed on 12 September 2019).
2. Lee, D. Google Self-Driving Car Hits a Bus. Available online: https://www.bbc.com/news/technology-35692845 (accessed on 5 November 2019).
3. Bevilacqua, M. Cyclist Killed by Tesla Car with Self-Driving Features. Available online: https://www.bicycling.com/news/a20034037/cyclist-killed-by-tesla-car-with-self-driving-features/ (accessed on 5 November 2019).
4. Halsey, A., III; Laris, M. Blind Man Sets Out Alone in Google's Driverless Car. Available online: https://www.washingtonpost.com/local/trafficandcommuting/blind-man-sets-out-alone-in-googles-driverless-car/2016/12/13/f523ef42-c13d-11e6-8422-eac61c0ef74d_story.html (accessed on 24 September 2019).
5. Li, K.; Wang, X.; Xu, Y.; Wang, J. Lane changing intention recognition based on speech recognition models. *Transp. Res. Part C Emerg. Technol.* **2016**, *69*, 497–514. [CrossRef]
6. Stamp, M. *A Revealing Introduction to Hidden Markov Models*; San Jose State University: San Jose, CA, USA, 2018; pp. 1–21.
7. Liang, J.; Ma, M.; Sadiq, M.; Yeung, K.-H. A filter model for intrusion detection system in Vehicle Ad Hoc Networks: A hidden Markov methodology. *Knowl.-Based Syst.* **2019**, *163*, 611–623. [CrossRef]
8. Saini, R.; Roy, P.P.; Dogra, D.P. A segmental HMM based trajectory classification using genetic algorithm. *Expert Syst. Appl.* **2018**, *93*, 169–181. [CrossRef]
9. Siddique, C.; Ban, X. (Jeff) State-dependent self-adaptive sampling (SAS) method for vehicle trajectory data. *Transp. Res. Part C Emerg. Technol.* **2019**, *100*, 224–237. [CrossRef]
10. Liu, P.; Kurt, A.; Ozguner, U. Synthesis of a behavior-guided controller for lead vehicles in automated vehicle convoys. *Mechatronics* **2018**, *50*, 366–376. [CrossRef]
11. Mingote, V.; Miguel, A.; Ortega, A.; Lleida, E. Supervector Extraction for Encoding Speaker and Phrase Information with Neural Networks for Text-Dependent Speaker Verification †. *Appl. Sci.* **2019**, *9*, 3295. [CrossRef]
12. Wang, X.; Cong, Z.; Fang, L. Determination of Real-time Vehicle Driving Status Using HMM. *Acta Autom. Sin.* **2014**, *39*, 2131–2142. [CrossRef]
13. Kato, J.; Watanabe, T.; Joga, S.; Liu, Y.; Hase, H. An HMM/MRF-Based Stochastic Framework for Robust Vehicle Tracking. *IEEE Trans. Intell. Transp. Syst.* **2004**, *5*, 142–154. [CrossRef]

14. Xiong, X.; Chen, L.; Liang, J. A New Framework of Vehicle Collision Prediction by Combining SVM and HMM. *IEEE Trans. Intell. Transp. Syst.* **2018**, *19*, 699–710. [CrossRef]
15. Jazayeri, A.; Cai, H.; Zheng, J.Y.; Tuceryan, M. Vehicle Detection and Tracking in Car Video Based on Motion Model. IEEE Trans. *Intell. Transp. Syst.* **2011**, *12*, 583–595. [CrossRef]
16. Choromański, W.; Grabarek, I. Driver with Varied Disability Level—Vehicle System: New Design Concept, Construction and Standardization of Interfaces. *Procedia Manuf.* **2015**, *3*, 3078–3084. [CrossRef]
17. Bennett, R.; Vijaygopal, R.; Kottasz, R. Attitudes towards autonomous vehicles among people with physical disabilities. *Transp. Res. Part A Policy Pract.* **2019**, *127*, 1–17. [CrossRef]
18. Bennett, R.; Vijaygopal, R.; Kottasz, R. Willingness of people with mental health disabilities to travel in driverless vehicles. *J. Transp. Health* **2019**, *12*, 1–12. [CrossRef]
19. Xu, X.W.; He, J.Y. Design of Intelligent Guide Vehicle for Blind People. *Appl. Mech. Mater.* **2013**, *268*, 1490–1493. [CrossRef]
20. Jeong, Y.; Son, S.; Lee, B. The Lightweight Autonomous Vehicle Self-Diagnosis (LAVS) Using Machine Learning Based on Sensors and Multi-Protocol IoT Gateway. *Sensors* **2019**, *19*, 2534. [CrossRef] [PubMed]
21. Jeong, Y.; Son, S.; Jeong, E.; Lee, B. An Integrated Self-Diagnosis System for an Autonomous Vehicle Based on an IoT Gateway and Deep Learning. *Appl. Sci.* **2018**, *8*, 1164. [CrossRef]
22. Kenneth, H.C. Lagrange Multipliers for Quadratic Forms with Linear Constraints. Available online: https://www.ece.k-state.edu/people/faculty/carpenter/document/lagrange.pdf (accessed on 5 October 2005).
23. Wang, Y.; Skerry-Ryan, R.J.; Stanton, D.; Wu, Y.; Weiss, R.J.; Jaitly, N.; Yang, Z.; Xiao, Y.; Chen, Z.; Bengio, S.; et al. Tacotron: Towards End-to-End Speech Synthesis. *arXiv* **2017**, arXiv:1703.10135.
24. Ito, K. The LJ Speech Dataset. Available online: https://keithito.com/LJ-Speech-Dataset/ (accessed on 3 September 2019).

© 2019 by the authors. Licensee MDPI, Basel, Switzerland. This article is an open access article distributed under the terms and conditions of the Creative Commons Attribution (CC BY) license (http://creativecommons.org/licenses/by/4.0/).

Article

Fuzzy Functional Dependencies as a Method of Choice for Fusion of AIS and OTHR Data

Medhat Abdel Rahman Mohamed Mostafa [1], Miljan Vucetic [2,*], Nikola Stojkovic [3], Nikola Lekić [2] and Aleksej Makarov [2]

- [1] School of Electrical Engineering and Computer Science, European University, 28 Carigradska St., 11000 Belgrade, Serbia; medhat.ma@yahoo.com
- [2] Vlatacom Institute, 5 Milutina Milankovica Blv., 11070 Belgrade, Serbia; nikola.lekic@vlatacom.com (N.L.); aleksej@vlatacom.com (A.M.)
- [3] School of Electrical Engineering, University of Belgrade, Serbia and Vlatacom Institute, 101801 Beograd, Serbia; nikola.stojkovic@vlatacom.com
- * Correspondence: miljan.vucetic@vlatacom.com; Tel.: +381-11-377-1100

Received: 2 October 2019; Accepted: 19 November 2019; Published: 26 November 2019

Abstract: Maritime situational awareness at over-the-horizon (OTH) distances in exclusive economic zones can be achieved by deploying networks of high-frequency OTH radars (HF-OTHR) in coastal countries along with exploiting automatic identification system (AIS) data. In some regions the reception of AIS messages can be unreliable and with high latency. This leads to difficulties in properly associating AIS data to OTHR tracks. Long history records about the previous whereabouts of vessels based on both OTHR tracks and AIS data can be maintained in order to increase the chances of fusion. If the quantity of data increases significantly, data cleaning can be done in order to minimize system requirements. This process is performed prior to fusing AIS data and observed OTHR tracks. In this paper, we use fuzzy functional dependencies (FFDs) in the context of data fusion from AIS and OTHR sources. The fuzzy logic approach has been shown to be a promising tool for handling data uncertainty from different sensors. The proposed method is experimentally evaluated for fusing AIS data and the target tracks provided by the OTHR installed in the Gulf of Guinea.

Keywords: HF-OTH radar; AIS; radar tracking; data fusion; fuzzy functional dependencies; maritime surveillance

1. Introduction

An exclusive economic zone (EEZ) is a 200 nmi (approximately 370 km)-wide area which spreads from territorial waters towards the open sea in which a coastal state has exclusive rights to exploit biological and mineral sea resources. The control of this zone posers technological, financial and organizational challenges. As parts of an integrated maritime surveillance (IMS) system, different types of electronic sensors and telecommunication systems can be used for monitoring the zone. However, the range of microwave and optical sensors depends on their working wavelengths and is limited due to atmospheric signal weakening and the Earth's curvature—distance to horizon. High-frequency (HF) radars operating at decameter wavelengths (3–30 MHz) use vertically polarized surface waves to detect and track targets beyond the horizon, enabling over-the-horizon (OTH) surveillance. Therefore, HF-OTH radars are well suited for EEZ surveillance [1,2].

For effective EEZ tracking, a fusion of data from multiple sensors is needed. In this case, term fusion means integration of data from at least two sensors. Most often, image of maritime situation beyond the horizon is provided by using fusion of data received from HF-OTHR and an automated identification system (AIS) that can be a land-based AIS (LAIS) or a satellite-based AIS (SAIS) [3].

Data fusion received from HF-OTHR and AIS provides extraction of paths which are not confirmed by AIS. This allows for detecting targets that are non-cooperative and which either pose a military threat or conduct illegal activities (such as smuggling) inside the EEZ [4,5].

In this paper we use real data from the HF-OTHR system installed in the Gulf of Guinea obtained with the pertaining detection and tracking software [6–9], and the current AIS data in order to perform data fusion. It is important to note that the Gulf of Guinea is among the most challenging environments for the targeted application. There are two major reasons for this:

1. HF noise levels in that area are among the highest in world [10], and
2. Absence of a strong regulatory institution (such as European Maritime Safety Agency—EMSA in Europe) sometimes leads to unpredictable behavior of participants in the maritime traffic. This causes a very high latency and questionable quality of AIS data.

Since our goal is fusion of the tracking data provided by different sensors for the same targets, we have employed the fuzzy functional dependency (FFD) concept in order to associate AIS data and OTHR tracks. Methods based on numerical statistics were used in the previous research on AIS and OTHR data fusion [8,11]. However, intensive traffic flows call for more reliable integration methods. Nowadays, artificial intelligence tools such as fuzzy approaches and neural networks have been applied to data fusion problems and resulted in higher fusion accuracy. A simple idea of the AIS and OTHR data fusion based on the fuzzy sets was brought forward in [12] and used in handling inaccuracy when computing association grades for different vectors. The fuzzy C-means clustering method was used in [13] for correlating tracks without merging them. Neural networks were proposed for improvement of AIS and OTHR data fusion [14,15], but the methods employed were rather complicated. In general, these approaches neglected ships which were either not equipped with AIS or unwilling to emit AIS information. Furthermore, the validity of the presented approaches was checked against research data without considering the latency of AIS messages.

In this work, we present a fuzzy-based method for addressing imprecision characterized by these types of sensors. The proposed approach offers better insight in numerical processing than neural networks. AIS and OTHR tracks are loaded into a relational database in order to perform advanced database operations such as data cleaning, filtering and record matching. Actually, FFDs have been used to define data integration constraints among two databases because they take similarity into account. Therefore, to address the development of fusion method, it is necessary to prepare data for processing and analyze different challenges appearing in real scenarios. This approach provides complete picture of maritime situation enabling the detection of possible threats to maritime nation's interests in EEZ, as well as better insight in events within it. The performance of the introduced method is experimentally tested and discussed. It is worth noting that this paper relies on concepts described in [11], but utilizes a different decision-making algorithm based on fuzzy concept.

This article is structured as follows. In Section 2 the research background is described. Section 3 presents methods of measuring the similarity between AIS and OTHR records. Section 4 gives the algorithm overview and implementation steps. Experimental results are discussed in Section 5. Finally, the paper ends with the conclusions presented in Section 6.

2. Background

In this section we considered functional dependencies (FDs), which are a basic concept for data fusion and data cleaning, between AIS records and OTHR tracks. Functional dependencies are one of the fundamentals of Codd's relational model used as a tool for the database design based on the normalization theory and redundancy elimination. An interpretation of FD, denoted by $X \to Y$, in the relation R defined over the set of attributes $attr(R)$ is the following: if two tuples t_1 and t_2 have the same value on the attribute X, then they also have the same value on attribute Y. It reads as X determines Y or Y is functionally dependent on X. But, in many practical problems data are not strictly equal. For example, OTHR by its nature involves some level of inaccuracy. In our case, it measures the range, azimuth and radial speed of targets with accuracy expressed in terms of the corresponding resolution cells. In this work, the range resolution cell of OTHR is 1.5 km, angular resolution is 10° and the radial velocity resolution is 0.32 m/s. The accuracy of measurement quantities is in the range of 0.5 to 1.5 respective resolution cells. Matching the AIS and OTHR data means that we have to find similar records. One of the challenging problems in our work is measuring similarities among AIS records and OTHR tracks considering their respective attributes. The main idea is to find the intensity of similarity in order to detect the same targets and possible threats. This assumption leads us to extending FDs to fuzzy functional dependencies. The equality relation used in the above canonical FDs is replaced by a similarity or proximity relation in FFDs as a measure of closeness [16]. The extension enables the specification of new application areas such as data fusion of AIS and OTHR. Today, FDs defined in various ways [17] are not used only for database design purposes, but also in other challenging applications such as data cleaning [18], record matching [19], query relaxation [20], and knowledge discovery [21].

In order to impose fuzzy data dependency, we introduce the definition of FFD $X \xrightarrow{\theta} Y$ [22]:

$$\min I \left(\approx (t_1(X), t_2(X)), \approx (t_1(Y), t_2(Y)) \right) \geq \theta \tag{1}$$

where $t_1, t_2 \in R$ (values of tuples t_1 and t_2 for attributes X and Y respectively in relation R), "$\approx$" is the closeness measure, I is a fuzzy implications operator, min denotes the "*and*" operator and $\theta \in [0, 1]$ is the strength of the dependency. The definition of FFD states that if $t_1(X)$ is similar to $t_2(X)$ then $t_1(Y)$ is also similar to $t_2(Y)$ with the strength of dependency θ, where X, Y are two sets of attributes in R, $X, Y \subseteq attr(R)$. Different similarity functions are employed to compare values of considered attributes. In the last two decades several definitions of FFDs have been introduced [23]. Recently, a new method has been developed for computing FFDs by [24].

In case of AIS and OTHR data fusion, t_1 and t_2 are records representing AIS object and OTHR track respectively. Record matching is performed by means of FFD, i.e., a fuzzy measure on [0, 1] is used to compute how similar the AIS object and an OTHR track are, considering common attributes from the corresponding databases and their aggregation. The first step is calculating closeness of the considered attributes using the defined similarity measures. The strength of dependency θ is a real number within the range [0, 1], describing the degree of similarity between an AIS object and an OTHR track. In order to detect the parameter θ, an aggregation of similarities per observed attributes is applied. The average aggregation function is suggested for calculating the strength of the dependency among AIS and OTHR records. Additionally, this function can provide flexibility in terms of weighting attributes when some of them are more important than others. Thus, in the next section, we will describe a step-by-step procedure for fusion of AIS and OTHR records.

An important aspect related to FD and FFDs is the presence of inference rules enabling the possibility of deriving the new dependencies from the existing ones [25]. This straightforwardly provides an efficient way to derive relations between other attributes in data analysis. These inference rules are based on Armstrong's axioms. Inference rules must be sound (rules generate valid dependencies) and complete (valid dependencies can be generated by only these rules). For example, we list the inference rules for FFDs (analogously to FDs):

1. Inclusive rule: If $X \xrightarrow{\theta_1} Y$ holds, and $\theta_1 \geq \theta_2$, then $X \xrightarrow{\theta_2} Y$ holds.
2. Reflexive rule: If $Y \subseteq X$, then $X \rightarrow Y$ holds.
3. Augmentation rule: $\{X \xrightarrow{\theta} Y\} \models XZ \xrightarrow{\theta} YZ$.
4. Transitivity rule: $\{X \xrightarrow{\theta_1} Y, Y \xrightarrow{\theta_2} Z\} \models X \xrightarrow{\min(\theta_1, \theta_2)} Z$.
5. Union rule: $\{X \xrightarrow{\theta_1} Y, X \xrightarrow{\theta_2} Z\} \models X \xrightarrow{\min(\theta_1, \theta_2)} YZ$.
6. Pseudotransitivity rule: $\{X \xrightarrow{\theta_1} Y, WY \xrightarrow{\theta_2} Z\} \models WX \xrightarrow{\min(\theta_1, \theta_2)} Z$.
7. Decomposition rule: If $X \xrightarrow{\theta} Y$ holds, and $Z \subseteq Y$, then $X \xrightarrow{\theta} Z$ holds.

3. Research Methodology

Our methodology described here answers the question as to how OTHR and AIS data are fused. Our main goal is to detect records referring to the same targets in a database. For that purpose, we explore the connection between FFDs and the problem of OTHR tracks and AIS data fusion. By matching records from two databases, we perform data analysis per common attributes in order to fuse related records. As an example, if tuple t_1 representing AIS record and tuple t_2 referring to OTHR track have similar values per common attributes, then they are candidates to represent the same vessel in the observed area. FFDs are used to quantify the closeness or remoteness of OTHR and AIS records. They specify a subset of tuples on which the dependency holds.

Initially, the source databases which need to be integrated have unique identifiers. The primary keys for OTHR tracks and AIS data are *ID_Number* and *Maritime Mobile Service Identity* (MMSI), respectively [26]. The use of primary keys eases matching as compared to a situation when alternative attributes need to be found in databases without unique identifiers. Obviously, there is FD between unique identifiers and other attributes in the source databases: *ID_Number* → *A*, and *MMSI* → *B*, where *A*, *B* are sets of attributes in source relations r_1 and r_2. However, in the case of OTHR and AIS data integration there is a FFD (*ID_Number*, *MMSI*) $\xrightarrow{\theta}$ *C*. where (*ID_Number*, *MMSI*) is a composite key (integration pair), θ parameter reflecting degree of similarity and *C* is set of common attributes in source relations: *Timestamp*, *Latitude*, *Longitude*, *Velocity* and *Ship Course*. This FFD is examined through the matching process between AIS and OTHR records in order to evaluate the closeness of observed tuples. Due to the inaccuracy and the inconsistency of data provided by OTHR and AIS platforms, we evaluate the closeness of records in order to detect FFDs which are possible candidates for associating related targets.

The parameter θ (the strength of fuzzy functional dependency) is calculated as follows:

$$\theta = \frac{1}{5}\left(c_{time} + c_{long} + c_{lat} + c_{vel} + c_{course}\right) \quad (2)$$

In fact, for each common attribute we calculate how close two attribute values are. Then, we aggregate and normalize the closeness values of common attributes. The records from OTHR and AIS source databases with the highest θ value are candidates for target association. In order to calculate the closeness values c for the attributes *Time* (c_{time}), *Longitude* (c_{long}), *Latitude* (c_{lat}), *Velocity* (c_{vel}) and *Course* (c_{course}) the following equations are used:

$$c_{vel} = 1 - \frac{|v_{AIS} - v_{OTHR}|}{v_{max}} \quad (3)$$

where v_{AIS} and v_{OTHR} represent the target's speed reported by AIS and OTHR networks respectively, while $v_{AIS} - v_{OTHR}$ is their difference. The variable v_{max} denotes the maximum speed of target communicated by OTHR and AIS systems, i.e., $v_{max} = \max(v_{AIS}^{max}, v_{OTHR}^{max})$. c_{vel} represents the speed matching value for the corresponding records from AIS and OTHR databases.

$$c_{course} = 1 - \frac{|C_{AIS} - C_{OTHR}|}{C_{max}} \quad (4)$$

where C_{AIS} and C_{OTHR} represent the target's course reported by AIS and OTHR networks respectively, $C_{AIS} - C_{OTHR}$ is their difference and C_{max} is maximum course provided for all the targets by OTHR and AIS databases, $C_{max} = \max(C_{AIS}^{max}, C_{OTHR}^{max})$. c_{course} is the matching coefficient for the course attribute between two attribute values representing records from AIS and OTHR.

$$c_{long} = 1 - \frac{|Long_{AIS} - Long_{OTHR}|}{Long_{max}} \quad (5)$$

$$c_{lat} = 1 - \frac{|Lat_{AIS} - Lat_{OTHR}|}{Lat_{max}} \qquad (6)$$

where $Long_{AIS}$ (Lat_{AIS}) and $Long_{OTHR}$ (Lat_{OTHR}) represent the target's longitude (latitude) reported by AIS and OTHR networks respectively, $Long_{AIS} - Long_{OTHR}$ ($Lat_{AIS} - Lat_{OTHR}$) is their difference and $Long_{max}$ (Lat_{max}) is maximum value of longitude (latitude) in OTHR and AIS databases. Longitude and latitude are expressed as decimal values (degrees). Closeness between AIS and OTHR records for longitude (latitude) position is shown as c_{long} (c_{lat}).

$$c_{time} = 1 - \frac{|t_{AIS} - t_{OTHR}|}{t_{interval}(sec)} \qquad (7)$$

where t_{AIS} and t_{OTHR} represent timestamps of data creation by AIS and OTHR sensors respectively and $t_{AIS} - t_{OTHR}$ is their difference. Timestamps t_{AIS} and t_{OTHR} are converted to seconds due to efficient computation. The interval of time ($t_{interval}$) represents the length of the observed time frame and it is also expressed in seconds. The interval is measured as time duration between two AIS messages.

Presented processing logic is triggered by newly received AIS message. We suppose that we operate with fairly precise AIS data. Furthermore, for every AIS record OTHR candidates are ranked by strength of dependency θ in the record matching process. This procedure is shown in the next Section.

4. Algorithm

In terms of data fusion from AIS and OTHR sensors, we present an algorithm based on aforementioned methodology. This algorithm with steps describing record matching is shown in Figure 1.

Algorithm initialization is triggered by the reception of AIS message. OTHR dataflow is periodic with repetition cycle of 33 s, while AIS system in this geographical area does not provide consistent message delivery. This is particularly the case with Satellite AIS transmissions where latency is even measured in hours. OTHR tracks representing the paths of targets must be stored in the database in order to perform record matching with AIS data. Sometimes, it can be a repository with a long history due to high latency of AIS message delivery. History of OTHR tracks helps in overcoming this issue. The raw messages of AIS and OTHR networks (XML and text file respectively) are converted and imported into MySQL databases. This approach allows efficient data manipulation, storing and flexible searching of data by time, position or target ID. Data is organized into tables according to OTHR and AIS sensors. Both tables have attributes related to target ID (*ID_Number* for OTHR and *MMSI* for AIS), position (*Longitude* and *Latitude*), *Course*, *Velocity* and *Timestamp* of data creation.

After the data is loaded and stored into databases, the first step is pre-processing of OTHR tracks and AIS measurements. This stage involves data filtering and validation for the purpose of completing the state of information. First, it includes data filtering by time. When a new AIS message is received, algorithm deletes old and unneeded OTHR tracks (older than the oldest AIS record and beyond the message time frame). Second, OTHR covers a static (fixed) area, while the AIS continuously changes its covering area. This pre-processing activity is focused on filtering AIS records that are outside the coverage area of OTHR. It is based on geographical coordinates of the data (latitude and longitude). Third, in order to avoid matching of possibly false targets with real AIS data, the Tracker Confidence attribute in OTHR database is used as a criterion for elimination of unconfirmed targets. Tracker Confidence attribute describes reliability of OTHR data [11]. Then data validation is also done over filtered records due to possibility of several detections for the same target in OTHR and AIS databases. Because of data validation, additional logic is applied for analyzed tracks (i.e., OTHR targets with velocity exceeding maximal possible speed of movement (>30 m/s) are excluded). Finally, algorithm performs the data uniformity task. This means that reported radial velocity and azimuth (angle radar—target relative to true north (TN)) registered with OTHR sensor cannot be used for fusion. For that purpose, we use approximations of target course and velocity of OTHR tracks in order to use the same units of measurement for record matching with AIS data. In addition, timestamps are converted to seconds due to closeness calculation. All aforementioned steps are performed over MySQL databases using data extraction queries. The advantage of storing data in relational form is simplicity of data manipulation, keeping of long history of OTHR tracks and further processing.

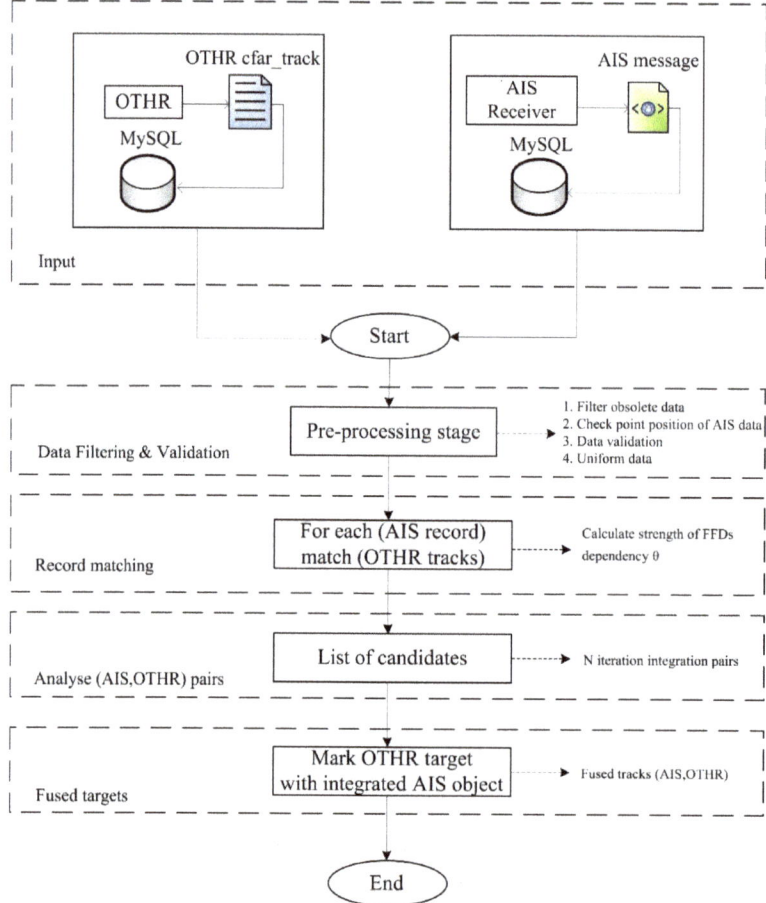

Figure 1. Algorithm for the fusion of automatic identification system (AIS) data and over-the-horizon radar (OTHR) tracks.

In the second step, processing logic based on FFDs is applied in the pre-processed tables. Actually, the algorithm searches for a set of possible associating candidates for each AIS point (record). The parameter θ is calculated using Equation (2). In this way, for every AIS point we compute a list of candidates for integration. Finally, for every AIS record OTHR integration candidates are ranked by the strength of the dependency θ. We determined experimentally the threshold value of the parameter θ for finding AIS-OTHR pairs. The criteria of $\theta > 0.80$ is applied.

Next step analyses results obtained in the previous activity: the association is done for a given pair of tracks based on the highest value of the parameter θ. The previous step is repeated for every new AIS message. If a pair of AIS and OTHR tracks within N cycles is found, then the algorithm confirms the association and passes the pair for fusion to it. Note that, despite the AIS point being treated as fused point, on every new appearance of AIS data, record matching and analyses of OTHR track candidates are repeated. This allows corrections in integration process and better resolution of difficulties and confusing situations (i.e., when multiple targets are close to each other).

In the last step, fused targets are marked as integrated pairs (ID_Number, MMSI). Association with an OTHR track is declared and AIS point is treated as the fused point.

5. Experiment and Discussion

In this section, the experimental results of the above proposed algorithm for data fusion between AIS and OTHR sensors are shown. As described in previous sections, the algorithm is triggered by reception of AIS message. Then the time frame of the AIS message is analyzed and data pre-processing is done in accordance with the steps for data filtering, validation and uniformity. It means that unnecessary (filtering by time) OTHR tracks are eliminated, AIS points outside the radar coverage area are not considered. Due to the absence of a strong regulatory institution, data provided by AIS can be of questionable quality (i.e., there are duplicate entries for the same targets). Before all aforementioned actions, we illustrate a picture of OTHR tracks and AIS targets as shown in Figure 2.

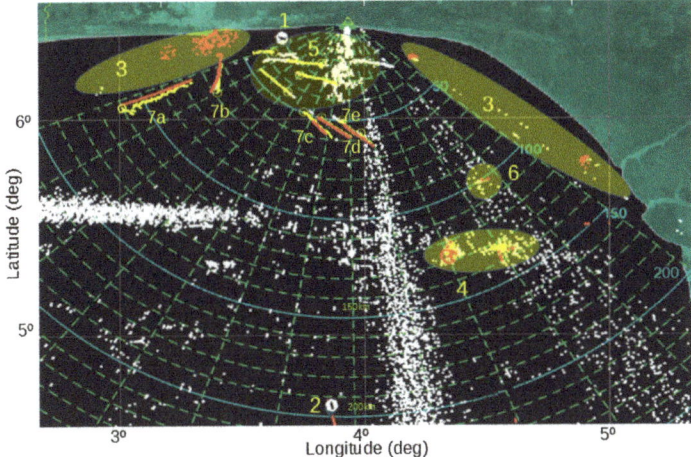

Figure 2. Maritime situation of the OTHR coverage area with reported AIS data in the Gulf of Guinea site (yellow—OTHR targets, red—AIS points, white—radar clutter).

In Figure 2, the following situations can be recognized:

1. Target leaving the radar coverage area;
2. Target entering the radar coverage area;
3. AIS data outside of the radar coverage area;
4. Multiple targets reported by AIS inside of one radar resolution cell. This situation is quite regular since resolution cell can be very large in comparison to the vessel size. In the presented scenario' two oil platforms with all corresponding vessels near them are shown. In practice, we mark these zones for a better understanding of maritime situation;
5. Vessels detected by OTHR, but not transmitting AIS data. Although this situation is rare in developed countries, in the Gulf of Guinee it is quite common and can present a security threat;
6. Vessels transmitting AIS data, but not detected by OTHR. This situation usually occurs when a smaller vessel delivers AIS data, although it may also present OTHR's missed target (there is no sensor with 100% detection probability);

7. Vessels labeled as 7a, 7b, 7c, 7d and 7e in the radar coverage area represent successful data fusion between AIS and OTHR sensors. Fusion parameters are calculated for them and shown in Figure 3.

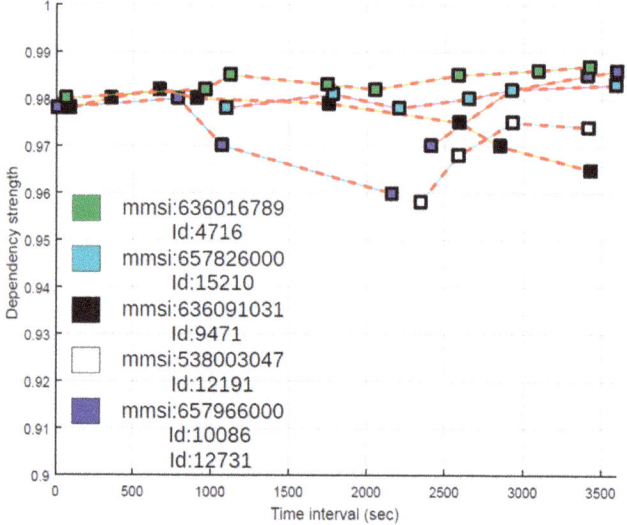

Figure 3. Results obtained for targets association over repeated iterations. Time interval from 16:33 h to 17:33 (3600 s).

The white points in Figure 2 are potential detections generated by OTHR. A large number of detections is a consequence of a complex environment in which OTHR operates, which contains high amounts of radar clutter. A high amount of clutter causes a large number of false detections and possibly missed detections of real radar targets. In our case, the sea swell is the main cause of radar clutter seen between 5 and 40 km away, at azimuth angles of ±10°. External radio sources such as Radio Frequency Interference (RFI) are also seen in two directions: −20° and −45°. A group of false detections comes from ionospheric interference, 10 km to 20 km wide, at the latitude of about 5.5°, covering azimuth angles from 10° to 60°. The clutter elimination (filtering) is performed during the pre-processing step.

As an example, a vessel labeled as 7a in the Figure 2. will be used to demonstrate the fusion process. In order to calculate the dependency strength θ, it is necessary to predict velocity and course of OTHR tracks using approximation and average measurements. OTHR target speed is approximated based on the difference in the distance registered by the radar and the period of time detection (Δt = 33 s) for the two neighboring positions. Projected OTHR target velocity is equal to the average value of the approximate speed measures over an observed time frame. Approximation of the course for OTHR ships is calculated by using WGS84 terrestrial reference system. To demonstrate the calculation of the dependency strength θ, let us consider the following AIS and OTHR records given in Tables 1 and 2 representing targets in Figure 2.

Table 1. AIS point.

MMSI	UTC Time	Latitude (deg)	Longitude (deg)	Course (deg)	Velocity (m/s)
636,091,031	16:48:10 (60,490 s)	6.142435	3.229788	247.1	9.36

Table 2. OTHR track.

ID_Number	UTC Time	Latitude (deg)	Longitude (deg)	Course (deg)	Velocity (m/s)
9471	16:48:05 (60,485 s)	6.128810	3.234660	248.68	9.77

Using Equations (3)–(7) we calculate closeness between attributes of AIS target and OTHR track inside a defined time frame as shown in Table 3:

Table 3. Closeness between AIS and OTHR attribute values.

c_{time}	c_{lat}	c_{long}	c_{course}	c_{vel}
0.9986	0.9979	0.999	0.9582	0.959

Finally, we compute the dependency strength of the FFD between records by using Equation (8):

$$\theta = \frac{c_{time} + c_{long} + c_{lat} + c_{vel} + c_{course}}{5} = \frac{4.9127}{5} = 0.9825 \quad (8)$$

Iterating through all OTHR records and matching with pointed AIS target, we get a pair (ID_Number, MMSI) for possible association (the highest value of parameter θ). In our case, AIS target with MMSI = 636,091,031 and OTHR track with ID_Number = 9471 are candidates for fusion. In order to increase the possibility of correct matching, we repeat the procedure described for newly received AIS message and calculate the dependency strength in this iteration as well. The results of matching for the fused pairs or targets within a time frame of 1 h (from 16:33 h to 17:33 h) are shown in Figure 3.

In the described experiment, there were 20+ vessels with AIS data and 13 OTHR targets. After AIS data outside of OTHR coverage area were filtered out, only 13 vessels reported by AIS remained for possible fusion. Results of data fusion for vessels labeled as 7a, 7b, 7c, 7d and 7e are depicted in Figure 3. Time axis provides more informative maritime picture. By using the MMSI code it is possible to obtain detailed information about the ship [27] as shown in Figure 4. It is important to note that cases 7b and 7c from Figure 2 are also represented as recognized targets and all the data are fused accordingly despite a significant AIS latency (7c) and later OTHR detection (7b). This is illustrated in Figure 3 where AIS message latency of a detected vessel is indicative (cyan squares) and first OTHR detections of vessel labeled as 7b were received long after the reception of AIS data (white squares). It is interesting to consider the fusion in the case of vessel labeled as 7e. The AIS target is fused with two different OTHR tracks due to the interruption of OTHR detection (blue squares in Figure 3). Since all objects around oil platforms are detected by OTHR as one, the OTHR track can be fused with only one AIS with the most likely attributes, while the other AIS will remain unfused. This should be marked for the operators of the maritime monitoring system.

AIS data that cannot be associated with any of the OTHR tracks will also stay unfused as well as OTHR tracks which cannot be associated with any AIS data (non-cooperative targets that pose potential threats). After all processing is done, complete operational picture in the OTHR coverage area is shown in Figure 5. It is worth mentioning that maritime situation presented in Figure 5 is formed after 4 h of collecting and processing data delivered by available sensors.

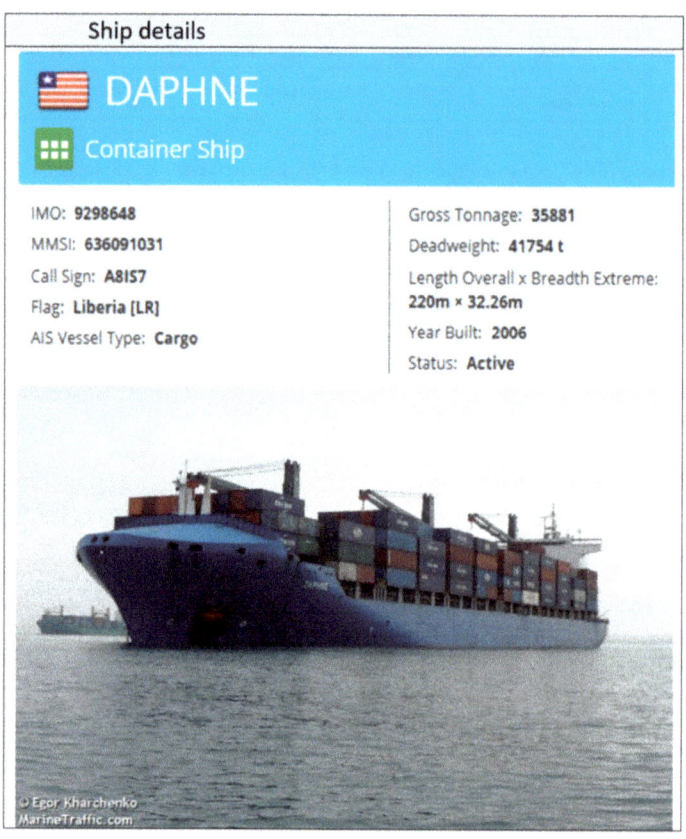

Figure 4. Fused AIS point with OTHR track—Dafne ex Emirates Asante [27].

There are critical and confusing situations such as when multiple targets are close to each other, which present a real challenge for any fusion procedure. In our work, we suggest an algorithm designed to autocorrect faulty AIS-OTHR fusion candidates through repeated iterations. This autocorrection happens when targets are moving away from each other. On the other hand, when there are multiple non-moving targets in small areas, usually in front of harbors and oil platforms, the finite resolution of OTHR prevents proper detection of vessels. Operators usually consider such OTHR detections as redundant or unnecessary. These areas are marked as illustrated in Figure 5 and often excluded from OTHR tracking. Furthermore, the correctness of fusion can be tested by estimating the AIS target's radar cross-section parameter, based on its position and orientation with respect to OTHR, as well as the ship size information, available from third-party sources. By comparing the measured Signal-to-Noise (SNR) of OTHR detection with pre-set experimental tables, it is possible to estimate the ship size class and perhaps to resolve to which AIS target the observed OTHR track is most likely to be associated. Thus, intensive research will be needed in future.

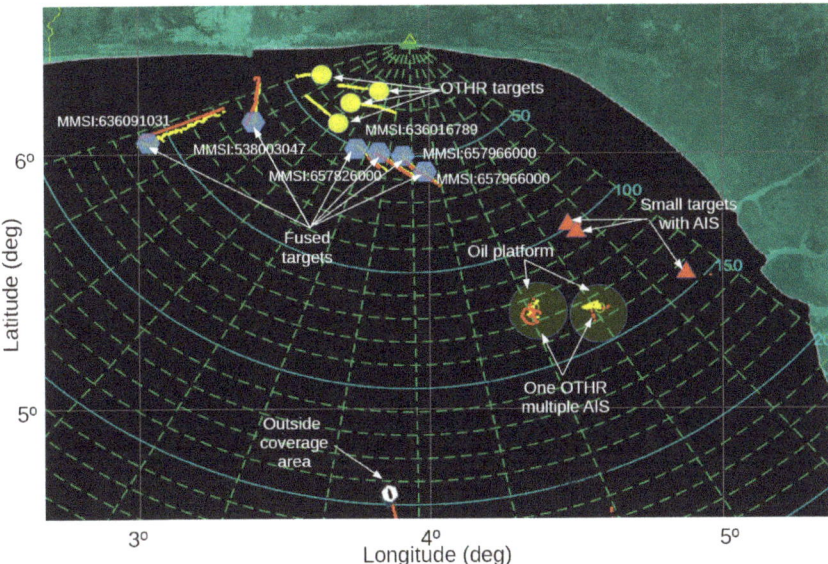

Figure 5. The results of experiment—complete operational picture for maritime surveillance (yellow—OTHR targets, red—AIS points, blue—fused data).

Figure 5 illustrates a complete maritime picture at OTH distances in the OTHR coverage area.

6. Conclusions

In this article, a method of fusing information obtained from AIS and OTHR sensors based on the fuzzy functional dependency approach has been presented. We describe a computational algorithm that focuses on the fuzzy decision making technique. It associates OTHR target tracks with available AIS ID records within the observed message time frame. Actually, the algorithm computes the record matching based on fuzzy dependency between AIS data and OTHR tracks. As shown, a number of iterations which take into account the similarity between database tuples, increases the odds for correct data fusion. This novel fuzzy-based method reveals the complementarity of two different sensors and enhances their integration. The integrated information provides a more informative maritime situational awareness. The reception of the AIS message is unsteady, especially in case of satellite transmission, causing difficulty in data fusion and analysis, as well as missing information. OTHR tracks not being fused with any AIS data can be identified as non-cooperative targets. Identification of these targets is a matter of future consideration. The performance of the proposed algorithm has been evaluated using real data. The experiment has shown promising results.

Future work is intended for applying this method to a larger-scale application problem (multiple OTHRs and multiple AIS sources covering a much larger geographical area). In addition, the application of decision-making theory is recognized as a possible direction for the method improvement.

Author Contributions: Conceptualization, M.A.R.M.M. and M.V.; methodology, M.V.; formal analysis, M.A.R.M.M.; investigation, N.S. and N.L. resources, N.S. and A.M.; data curation, M.V.; writing—original draft preparation, M.A.R.M.M. and M.V.; writing—review and editing, M.V. and A.M.; visualization, N.L.; project administration, N.L.

Funding: This research was funded by Vlatacom Institute. The APC was funded by Vlatacom Institute.

Conflicts of Interest: The authors declare no conflict of interest

References

1. Sevgi, L.D.; Ponsford, A.M. An HF Radar Base Integrated Maritime Surveillance System. In Proceedings of the 3rd International Multiconference IMACS/IEEE CSCC'99, Athens, Greece, 4–8 July 1999; pp. 5801–5806.
2. Anderson, S.J. Optimizing HF Radar Siting for Surveillance and Remote Sensing in the Strait of Malacca. *IEEE Trans. Geosci. Remote Sens.* **2013**, *51*, 1805–1816. [CrossRef]
3. Braca, P.; Maresca, S.; Grasso, R.; Bryan, K.; Hortsmann, J. Maritime surveillance with multiple over-the-horizon HFSW radars: An overview of recent experimentation. *IEEE Aerosp. Electron. Syst. Mag.* **2015**, *30*, 4–18. [CrossRef]
4. Ding, Z.; Kannappan, G.; Benameur, K.; Kirubarajan, T.; Farooq, M. Wide area integrated maritime surveillance: An updated architecture with data fusion. In Proceedings of the 6th International Conference of Information Fusion–FUSION 2003, Cairns, Australia, 8–11 July 2003; pp. 1324–1333.
5. Cimino, G.; Arcieri, G.; Horn, S.; Bryan, K. *Technical Report CMRE-FR-2014-017, Sensor Data Management to Achieve Information Superiority in Maritime Situational Awareness*; NATO Science and Technology Organization Centre for Maritime Research and Experimentation: La Spezia, Italy, 2014.
6. Dzvonkovskaya, A.; Gurgel, K.-W.; Rohling, H.; Schlick, T. Low power high frequency surface wave radar application for ship detection and tracking. In Proceedings of the Radar 2008 Conference, Adelaide, Australia, 16–18 April 2008; pp. 654–659.
7. Dzvonkovskaya, A.; Rohling, H. HF Radar Performance Analysis Based on AIS Ship Information. In Proceedings of the IEEE Radar Conference, Washington, DC, USA, 10–14 May 2010; pp. 1239–1244.
8. Nikolic, D.; Stojkovic, N.; Popovic, Z.; Tosic, N.; Lekic, N.; Stankovic, Z.; Doncov, N. Maritime Over the Horizon Sensor Integration: HFSWR Data Fusion Algorithm. *Remote Sens.* **2019**, *11*, 852. [CrossRef]
9. Stojković, N.; Nikolić, D.; Džolić, B.; Tošić, N.; Orlić, V.; Lekić, N.; Todorović, B. An Implementation of Tracking Algorithm for Over-The-Horizon Surface Wave Radar. In Proceedings of the Telfor 2016, Belgrade, Serbia, 22–23 November 2016.
10. Radio Noise. *Recommendation ITU-R P. 372-13*; CCIR: Geneva, Switzerland, 2016.
11. Nikolic, D.; Stojkovic, N.; Lekic, N. Maritime over the Horizon Sensor Integration: High Frequency Surface-Wave-Radar and Automatic Identification System Data Integration Algorithm. *Sensors* **2018**, *11*, 1147. [CrossRef]
12. Chang, L.; Xiaofei, S. Study of Data Fusion of AIS and Radar. In Proceedings of the International Conference of Soft Computing and Pattern Recognition, Malacca, Malaysia, 4–7 December 2009; pp. 674–677. [CrossRef]
13. Jidong, S.; Xiaoming, L. Fusion of Radar and AIS Data. In Proceedings of the 7th International Conference on Signal Processing ICSP'04, Bejing, China, 31 August–4 September 2004; pp. 2604–2607. [CrossRef]
14. Xiaorui, H.; Changchuan, L. A Preliminary Study on Targets Association Algorithm of Radar and AIS Using BP Neural Network. *Procedia Eng.* **2011**, *15*, 1441–1445. [CrossRef]
15. Jiachun, Z.; Zongheng, C. A Study of Radar and AIS Object Data Fusion Based on AFS_RBF's Neural Network. *J. Jimei Univ. Nat. Science* **2005**, *10*, 216–220.
16. Shenoi, S.; Melton, A. Proximity relations in the fuzzy relational database model. *Fuzzy Sets Syst.* **1989**, *100*, 51–62. [CrossRef]
17. Caruccio, L.; Deufemia, V.; Polese, G. Relaxed Functional Dependencies—A Survey of Approaches. *IEEE Trans. Knowl. Data Eng.* **2016**, *28*, 147–165. [CrossRef]
18. Bohannon, P.; Fan, W.; Geerts, F.; Jia, X.; Kementsietsidis, A. Conditional Functional Dependencies for Data Cleaning. In Proceedings of the IEEE 23rd International Conference on Data Engineering, Istanbul, Turkey, 11–15 April 2007; pp. 746–755.
19. Fan, W.; Gao, H.; Jia, X.; Li, J.; Ma, S. Dynamic constraints for record matching. *VLDB J.* **2011**, *20*, 495–520. [CrossRef]
20. Nambiar, U.; Kambhampati, S. Mining approximate functional dependencies and concept similarities to answer imprecise queries. In Proceedings of the 7th International Workshop Web Databases, Paris, France, 17–18 June 2004; pp. 73–78.
21. Hudec, M.; Vučetić, M.; Vujošević, M. Comparison of linguistic summaries and fuzzy functional dependencies related to data mining. In *Biologically-Inspired Techniques for Knowledge Discovery and Data Mining*; Alam, S., Dobbie, G., Koh, Y.S., Rehman, S., Eds.; IGI Global: Hershey, PA, USA, 2014; pp. 174–203, ISBN 1466660783.

22. Chen, G.; Kerre, E.E.; Vandenbulcke, J. Normalization based on fuzzy functional dependency in a fuzzy relational data model. *Inf. Syst.* **1996**, *21*, 299–310. [CrossRef]
23. Vučetić, M.; Vujošević, M. A literature overview of functional dependencies in fuzzy relational database models. *Tech. Technol. Educ. Manag.* **2012**, *7*, 1593–1604.
24. Vucetic, M.; Hudec, M.; Vujošević, M. A new method for computing fuzzy functional dependencies in relational database systems. *Expert Syst. Appl.* **2013**, *40*, 2738–2745. [CrossRef]
25. Sözat, M.I.; Yazici, A. A complete axiomatization for fuzzy functional and multivalued dependencies in fuzzy database relations. *Fuzzy Sets Syst.* **2001**, *117*, 161–181. [CrossRef]
26. ITU Recommendation. *M.585–7. Assignment and Use of Identities in the Maritime Mobile Service*; ITU: Geneva, Switzerland, March 2015.
27. Ships Database. Available online: https://www.marinetraffic.com/sr/ais/details/ships/shipid:757561 (accessed on 2 October 2019).

© 2019 by the authors. Licensee MDPI, Basel, Switzerland. This article is an open access article distributed under the terms and conditions of the Creative Commons Attribution (CC BY) license (http://creativecommons.org/licenses/by/4.0/).

Article

A Clamping Force Estimation Method Based on a Joint Torque Disturbance Observer Using PSO-BPNN for Cable-Driven Surgical Robot End-Effectors

Zhengyu Wang [1,2], Daoming Wang [1,*], Bing Chen [1], Lingtao Yu [3], Jun Qian [1] and Bin Zi [1]

1. School of Mechanical Engineering, Hefei University of Technology, Hefei 230009, China; wangzhengyu_hfut@hfut.edu.cn (Z.W.); chbing@hfut.edu.cn (B.C.); qianjun@hfut.edu.cn (J.Q.); binzi@hfut.edu.cn (B.Z.)
2. Tianjin Key Laboratory of Aerospace Intelligent Equipment Technology, Tianjin Institute of Aerospace Mechanical and Electrical Equipment, Tianjin 300301, China
3. College of Mechanical and Electrical Engineering, Harbin Engineering University, Harbin 150001, China; yulingtao@hrbeu.edu.cn
* Correspondence: denniswang@hfut.edu.cn; Tel.: +86-0551-62901326

Received: 28 October 2019; Accepted: 29 November 2019; Published: 1 December 2019

Abstract: The ability to sense external force is an important technique for force feedback, haptics and safe interaction control in minimally-invasive surgical robots (MISRs). Moreover, this ability plays a significant role in the restricting refined surgical operations. The wrist joints of surgical robot end-effectors are usually actuated by several long-distance wire cables. Its two forceps are each actuated by two cables. The scope of force sensing includes multidimensional external force and one-dimensional clamping force. This paper focuses on one-dimensional clamping force sensing method that do not require any internal force sensor integrated in the end-effector's forceps. A new clamping force estimation method is proposed based on a joint torque disturbance observer (JTDO) for a cable-driven surgical robot end-effector. The JTDO essentially considers the variations in cable tension between the actual cable tension and the estimated cable tension using a Particle Swarm Optimization Back Propagation Neural Network (PSO-BPNN) under free motion. Furthermore, a clamping force estimator is proposed based on the forceps' JTDO and their mechanical relations. According to comparative analyses in experimental studies, the detection resolutions of collision force and clamping force were 0.11 N. The experimental results verify the feasibility and effectiveness of the proposed clamping force sensing method.

Keywords: surgical robot end-effector; clamping force estimation; joint torque disturbance observer; PSO-BPNN; cable tension measurement

---

1. Introduction

In recent years, more and more researchers, companies and hospitals have paid attention to the development, commercial aspects, and application of minimally-invasive surgical robot (MISR) techniques. The advantages of robot-assisted minimally-invasive surgery (MIS) include positioning accuracy, easier realization of MIS, an improved success rate, a reduction in pain, and a reduction in recovery time [1,2]. Many MISR systems including the da Vinci surgical system [3], the DLR MIRO [4], the Raven system [5] have been shown to be valid and feasible and to provide advancements in MIS operations. The capacity to sense force is an important technique for force feedback, haptics, and safe interaction control in MISRs. This ability can also help to restrict refined operations and improve operational safety [6]. Moreover, this ability can help surgeons to determine tissue hardness and evaluate the anatomical and histological properties of object organs in order to perform better manipulation in MIS [7,8]. Thus, the study of force sensing is an important research direction of MISR.

The wrist joints of a surgical robot end-effector usually have multiple degrees of freedom to ensure good flexibility and dexterity [9]. Cable-driven actuators have been widely used in rehabilitation and medical robots [10–12]. Furthermore, flexible continuum instruments present good application prospects. Hwang and Kwon proposed a novel constrained strong continuum manipulator by using auxiliary links attached to the main continuum links [13]. Li et al. developed a flexible endoscope based on the tendon-driven continuum mechanism [14]. The wrist joints of an end-effector are usually actuated by several long-distance cables with a small diameter. Each of its two forceps are actuated by two cables. The scope of force sensing includes multidimensional external force (including three-dimensional force and three-dimensional torque) and one-dimensional clamping force. Use of a surgical robot end-effector with the ability to sense force is an important and a feasible way to realize feedback from an external force or a clamping force.

Two methods for sensing external forces have been studied by MISR system researchers. One is the direct sensing method, which integrates force sensors into the surgical instrument's tip. Some surgical tools have been designed for an MISR system with the ability to sense force [15–17]. Li et al. designed and developed a three-axis force sensor using a resistance-based sensing method [18]. Yu et al. developed a six-dimensional force/torque sensor with a double-crossbeam structure for a surgical robot end-effector [19]. Lim et al. proposed a kind of forceps with an optical fiber Bragg grating sensor integrated into it [20]. Kim et al. developed a surgical robot with a multi-axis force sensing instrument [21,22]. Radó et al. developed a surgery robot with a three-axis force sensor integrated into it to provide force feedback [21]. However, because wrist joints need to be designed to be narrow and small in size in order to meet the requirements of surgical robot end-effectors, and considering the limiting factors of the disinfection method, the material hemolysis effect, and the economic cost, many challenges and difficulties remain with the application of these force sensing methods by integrating internal force sensors into the tip of surgical forceps [23,24].

The other method for sensing external forces is the indirect sensing method, which uses system information including current, torque, tension, displacement, pressure, and a visual image of the laparoscope. The design of an external force disturbance observer or a similar estimation method has attracted a great deal of attention. Zhao and Nelson developed a cable-driven decoupled surgical robot end-effector and proposed a method for estimating the three-axis force using a motor current [25]. Because the friction and elasticity of cables are omitted in the overall model, as well as the filtering processing of the motor current, so the estimated accuracy was limited with relatively big errors. Li et al. [26] and Haraguchi et al. [27] both proposed external force disturbance observers for the force estimating the force acting at the forceps of pneumatically driven MISR. These studies were purely based on the system dynamic model and the cylinder pressure disturbance observers. The results showed good dynamic performance with acceptable estimated accuracy. Xue et al. proposed a tension sensor using fiber gratings for estimating the grasping force in a laparoscope surgical robot [28]. The property index of the sensor showed good performance. The hysteresis characteristics of the sensor and the friction between the cable and beams affect the resolution of the proposed sensor. They also proposed a method for estimating the grasping force based on a model of cable-pulley systems considering its tension transmission characteristics [29]. These two clamping force estimation methods were deeply affected by the model precision of cable-pulley transmission system. Li and Hannaford [30] proposed a Gaussian process regression for predicting the clamping force of a cable-driven surgical robot end-effector for the Raven system. The estimated accuracy was still the main problem for these clamping force estimation methods using the motor currents. In addition, artificial neural networks and deep learning algorithms provide us with another way to realize sensorless force sensing in MISRs [31–33]. Yu et al. proposed a bidimensional external force and clamping force sensing method based on changes in cable tension for a surgical micromanipulator [34]. The results showed acceptable accuracy. However, the comprehensive resistance of the cable tension was estimated and limited by the BP Neural network under free motion. This work was the preliminary study of the clamping force sensing compared with the new estimation method in this paper. Hwang and Lim proposed a force estimation method based on a deep learning method that utilize sequential images of an object's shape changed by an external force [35]. It was difficult to learn and predict the interaction force using such many sequential and variational images in real time, due to the camera movement during the surgery operation. Huang et al. proposed a method for clamping force estimation based on a neural network for a cable-driven surgical robot [36]. The results showed the training process and training errors, but not considered the generalization ability verification. Marban et al. proposed a force estimation model for robotic surgery based on convolutional neural networks and long-short term memory networks [37]. The camera and organs were static while the surgical instrument was in motion. The real-time estimation is still a big challenge during the dynamic process of the real surgery.

To summarize, the property that these above proposed indirect sensing methods have in common is the combination or segregation of the disturbance observers and neural networks or learning methods. The accuracy, rapidity and robustness of the clamping force estimation are still the main challenges for the cable-driven surgical robot end-effector without the forceps' internal force sensors. The motivation and contribution of this study are trying to find a new way of the clamping force estimation, which is considering the cable-driven system dynamics, the cable tension measurements and estimation, and the joint torque disturbance observer of the forceps in real time, such that aims to achieve a good comprehensive performance with low-cost and easy realization.

This paper focuses on one-dimensional clamping force sensing method that do not require any internal force sensors to be integrated into the wrist joints of a surgical robot end-effector. The main contribution of this study is the development of a novel method for estimating the clamping force of a forceps based on a joint torque disturbance observer, which essentially considers the variations in cable tension between the actual cable tension and the real-time estimated cable tension using a Particle Swarm Optimization Back Propagation Neural Network (PSO-BPNN) under free motion of the end-effector. A clamping force estimator is proposed based on the JTDO and the mechanical relations in the forceps. The main advantages of this method are the combination of cable-driven system dynamics and joint torque disturbance observer using PSO-BPNN, for improving the comprehensive performance. We verify the estimation method through a series of experiments with an equivalent experimental system.

2. Methods

2.1. Description of the 3-Degrees of Freedom (3-DoF) Cable-Driven Surgical Robot End-Effector

A surgical robot end-effector is a key device in a surgical robotic system. Figure 1a shows the experimental prototype of a 3-DoF surgical robot end-effector, which is actuated by three cable-driven actuators with six cables. The external diameter of the wrist joints and the slender shaft is limited to 8 mm.

Figure 1b shows the principle diagram of the configurational characteristics and the cable-driven systems. This 3-DoF forceps of a surgical robot end-effector has one yaw joint and two pitch joints, which are all actuated by cables. Meanwhile, the 3-DoF consist of yaw, pitch, and opening & closing. The opening & closing and pitch movement are compound motions between forceps A and B with different combinations of rotation directions. A multi-axis motion is the result of a 3-DoF compound action. The wrist joints of the end-effector are actuated by three long-distance cable-driven modules with a cable of approximately 500 mm in length inside of the slender shaft. The cable-pulley systems form the transition bridge between the wrist joints and three servo motors. Obviously, clamping force estimation focuses on the force conditions of forceps A and B for the 3-DoF cable-driven surgical robot end-effector. Furthermore, the driving torques of forceps A and B are directly related to the driving cable tensions. This means that the relationship between the driving torques and the cable tensions can be used to estimate the clamping force.

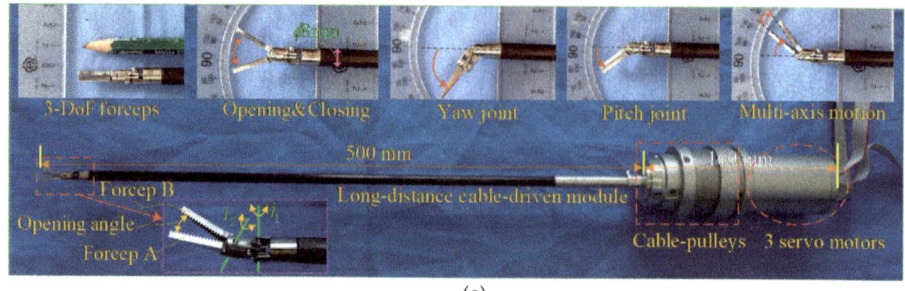
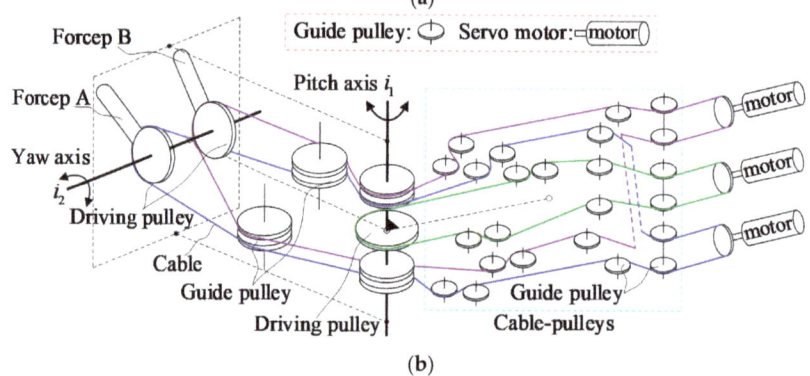

Figure 1. 3-DoF cable-driven surgical robot end-effector. (**a**) Experimental prototype; (**b**) Principle diagram.

2.2. Equivalent Experimental System for the Surgical Robot End-Effector

In order to study external force and clamping force sensing methods, we designed and built an equivalent experimental prototype for the 3-DoF cable-driven surgical robot end-effector.

As shown in Figure 2a, a clamping force sensing study can be performed by using this equivalent experimental system. The wrist joints are the same as shown in Figure 1. Their cable-pulley systems are expanded into a horizontal plate form, in order to achieve convenient installation and an easy layout. The equivalent experimental prototype consists of 3-DoF wrist joints, one slender shaft, cable-pulley systems, a cable tension detection module, motor driving system, a data acquisition system, a computer, and control software.

Overall, the difference between Figures 1 and 2 are in the motor driving system. Figure 1 shows three servo motors for driving six cables. Figure 2 shows six linear stepping motors for driving six cables. Figure 2b shows a block diagram of the system's composition. The nominal diameter of the driving wire cable is 0.45 mm. The measuring range of the tension sensor is 0–50 N. The max subdivs value of the linear stepping motor driver is 512. The motion control card can drive six motors. Six tension sensors are integrated between the motor driving system and the cable-pulley systems. This means that the cable tension of the input side can be measured. For detecting the actual clamping force, two pressure sensors (FlexiForce A201, Tekscan®, South Boston, MA, USA) are installed on the two sides of the triangular block, which is the grasping object for forceps A and B. In terms of overall function and robotic mechanisms, the equivalent experimental system (shown in Figure 2) is equal to the 3-DoF cable-driven surgical robot end-effector (shown in Figure 1).

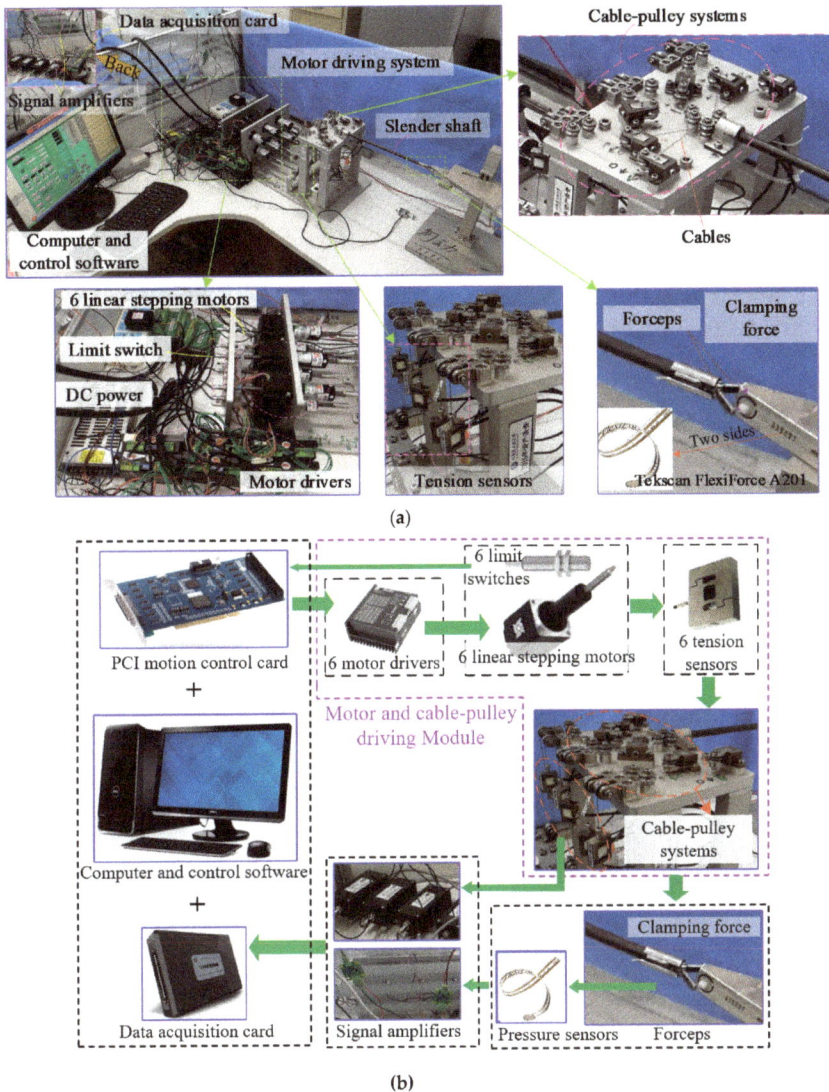

Figure 2. Experimental set-up. (**a**) Equivalent experimental prototype; (**b**) System composition block diagram.

As mentioned in the descriptions above, the equivalent experimental prototype is designed equivalent with the 3-DoF cable driven surgical robot end-effector. Their mechanical structures are the same on the aspect of mechanism principle. Moreover, they have the same 3-DoF wrist joints and forceps. The other main reason of equivalence principle is that, their kinematic mapping methods between the motor position and joint angle are the same principles only with different parameters. This means that the equivalent experimental prototype has the function and capability to represent the kinematics, dynamics, external forces, clamping force and kinematic mapping of the 3-DoF cable driven surgical robot end-effector. In the experiments on clamping force, the pitch joint lacked convenience. The deeply reason is that the manufacturing and assembly errors of the pitch joint greatly influence the decoupling of the wrist joints. Therefore, in order to verify the feasibility and effectiveness of the clamping force sensing method that we proposed in this paper, we discarded the pitch joint as shown in Figure 2. This means that the experiments of clamping force estimation were carried out with a pitch joint angle of 0 degrees and a specific opening angle of 65 degrees, which is a limitation of the experimental mechanical structure.

2.3. Strategy for Estimating the Clamping Force of the Surgical Robot End-Effector

2.3.1. Modeling the System Dynamics of the Forceps

As mentioned in previous sections, the mechanical model between the joint torques and the cable tensions is the key to establishing the overall dynamics model, which provides a model for calculating the clamping force. Figure 3 shows a simplified dynamic model of the cable-driven surgical robot end-effector. Because the initial cable tensions were preadjusted to suitable values during the adjustment process, the elasticity of cables can be neglected in the modeling process. Moreover, the friction of the cable-pulley system is temporarily hided in the cable tension losses F_f of the overall dynamics, as well as the nonlinear characteristics of cables. These two factors are the uncertain models. But they can be estimated by the PSO-BPNN. In order to study the problem of estimating the clamping force of the cable-driven surgical robot end-effector, the modelling, analysis and experiment were all carried out at the system's zero position without considering the coupling problem and manufacturing errors of the wrist joints' motion. The main reason is that the experiment condition is limited at the system's zero position according to the measurement of the clamping force using two flexible pressure sensors. Under these circumstances, the feasibility and effectiveness of the estimation method can be verified.

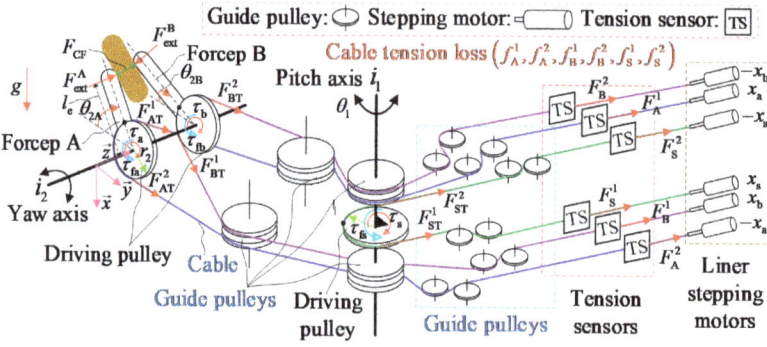

Figure 3. A simplified dynamic model of the cable-driven surgical robot end-effector. linear.

The subscripts A/a, B/b, S/s in the following formulas denote the relations to yaw joint A, yaw joint B, and pitch joint S, respectively. F_A^1, F_B^1, and F_S^1 denote the respective actual measured values of the motor driving cables. F_A^2, F_B^2, and F_S^2 denote the respective actual measured values of the motor returning cables. F_{AT}^1, F_{BT}^1, and F_{ST}^1 denote the respective driving cable tensions. F_{AT}^2, F_{BT}^2, and F_{ST}^2 denote the respective returning cable tensions. f_A^1, f_B^1, and f_S^1 denote the respective driving cable tension losses. f_A^2, f_B^2, and f_S^2 denote the respective returning cable tension losses. The cable tension losses include the frictions of the cable-pulley systems, the nonlinear characteristics of cables, and other uncertain items. τ_a, τ_b, and τ_s denote the respective joint driving torques. τ_{fa}, τ_{fb}, and τ_{fs} denote the respective joint friction torques. x_a, x_b, and x_s denote the respective motor displacements of driving cables. $-x_a$, $-x_b$, and $-x_s$ denote the respective motor displacements of returning cables. θ_{2A}, θ_{2B}, and θ_1 denote the respective joint angles of yaw joint A, yaw joint B and pitch joint S. Meanwhile, the opening angle of forceps A and B can be calculated as $(\theta_{2A} + \theta_{2B})$, according to the initial position in Figure 2. r_1 and r_2 denote the effective drive radius of yaw joint A, yaw joint B and pitch joint S, respectively.

Suppose that F_M is the set of actual measured values of the motor driving cable tensions for forceps:

$$F_M = \begin{bmatrix} F_A^1 & F_A^2 & F_B^1 & F_B^2 \end{bmatrix}^T \tag{1}$$

Suppose that F_T is the set of cable tensions for actuating forceps A and B:

$$F_T = \begin{bmatrix} F_{AT}^1 & F_{AT}^2 & F_{BT}^1 & F_{BT}^2 \end{bmatrix}^T \tag{2}$$

Suppose that F_f is the set of cable tension losses for actuating forceps A and B:

$$F_f = \begin{bmatrix} f_A^1 & f_A^2 & f_B^1 & f_B^2 \end{bmatrix}^T \tag{3}$$

Obviously, from Equations (1) to (3), the following relationship can be obtained:

$$F_M = F_T + F_f \tag{4}$$

The driving torques τ_{ab} of forceps A and B can be calculated as:

$$\tau_{ab} = \begin{bmatrix} \tau_a \\ \tau_b \end{bmatrix} = r_2 \begin{bmatrix} F_{AT}^1 - F_{AT}^2 \\ F_{BT}^1 - F_{BT}^2 \end{bmatrix} = r_2 \begin{bmatrix} (F_A^1 - f_A^1) - (F_A^2 + f_A^2) \\ (F_B^1 - f_B^1) - (F_B^2 + f_B^2) \end{bmatrix} \tag{5}$$

where, τ_a and τ_b denote the driving torques of forceps A and B.

Suppose that $\hat{F}_E$ is the set of estimated values from the cable tension sensors under free motion of the forceps, which means that no clamping force or external force is applied to the wrist joints of the surgical robot end-effector. An artificial neural network model can be employed for the estimated values of the cable tension sensors' tensions $\hat{F}_E$ under free motion. Suppose that $\hat{F}_{fE}$ is the set of cable tension losses for actuating forceps A and B under free motion:

$$\hat{F}_E = \begin{bmatrix} \hat{F}_A^1 & \hat{F}_A^2 & \hat{F}_B^1 & \hat{F}_B^2 \end{bmatrix}^T \tag{6}$$

$$\hat{F}_{fE} = \begin{bmatrix} \hat{f}_{AE}^1 & \hat{f}_{AE}^2 & \hat{f}_{BE}^1 & \hat{f}_{BE}^2 \end{bmatrix}^T \tag{7}$$

Therefore, the set of cable tension disturbances from the tension sensor measurement modules can be defined as $\hat{F}_D$:

$$\hat{F}_D = \begin{bmatrix} \hat{F}_{AD}^1 & \hat{F}_{AD}^2 & \hat{F}_{BD}^1 & \hat{F}_{BD}^2 \end{bmatrix}^T = F_M - \hat{F}_E \tag{8}$$

The simplified dynamic model of forceps A and B can be given as:

$$\begin{aligned} \tau_{ab} &= \begin{bmatrix} \tau_a \\ \tau_b \end{bmatrix} = \begin{bmatrix} J_A \ddot{\theta}_{2A} + g(\theta_{2A}) + \tau_{fa} + F_{ext}^A l_{eA} \\ J_B \ddot{\theta}_{2B} + g(\theta_{2B}) + \tau_{fb} + F_{ext}^B l_{eB} \end{bmatrix} \\ &= r_2 \begin{bmatrix} F_{AT}^1 - F_{AT}^2 \\ F_{BT}^1 - F_{BT}^2 \end{bmatrix} \\ &= r_2 \begin{bmatrix} (F_A^1 - F_A^2) - (f_A^1 + f_A^2) \\ (F_B^1 - F_B^2) - (f_B^1 + f_B^2) \end{bmatrix} \end{aligned} \tag{9}$$

where, J_A and J_B, τ_{fa} and τ_{fb}, $\theta_{2A} = x_a/r_2$ and $\theta_{2B} = x_b/r_2$, l_{eA} and l_{eB}, and F_{ext}^A and F_{ext}^B are the equivalent rotating inertias, the joint frictions, the joint angles, the arms of external force, and the external force of the joints of forceps A and B, respectively. We neglected the forceps' gravities since they are installed in the horizontal plane.

Equation (9) can be revised as Equation (10) when there is no external force or clamping force under free motion:

$$\hat{\tau}_{ab} = \begin{bmatrix} \hat{\tau}_a \\ \hat{\tau}_b \end{bmatrix} = \begin{bmatrix} J_A \ddot{\theta}_{2A} + g(\theta_{2A}) + \hat{\tau}_{fa} \\ J_B \ddot{\theta}_{2B} + g(\theta_{2B}) + \hat{\tau}_{fb} \end{bmatrix} \\ = r_2 \begin{bmatrix} (\hat{F}_A^1 - \hat{F}_A^2) - (\hat{f}_{AE}^1 + \hat{f}_{AE}^2) \\ (\hat{F}_B^1 - \hat{F}_B^2) - (\hat{f}_{BE}^1 + \hat{f}_{BE}^2) \end{bmatrix} \quad (10)$$

Considering that linear stepping motors were chosen to be the drive units for the wrist joints, the acceleration and deceleration modes were set to the 'S' type in the motion control card, their acceleration and deceleration time were set to be 0.02 s, and the constant velocity was set to be variable. Because the inertias of the wrist joints are very small, their inertial forces only have marginal effects on the wrist joints' system dynamics, so the inertia forces of the wrist joints were ignored in this paper. If the joints' driving torques can be estimated under free motion, then Equations (9) and (10) can be simplified as Equations (11) and (12), respectively:

$$\tau_{ab} = \begin{bmatrix} \tau_a \\ \tau_b \end{bmatrix} = \begin{bmatrix} \tau_{fa} + F_{ext}^A l_{eA} \\ \tau_{fb} + F_{ext}^B l_{eB} \end{bmatrix} \\ = r_2 \begin{bmatrix} (F_A^1 - F_A^2) - (f_A^1 + f_A^2) \\ (F_B^1 - F_B^2) - (f_B^1 + f_B^2) \end{bmatrix} \quad (11)$$

$$\hat{\tau}_{ab} = \begin{bmatrix} \hat{\tau}_a \\ \hat{\tau}_b \end{bmatrix} = \begin{bmatrix} \hat{\tau}_{fa} \\ \hat{\tau}_{fb} \end{bmatrix} \\ = r_2 \begin{bmatrix} (\hat{F}_A^1 - \hat{F}_A^2) - (\hat{f}_{AE}^1 + \hat{f}_{AE}^2) \\ (\hat{F}_B^1 - \hat{F}_B^2) - (\hat{f}_{BE}^1 + \hat{f}_{BE}^2) \end{bmatrix} \quad (12)$$

Combining Equations (11) and (12), the joint torque disturbance $\hat{\tau}_D = \begin{bmatrix} \hat{\tau}_{aD} & \hat{\tau}_{bD} \end{bmatrix}^T$ of forceps A and B can be given as:

$$\hat{\tau}_D = \tau_{ab} - \hat{\tau}_{ab} = \begin{bmatrix} \tau_{fa} - \hat{\tau}_{fa} + F_{ext}^A l_{eA} \\ \tau_{fb} - \hat{\tau}_{fb} + F_{ext}^B l_{eB} \end{bmatrix} \\ = r_2 \begin{bmatrix} (\hat{F}_{AD}^1 - \hat{F}_{AD}^2) - (f_A^1 + f_A^2) + (\hat{f}_{AE}^1 + \hat{f}_{AE}^2) \\ (\hat{F}_{BD}^1 - \hat{F}_{BD}^2) - (f_B^1 + f_B^2) + (\hat{f}_{BE}^1 + \hat{f}_{BE}^2) \end{bmatrix} \quad (13)$$

Under the circumstances of free motion and no clamping force, the joint frictions and cable tension losses of forceps A and B satisfy the condition approximated below:

$$\begin{bmatrix} (f_A^1 + f_A^2) & (f_B^1 + f_B^2) & \tau_{fa} & \tau_{fb} \end{bmatrix}^T \approx \begin{bmatrix} (\hat{f}_A^1 + \hat{f}_A^2) & (\hat{f}_B^1 + \hat{f}_B^2) & \hat{\tau}_{fa} & \hat{\tau}_{fb} \end{bmatrix}^T \quad (14)$$

Therefore, the joint torque disturbance $\hat{\tau}_D$ of Equation (13) can be further simplified as:

$$\hat{\tau}_D = \begin{bmatrix} \hat{\tau}_{aD} \\ \hat{\tau}_{bD} \end{bmatrix} \approx r_2 \begin{bmatrix} \hat{F}_{AD}^1 - \hat{F}_{AD}^2 \\ \hat{F}_{BD}^1 - \hat{F}_{BD}^2 \end{bmatrix} = \begin{bmatrix} F_{ext}^A l_{eA} \\ F_{ext}^B l_{eB} \end{bmatrix} \quad (15)$$

Furthermore, the estimated external force $\hat{F}_{ext}$ of forceps A and B can be given by Equation (16):

$$\hat{F}_{ext} = \begin{bmatrix} \hat{F}_{ext}^A & \hat{F}_{ext}^B \end{bmatrix}^T \approx \begin{bmatrix} \dfrac{\hat{\tau}_{aD}}{l_{eA}} & \dfrac{\hat{\tau}_{bD}}{l_{eB}} \end{bmatrix}^T \quad (16)$$

Finally, the estimated clamping force of forceps A and B can be calculated by Equation (17):

$$\hat{F}_{CF} \approx \frac{1}{2}\left(\left|\hat{F}_{ext}^{A}\right| + \left|\hat{F}_{ext}^{B}\right|\right) \quad (17)$$

2.3.2. Strategy for Estimating the Clamping Force

According to the process for modeling the forceps' clamping force, a strategy for estimating the clamping force of the forceps is given as shown in Figure 4. Essentially, the block diagram of this estimation strategy is a kind of open-loop form, which is constructed according to the open-loop control strategy of the joints' motion using the feedback of the kinematic mapping relation.

Firstly, the desired opening angle or grasping angle θ_C is input into the control software; then, it is converted to yaw joint angle θ_Y according to the kinematic relation.

Secondly, angle θ_Y is input into the driving system according to the mapping relation of the joint angle to the cable and motor displacement; then, the calculated motor displacement x_m and setting motor velocity v_m are input into the driving system of the linear stepping motors.

Thirdly, the motors pull or push the cable loops with the desired displacement while the force sensors measure the driving and returning cable tensions F_M. Meanwhile, the wrist joints are actuated by the difference in the cable tensions in the cable-pulley systems. Forceps A and B are actuated by the difference in tension with the result of an opening or a closing action on the two flexible pressure sensors, which provide comparisons of the measured joint external forces F_{ext}^{A} and F_{ext}^{B}, and the measured clamping force F_{CF}.

Fourthly, the motor displacement x_m and the motor velocity v_m of forceps A and B are input into the cable tension estimation model based on PSO-BPNN under free motion in order to estimate the tensions $\hat{F}_E$. Therefore, the cable tension disturbance $\hat{F}_D$ can be calculated by inputting the actual cable tensions F_M and the estimated cable tensions $\hat{F}_E$. The yaw joint torque disturbance $\hat{\tau}_D$ can be obtained using the cable tension disturbance $\hat{F}_D$.

Finally, the external force $\hat{F}_{ext}$ of the yaw joints can be estimated by Equation (16), and the clamping force $\hat{F}_{CF}$ can be calculated using the clamping force estimator. To sum up, the strategy for estimating the clamping force is based on the measured motor driving cable tensions, the cable tension model based on PSO-BPNN under free motion, the joint torque disturbance observer, and the clamping force estimator that does not require internal force sensors to be integrated into the end tips of the wrist joints.

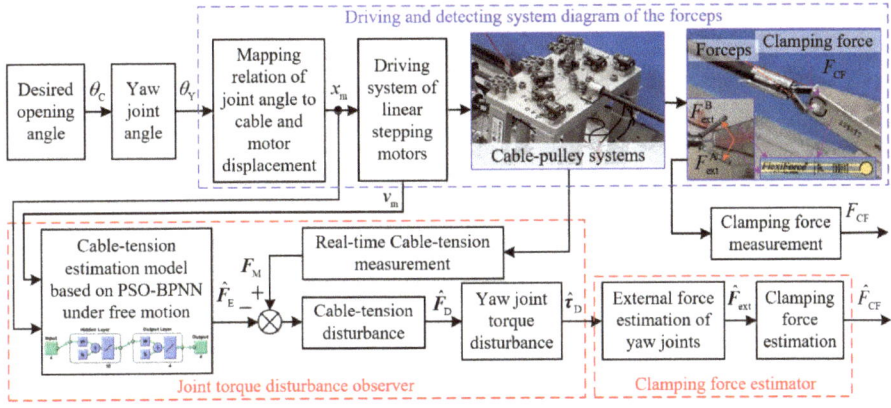

Figure 4. Strategy for estimating the clamping force of the forceps.

2.3.3. Joint Torque Disturbance Observer Using PSO-BPNN

The purpose of designing the yaw joint torque disturbance observer shown in Equation (15) is to estimate or predict the variation in the motor's driving cable tension as compared with the same motor's displacement and velocity under free motion. This means that the forceps' external force or clamping force affects the motor's driving cable tension with respect to the condition of free motion or no clamping force. The core requirement for this joint torque disturbance observer is to build a high-accuracy model for predicting the cable tension under free motion with the same motor displacement and velocity. The parameters estimations with evolutionary algorithms using tweezers show its superiority in a coupled nonlinear dynamics [38]. In this paper, the PSO-BPNN was employed to fit an artificial neural network model for estimating or predicting the motor's driving cable tension under free motion.

The artificial neural networks can approximate any function to an arbitrary degree of accuracy due to its high learning capability and parallel computing nature [39]. The BPNN is a supervised artificial neural network, and it is widely used for nonlinear and non-convex function approximation [40]. However, its main disadvantages include slow learning speed, a propensity to easily fall into a local minimum, a limited number of network layers, and overfitting [41]. The traditional BPNN can be improved by global optimization of PSO to solve the problems of oscillation, slow convergence and local extremum in the training process [42].

In this paper, the PSO-BPNN is employed to fit an artificial neural network model for estimating or predicting the motor driving cable tension under free motion. PSO-BPNN [43] is a combination of Particle Swarm Optimization and a BP neural network. Because the PSO algorithm is based on a heuristic learning algorithm, it can search different regions of the solution space at the same time, avoid falling into local minima and realize global optimization. In each iteration of the optimal solution, the particle updates itself by tracking two extreme values. The first one is called individual extreme value, which is the optimal solution found by the particle itself. The other one is called global extreme value, which is the optimal solution found by the whole population. When the two optimal extreme values are found, the particle updates its velocity and position according to the following formula [44]:

$$\begin{cases} v_{id} = wv_{id} + c_1u_1(p_{id} - x_{id}) + c_2u_2(p_{gd} - x_{id}) \\ x_{id} = x_{id} + v_{id} \end{cases} \quad (18)$$

where, x_{id} and v_{id} are the position and velocity of the i-th particle, respectively; p_{id} is the optimal position that the i-th particle searches for; p_{gd} is the optimal position searched by the whole particle swarm; c_1 and c_2 are learning factors; u_1 and u_2 are uniform random numbers within $[-1,1]$; w is the inertia weight.

The PSO algorithm was used to train the BP neural network, and the weight and threshold of each neuron were taken to be a particle's iterative optimization of the solution space. The specific steps for PSO of the BP neural network algorithm are introduced in [45,46]. The specific steps for PSO of the BP neural network algorithm are as follows:

(a) Determine the topological structure of the BP neural network and set the number of neurons in each layer of the BP neural network. The particle population is initialized, and the velocity and position of each particle are randomly set. The main operating parameters of particle swarm optimization are shown in Table 1.
(b) Calculate the fitness value Fit (i) of each particle;
(c) Compare the fitness value Fit (i) of each particle with the individual extreme value. If Fit (i) > pbest (i), replace pbest (i) with Fit (i).
(d) Compare the fitness value Fit (i) of each particle with the global extreme value gbest (i). If Fit (i) > gbest (i), replace gbest (i) with Fit (i).
(e) Update the position and velocity of each particle according to Equation (18);

(f) If the condition is satisfied (the error is sufficiently small or the number of cycles has reached its maximum), exit; otherwise, return to the second step (b);
(g) The global extreme value gbest (i) from the PSO algorithm is used as the weight and threshold for the BP neural network and to train the neural network with training samples;
(h) The generalization ability of the PSO-BPNN can be tested by simulation with the test samples.

Table 1. The main parameters of PSO-BPNN.

Parameters	Values
Swarm size	10
c_1, c_2	1.49445
Max iteration	30
w	1
Number of the input layer nodes	4
Number of the hidden layer nodes	10
Number of the output layer nodes	4
Number of neural network training	100
Learning rate of neural network	0.005
Percentage of training data	90%
Percentage of testing data	10%

To sum up, the PSO-BPNN plays an important role in the online estimation of motor cable tensions under free motion. The estimated values are input into the joint torque disturbance observer as contrasting values. The clamping force estimator can be established on this basis.

2.3.4. The Clamping Force Estimator

As shown in Figure 4 and described in Equations (15) and (16), the external forces $\hat{F}_{ext}$ of forceps A and B can be estimated using the joint torque disturbance observer and the contact force arm. Equation (17) shows the definition and an estimation of the comprehensive clamping force $\hat{F}_{CF}$ of the two forceps A and B. The clamping force estimator proposed in this paper essentially considers and combines the system's dynamic characteristics and an artificial neural network. Moreover, the general accuracy of the trained PSO-BPNN model determines the accuracy of the method for estimating the clamping force.

3. Results and Discussion

In order to validate and evaluate the comprehensive performance of the proposed method for estimating clamping force based on a joint torque disturbance observer using PSO-BPNN, a series of experiments were carried out. In this section, we provide the results from the cable tension estimation model based on the PSO-BPNN under free motion, and the experimental results of clamping force estimation considering collision detection and a clamping action.

3.1. Training and Testing Results from the Cable Tension Estimation Model Based on the PSO-BPNN under Free Motion

A series of experiments were carried out to explore the potential relations between the motor's displacement and velocity and the measured cable tension value under free motion. The displacement corresponding to one pulse of the linear stepping motor is 0.00127 mm. The acceleration and deceleration modes are selected as "S" type in the motion control card. The acceleration and deceleration time are set to 0.02s, and the constant motion velocity is set to be variable. Five kinds of motors' velocities (0.5, 0.6, 0.7, 0.8 and 1.0 mm/s) with increasing ladder are planned for the joint motion with the forceps' opening angle ranging from zero to maximum angle. The maximum opening angle of the forceps A and B is 130 degrees. This means that forceps A and B are actuated with joint angle range [0,65] degrees on the opposite direction, respectively. The input data were the displacement and velocity of the two linear stepping motors for pulling the driving cables of forceps A and B. The output data were the cable tension values that were measured through four tension sensors. Table 1 shows the main parameters of the PSO-BPNN.

The training data comprised 90% of the randomized input data and their related output data. The remaining 10% of the randomized input and output data were used as the testing data for examining the generalization ability and prediction accuracy of the PSO-BPNN. The number of training data was 4892, and the number of testing data was 544.

The training and testing outcome indicators of the PSO-BPNN model are shown as Figure 5. We can see that the trained PSO-BPNN model has good comprehensive performance. Figure 6 shows the estimation results when the testing data were input into the PSO-BPNN model. The mean square errors of the testing data were 0.2171, 0.2918, 0.2459, and 0.2379 N with respect to the estimation ability of $\hat{F}_E$, which are the motor cable tensions for actuating forceps A and B under free motion.

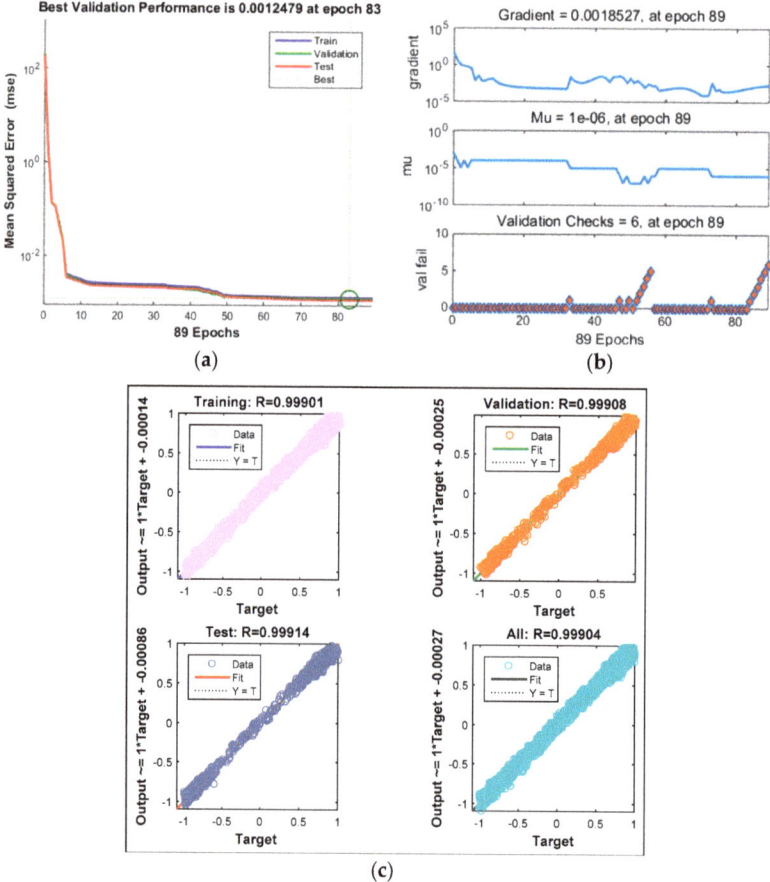

Figure 5. The training and testing outcome indicators of the PSO-BPNN model. (**a**) Mean squared error; (**b**) Gradient, mu and validation checks; (**c**) Regression results.

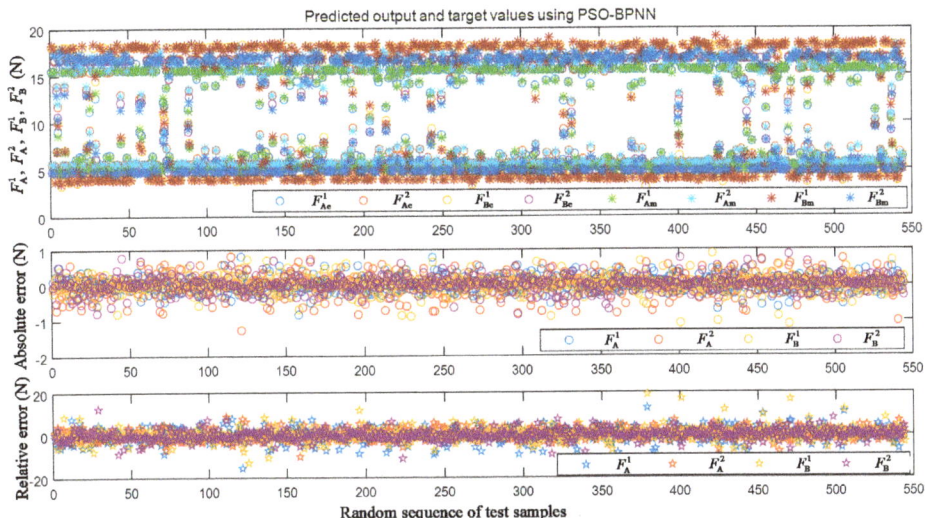

Figure 6. The estimation results of the testing data using the PSO-BPNN model.

3.2. Experimental Results of Clamping Force Estimation

As shown in Figures 2a and 4, the two forceps A and B were actuated from an opening angle of 120 degrees to the clamping position limited by the contact object. To test the comprehensive performance of the proposed method for clamping force estimation, the continuous and stair-loading types of clamping motion experiments were carried out, as shown in Figures 7–9. "No loading region" means that the forceps perform the clamping motion without contacting the object. "Start loading" means that the forceps come into contact with the object for the first time. "Stop loading" means that the forceps come into contact with the object for the final time. "Loading region" means that the clamping force increases the area between the start and stop loading moments. "Collision detection" means that the clamping force can be estimated with a set threshold value, which is called the collision detection resolution. "Constant loading region" means that the clamping motion had stopped due to constant external forces and the clamping force of forceps A and B.

Figure 7 shows the experimental results of the continuous clamping force estimation with a collision analysis. The collision detection threshold was 0.11 N for the estimated external forces of forceps A and B. Over the whole region, the overall root mean square errors (RMSEs) of the estimated external forces were 0.1010 N and 0.4035 N for forceps A and B, respectively; and the overall RMSE of the clamping force was 0.2321 N. The overall average errors in the estimated external forces were −0.0088 N and −0.3042 N for forceps A and B, respectively; the overall average error in the estimated clamping force was −0.1751 N; and the errors in the estimated clamping force can be summarized in the interval of [−0.3482, 0.1718] N. In the constant loading region, the estimation accuracies of the external forces $\hat{F}_{ext}^A$ and $\hat{F}_{ext}^B$ and the clamping force $\hat{F}_{CF}$ were 93.83%, 80.13%, and 86.91%, respectively.

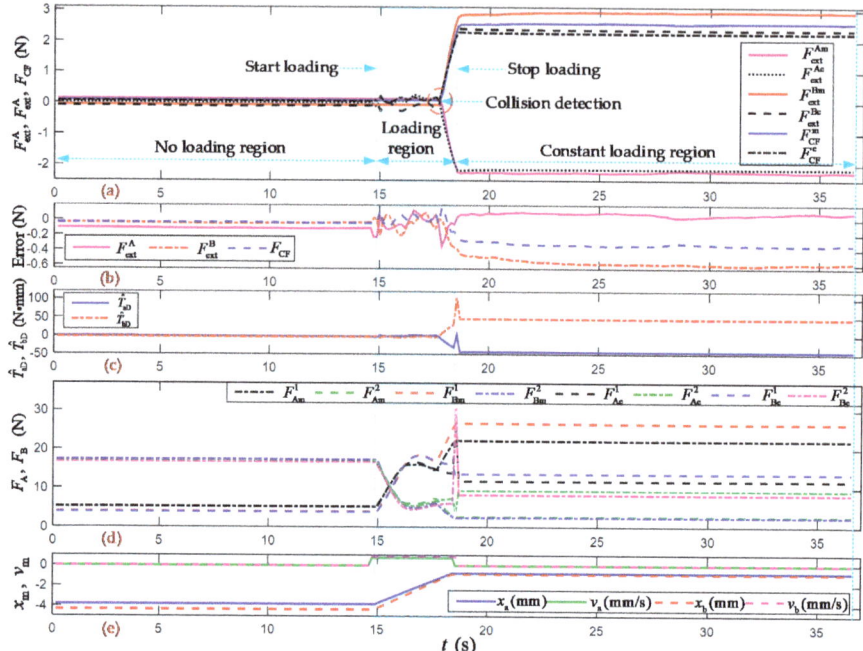

Figure 7. Experimental results of the continuous clamping force estimation with a collision analysis. (**a**) Measured and estimated values of the external forces and clamping force of forceps A and B. The superscripts m and e denote the measured and estimated values, respectively; (**b**) Errors in the estimated external forces and clamping force; (**c**) Joint torque disturbances of forceps A and B; (**d**) Measured and estimated values of the cable tensions of forceps A and B. The superscripts m and e denote the measured and estimated values, respectively; (**e**) Displacements and velocities of the linear stepping motors for pulling the driving cables of forceps A and B.

Figures 8 and 9 show the two groups of experimental results of the incremental clamping force estimation. At the end of the no loading region, five stepwise incremental loading motions were performed in the clamping force experiments. The first contact position was set at the opening angle (65 degrees). The estimated cable tensions $\hat{F}_A^1$, $\hat{F}_A^2$, $\hat{F}_B^1$, and $\hat{F}_B^2$ from the trained PSO-BPNN model were 14.8259 N, 6.7541 N, 17.1017 N, and 5.7025 N, respectively. The estimated external forces $\hat{F}_{ext}^A$ and $\hat{F}_{ext}^B$ are compensated for with their overall average errors of −0.1395 N and −0.1404 N, respectively, from the first group.

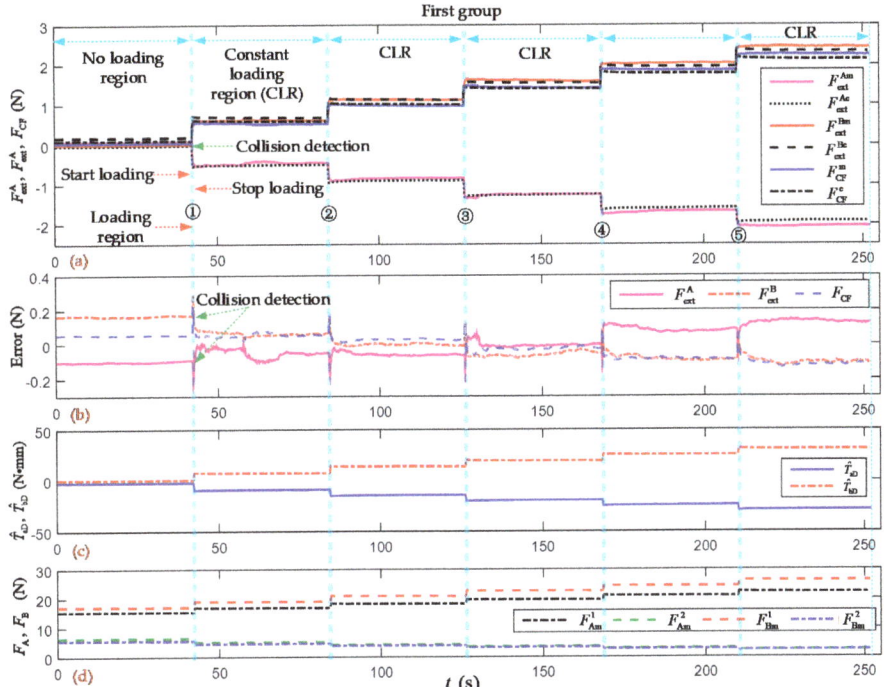

Figure 8. Experimental results of the incremental clamping force estimation (first group). (**a**) Measured and estimated values of the external forces and clamping force of forceps A and B. The superscripts m and e denote the measured and estimated values; (**b**) Errors in the estimated external forces and clamping force; (**c**) Joint torque disturbances of forceps A and B; (**d**) Measured values of the cable tensions of forceps A and B. The superscript m denotes the measured values.

Over the whole region of the first group of experiments as shown in Figure 8, the overall RMSE of the estimated external forces $\hat{F}_{ext}^A$ and $\hat{F}_{ext}^B$ were 0.0825 N and 0.0956 N, respectively; and the overall RMSE of the clamping force $\hat{F}_{CF}$ was 0.070 N. The overall average errors in the estimated external forces were 0.00002 N and −0.00004 N, respectively. The overall average error in the estimated clamping force was −0.013 N, and the errors in the estimated clamping force can be summarized in the interval of [−0.1343, 0.2256] N.

Over the whole region of the second group of experiments as shown in Figure 9, the overall RMSE of the estimated external forces $\hat{F}_{ext}^A$ and $\hat{F}_{ext}^B$ was 0.1316 N and 0.1087 N, respectively. The overall RMSE of the clamping force $\hat{F}_{CF}$ was 0.080 N. The overall average errors in the estimated external forces were −0.1223 N and 0.0628 N, respectively; and the overall average error in the estimated clamping force was 0.0736 N. The errors in the estimated clamping force can be summarized in the interval of [−0.0100, 0.2794] N.

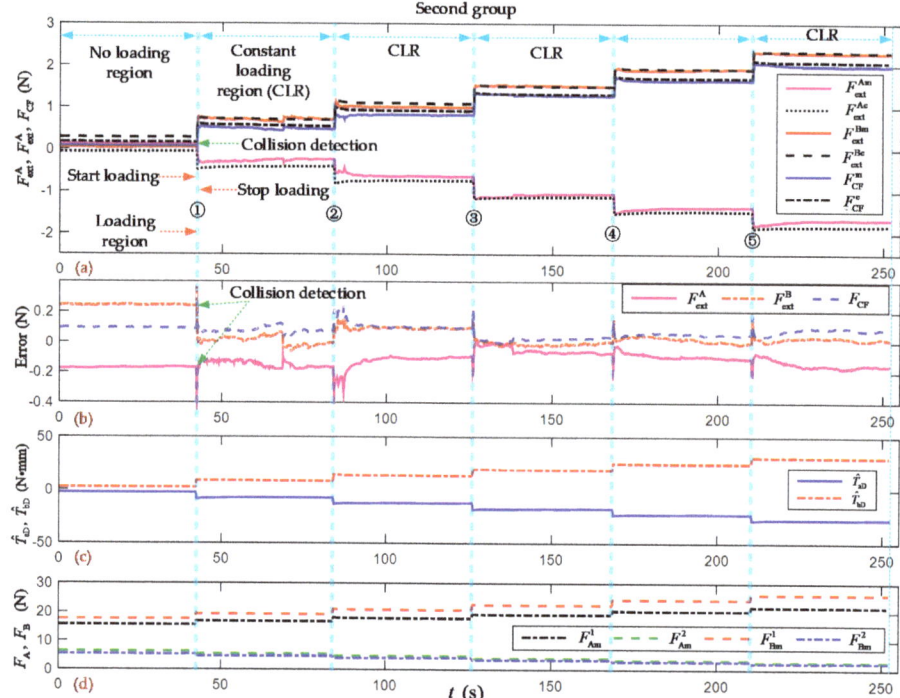

Figure 9. Experimental results of the incremental clamping force estimation (second group). (**a**) Measured and estimated values of the external forces and clamping force of forceps A and B. The superscripts *m* and *e* denote the measured and estimated values, respectively; (**b**) Errors in the estimated external forces and clamping force; (**c**) Joint torque disturbances of forceps A and B; (**d**) Measured values of the cable tensions of forceps A and B. The superscript *m* denotes the measured values.

3.3. Analysis and Discussion

When training a PSO-BPNN model, the estimation accuracy of the trained neural network can be limited by multiple effects, including the model parameter settings and the training data size. Because the five different velocities of the yaw joint motion were quantitative, the estimation accuracy of the trained PSO-BPNN model was only affected by the training data size. So, the estimation accuracy of the trained PSO-BPNN model could be increased by collecting more training data with a higher number of velocity bands. However, our trained PSO-BPNN model showed sufficiently good comprehensive performance to be employed in the estimation of the cable tension of forceps A and B.

Figures 7–9 show the results of the clamping force estimation in the continuous and stair-loading types of clamping motion experiments. In the curves of the estimation errors and the joint torque disturbance, some relatively large errors arise at the beginning of the start loading region and the end of the stop loading region. The reasons for these errors include, on the one hand, the effects of ignoring the nonlinear elasticity of and the creep and hysteresis in the wire cables. These nonlinear parameters and factors cannot be totally included in the system dynamic model and PSO-BPNN model. On the other hand, the measured errors in the two flexible pressure sensors cannot be ignored when they are pressed at the beginning and at the end of the clamping action.

To sum up, the collision detection threshold was found to reach 0.11 N, and the clamping force estimation resolution was found to be the same as the collision detection threshold. The estimation accuracy for the static clamping force was greater than 86%. In this study, the proposed method for clamping force estimation was shown to perform well. The two groups of incremental clamping force estimation experiments demonstrated the more complete comprehensive performance of the proposed method, with average errors in the interval of [−0.0722, 0.2525] N. Meanwhile, the experiments showed that this method can provide effectively detection range in the interval of [0,2] N. Moreover, the length of forceps direct influences the arms of external force, this means that the effectively detection range can be extended by reducing the overall length of the forceps within reasonable limits. Overall, the experimental results mean that the proposed method for clamping force estimation has the potential to be used in the cable-driven end-effector of a surgical robotic system for MIS.

4. Conclusions

In this study, we proposed a method for clamping force estimation based on a joint torque disturbance observer using PSO-BPNN for a cable-driven surgical robot end-effector. This estimation method considers both the cable-driven end-effector's system dynamics and the estimated cable tension using the PSO-BPNN under free motion. Moreover, the clamping force can be estimated by only using known information about the motor's displacement and velocity, and the measured cable tension value, without the need for internal force sensors to be integrated into the wrist joints of the surgical robot end-effector. The PSO-BPNN-based joint torque disturbance observer performed well in the disturbance estimation. The experimental results showed that the proposed method for clamping force estimation has good comprehensive performance. Our future work will focus on solving the decoupling precision problem in multi-DoF wrist joints, and the engineering and application of this method.

Author Contributions: Conceptualization and methodology, Z.W.; software J.Q. and B.C.; validation, Z.W. and D.W.; formal analysis, Z.W., L.Y. and B.C.; investigation, Z.W., B.C and L.Y.; resources, B.Z.; writing—original draft preparation, Z.W. and J.Q.; writing—review and editing, Z.W., D.W., L.Y. and B.Z.

Funding: This research was funded by National Natural Science Foundation of China, grant number 51805129 and 91748109; the Fundamental Research Funds for the Central Universities, grant number JZ2019HGTB0078 and JZ2019HGTB0084; the Open Project Program of Tianjin Key Laboratory of Aerospace Intelligent Equipment Technology, Tianjin Institute of Aerospace Mechanical and Electrical Equipment, grant number TJYHZN2019KT006.

Acknowledgments: Wenjie Wang and Jing Yang are acknowledged for their support in the process of experiments.

Conflicts of Interest: The authors declare no conflict of interest.

References

1. Marcus, H.; Nandi, D.; Darzi, A.; Yang, G. Surgical Robotics Through a Keyhole: From Today's Translational Barriers to Tomorrow's 'Disappearing' Robots. *IEEE Trans. Bio-Med. Eng.* **2013**, *60*, 674–681. [CrossRef] [PubMed]
2. Wang, Z.; Zi, B.; Ding, H.; You, W.; Yu, L. Hybrid grey prediction model-based autotracking algorithm for the laparoscopic visual window of surgical robot. *Mech. Mach. Theory* **2018**, *123*, 107–123. [CrossRef]
3. Mohareri, O.; Ramezani, M.; Adebar, T.; Abolmaesumi, P.; Salcudean, S. Automatic Localization of the da Vinci Surgical Instrument Tips in 3-D Transrectal Ultrasound. *IEEE Trans. Bio-Med. Eng.* **2013**, *60*, 2663–2672. [CrossRef] [PubMed]
4. Blake, H.; Jacob, R.; Diana, W.; King, H.; Roan, P.; Cheng, L.; Glozman, D.; Ma, J.; Kosari, S.; White, L. Raven-II: An Open Platform for Surgical Robotics Research. *IEEE Trans. Bio-Med. Eng.* **2013**, *60*, 954–959.
5. Hagn, U.; Nickl, M.; Jörg, S.; Passig, G.; Bahls, T.; Nothhelfer, A.; Hacker, F.; Le-Tien, L.; Albu-Schäffer, A.; Konietschke, R.; et al. The DLR MIRO: A versatile lightweight robot for surgical applications. *Ind. Robot* **2008**, *35*, 324–336. [CrossRef]
6. Kim, U.; Lee, D.H.; Yoon, W.J.; Hannaford, B.; Choi, H.R. Force sensor integrated surgical forceps for minimally invasive robotic surgery. *IEEE Trans. Robot.* **2015**, *31*, 1214–1224. [CrossRef]
7. Gonenc, B.; Chamani, A.; Handa, J.; Gehlbach, P.; Taylor, R.; Iordachita, I. 3-DOF Force-Sensing Motorized Micro-Forceps for Robot-Assisted Vitreoretinal Surgery. *IEEE Sens. J.* **2017**, *17*, 3526–3541. [CrossRef]

8. Aviles, A.; Alsaleh, S.; Hahn, J.; Casals, A. Towards retrieving force feedback in robotic-assisted surgery: A supervised neuro-recurrent-vision approach. *IEEE Trans. Haptics* **2016**, *10*, 431–443. [CrossRef]
9. Wang, Z.; Zi, B.; Wang, D.; Qian, J.; You, W.; Yu, L. External force self-sensing based on cable-tension disturbance observer for surgical instrument. *IEEE Sens. J.* **2019**, *19*, 5274–5284. [CrossRef]
10. Chen, B.; Zi, B.; Wang, Z.; Qin, L.; Liao, W. Knee exoskeletons for gait rehabilitation and human performance augmentation: A state-of-the-art. *Mech. Mach. Theory* **2019**, *134*, 499–511. [CrossRef]
11. Chen, Q.; Zi, B.; Sun, Z.; Li, Y.; Xu, Q. Design and Development of a New Cable-Driven Parallel Robot for Waist Rehabilitation. *IEEE/ASME Trans. Mech.* **2019**, *24*, 1497–1507. [CrossRef]
12. Wu, L.; Crawford, R.; Roberts, J. Dexterity analysis of three 6-DOF continuum robots combining concentric tube mechanisms and cable-driven mechanisms. *IEEE Robot. Autom. Lett.* **2016**, *2*, 514–521. [CrossRef]
13. Hwang, M.; Kwon, D.S. Strong continuum manipulator for flexible endoscopic surgery. *IEEE/ASME Trans. Mech.* **2019**, *24*, 2193–2203. [CrossRef]
14. Ma, X.; Song, C.; Chiu, P.W.; Li, Z. Autonomous Flexible Endoscope for Minimally Invasive Surgery with Enhanced Safety. *IEEE Robot. Autom. Lett.* **2019**, *4*, 2607–2613. [CrossRef]
15. Kuebler, B.; Seibold, U.; Hirzinger, G. Development of actuated and sensor integrated forceps for minimally invasive surgery. *Int. J. Med. Robot. Comp.* **2006**, *1*, 96–107. [CrossRef] [PubMed]
16. Trejos, A.; Escoto, A.; Naish, M.; Patel, R. Design and Evaluation of a Sterilizable Force Sensing Instrument for Minimally Invasive Surgery. *IEEE Sens. J.* **2017**, *17*, 3983–3993. [CrossRef]
17. Yu, L.; Yan, Y.; Yu, X.; Xia, Y. Design and Realization of Forceps With 3-D Force Sensing Capability for Robot-Assisted Surgical System. *IEEE Sens. J.* **2018**, *18*, 8924–8932. [CrossRef]
18. Li, K.; Pan, B.; Zhan, J.; Gao, W.; Fu, Y.; Wang, S. Design and performance evaluation of a 3-axis force sensor for MIS palpation. *Sens. Rev.* **2015**, *35*, 219–228. [CrossRef]
19. Yu, H.; Jiang, J.; Xie, L.; Liu, L.; Shi, Y.; Cai, P. Design and static calibration of a six-dimensional force/torque sensor for minimally invasive surgery. *Minim. Invasive Ther. Allied Technol.* **2014**, *23*, 136–143. [CrossRef]
20. Lim, S.; Lee, H.; Park, J. Grip force measurement of forceps with fibre Bragg grating sensors. *Electron. Lett.* **2014**, *50*, 733–735. [CrossRef]
21. Kim, U.; Lee, D.; Kim, Y.; Seok, D.Y.; So, J.; Choi, H. S-surge: Novel portable surgical robot with multiaxis force-sensing capability for minimally invasive surgery. *IEEE/ASME Trans. Mech.* **2017**, *22*, 1717–1727. [CrossRef]
22. Kim, U.; Kim, Y.B.; So, J.; Seok, D.Y.; Choi, H.R. Sensorized surgical forceps for robotic-assisted minimally invasive surgery. *IEEE Trans. Ind. Electron.* **2018**, *65*, 9604–9613. [CrossRef]
23. Radó, J.; Dücső, C.; Földesy, P.; Szebényi, G.; Nawrat, Z.; Rohr, K.; Fürjes, P. 3D force sensors for laparoscopic surgery tool. *Microsyst. Technol.* **2018**, *24*, 519–525. [CrossRef]
24. Overtoom, E.; Horeman, T.; Jansen, F.; Dankelman, J.; Schreuder, H. Haptic feedback, force feedback, and force-sensing in simulation training for laparoscopy: A systematic overview. *J. Surg. Educ.* **2019**, *76*, 242–261. [CrossRef] [PubMed]
25. Zhao, B.; Nelson, C. Estimating Tool-Tissue Forces Using a 3-Degree-of-Freedom Robotic Surgical Tool. *J. Mech. Robot.* **2016**, *8*, 051015. [CrossRef]
26. Li, H.; Kawashima, K.; Tadano, K.; Ganguly, S.; Nakano, S. Achieving haptic perception in forceps' manipulator using pneumatic artificial muscle. *IEEE/ASME Trans. Mech.* **2013**, *18*, 74–85. [CrossRef]
27. Haraguchi, D.; Kanno, T.; Tadano, K.; Kawashima, K. A pneumatically driven surgical manipulator with a flexible distal joint capable of force sensing. *IEEE/ASME Trans. Mech.* **2015**, *20*, 2950–2961. [CrossRef]
28. Xue, R.; Ren, B.; Huang, J.; Yan, Z.; Du, Z. Design and Evaluation of FBG-Based Tension Sensor in Laparoscope Surgical Robots. *Sensors* **2018**, *18*, 2067. [CrossRef]
29. Xue, R.; Du, Z.; Yan, Z.; Ren, B. An estimation method of grasping force for laparoscope surgical robot based on the model of a cable-pulley system. *Mech. Mach. Theory* **2019**, *134*, 440–454. [CrossRef]
30. Li, Y.; Hannaford, B. Gaussian process regression for sensorless grip force estimation of cable-driven elongated surgical instruments. *IEEE Robot. Autom. Lett.* **2017**, *2*, 1312–1319. [CrossRef]
31. Liang, Y.; Du, Z.; Wang, W.; Yan, Z.; Sun, L. An improved scheme for eliminating the coupled motion of surgical instruments used in laparoscopic surgical robots. *Robot. Auton. Syst.* **2019**, *112*, 49–59. [CrossRef]
32. Gessert, N.; Beringhoff, J.; Otte, C.; Schlaefer, A. Force estimation from OCT volumes using 3D CNNs. *Int. J. Comput. Ass. Rad.* **2018**, *13*, 1073–1082. [CrossRef]

33. Li, X.; Cao, L.; Tiong, A.; Phan, P.; Phee, S. Distal-end force prediction of tendon-sheath mechanisms for flexible endoscopic surgical robots using deep learning. *Mech. Mach. Theory* **2019**, *134*, 323–337. [CrossRef]
34. Yu, L.; Wang, W.; Zhang, F. External force sensing based on cable tension changes in minimally invasive surgical micromanipulators. *IEEE Access* **2018**, *6*, 5362–5373. [CrossRef]
35. Hwang, W.; Lim, S.C. Inferring Interaction Force from Visual Information without Using Physical Force Sensors. *Sensors* **2017**, *17*, 2455. [CrossRef]
36. Huang, J.; Yan, Z.; Xue, R. Grip Force Estimation of Laparoscope Surgical Robot based on Neural Network Optimized by Genetic Algorithm. In Proceedings of the 3rd International Conference on Robotics, Control and Automation, Chengdu, China, 11–13 August 2018; pp. 95–100.
37. Marban, A.; Srinivasan, V.; Samek, W.; Fernández, J.; Casals, A. A recurrent convolutional neural network approach for sensorless force estimation in robotic surgery. *Biomed. Signal Proces.* **2019**, *50*, 134–150. [CrossRef]
38. Verotti, M.; Di Giamberardino, P.; Belfiore, N.P.; Giannini, O. A genetic algorithm-based method for the mechanical characterization of biosamples using a MEMS microgripper: Numerical simulations. *J. Mech. Behav. Biomed.* **2019**, *96*, 88–95. [CrossRef] [PubMed]
39. Yeh, W.C. A squeezed artificial neural network for the symbolic network reliability functions of binary-state networks. *IEEE Trans. Neural Netw. Learn. Syst.* **2016**, *28*, 2822–2825. [CrossRef] [PubMed]
40. Jethmalani, C.R.; Simon, S.P.; Sundareswaran, K.; Nayak, P.S.R.; Padhy, N.P. Auxiliary hybrid PSO-BPNN-based transmission system loss estimation in generation scheduling. *IEEE Trans. Ind. Inform.* **2016**, *13*, 1692–1703. [CrossRef]
41. Liu, Q.; Brigham, K.; Rao, N.S. Estimation and fusion for tracking over long-haul links using artificial neural networks. *IEEE Trans. Signal Inf. Process. Netw.* **2017**, *3*, 760–770. [CrossRef]
42. Hou, C.; Yu, X.; Cao, Y.; Lai, C.; Cao, Y. Prediction of synchronous closing time of permanent magnetic actuator for vacuum circuit breaker based on PSO-BP. *IEEE Trans. Dielectr. Electr. Insul.* **2017**, *24*, 3321–3326. [CrossRef]
43. Liu, C.; Ding, W.; Li, Z.; Yang, C. Prediction of high-speed grinding temperature of titanium matrix composites using BP neural network based on PSO algorithm. *Int. J. Adv. Manuf. Technol.* **2017**, *89*, 2277–2285. [CrossRef]
44. Zhang, C.; Chen, Z.; Mei, Q.; Duan, J. Application of particle swarm optimization combined with response surface methodology to transverse flux permanent magnet motor optimization. *IEEE Trans. Magn.* **2017**, *53*, 1–7. [CrossRef]
45. Ren, C.; An, N.; Wang, J.; Li, L.; Hu, B.; Shang, D. Optimal parameters selection for BP neural network based on particle swarm optimization: A case study of wind speed forecasting. *Knowl. Based syst.* **2014**, *56*, 226–239. [CrossRef]
46. Mohamad, E.; Armaghani, D.; Momeni, E.; Yazdavar, A.; Ebrahimi, M. Rock strength estimation: A PSO-based BP approach. *Neural Comput. Appl.* **2018**, *30*, 1635–1646. [CrossRef]

© 2019 by the authors. Licensee MDPI, Basel, Switzerland. This article is an open access article distributed under the terms and conditions of the Creative Commons Attribution (CC BY) license (http://creativecommons.org/licenses/by/4.0/).

Article

Real-Time Queue Length Detection with Roadside LiDAR Data

Jianqing Wu [1,2], Hao Xu [2], Yongsheng Zhang [2], Yuan Tian [2] and Xiuguang Song [1,*]

1. School of Qilu Transportation, Shandong University, Jinan 250061, China; jianqingwusdu@sdu.edu.cn
2. Department of Civil and Environmental Engineering, University of Nevada, Reno, NV 89557, USA; haox@unr.edu (H.X.); yongshengz@unr.edu (Y.Z.); yuantian@nevada.unr.edu (Y.T.)
* Correspondence: songxiuguang@sdu.edu.cn; Tel.: +86-15865275372

Received: 24 March 2020; Accepted: 17 April 2020; Published: 20 April 2020

Abstract: Real-time queue length information is an important input for many traffic applications. This paper presents a novel method for real-time queue length detection with roadside LiDAR data. Vehicles on the road were continuously tracked with the LiDAR data processing procedures (including background filtering, point clustering, object classification, lane identification and object association). A detailed method to identify the vehicle at the end of the queue considering the occlusion issue and package loss issue was documented in this study. The proposed method can provide real-time queue length information. The performance of the proposed queue length detection method was evaluated with the ground-truth data collected from three sites in Reno, Nevada. Results show the proposed method can achieve an average of 98% accuracy at the six investigated sites. The errors in the queue length detection were also diagnosed.

Keywords: queue length; roadside sensor; vehicle detection

---

1. Introduction

Queue length has been used in many transportation areas, including but not limited to performance evaluation at signalized intersections, adaptive signal control, adaptive ramp metering and travel route selection [1]. Some applications such as the optimal signal control and travel route selection require real-time queue length information [2]. Queue length can either be estimated or directly detected. The typical queue estimation methods include the input-output method and the shockwave method. The input-output method uses advanced detector actuation, parametric data (headway, storage capacity, etc.) and phase change information (for signalized intersections) to estimate the queue length. The input-output method has the assumptions that the vehicles stay in the same lane after passing the advanced detector and the vehicles follow the first-in-first-out (FIFO) principle [3]. However, the accuracy of the input-output approach is limited by the detector's counting error [4]. Lee et al. proposed a singular-point correction method to eliminate the accumulated counting error over time [5], but the singular-point correction method required the proper calibration based on the different features of field data, which limited the transferability of the method. Liu et al. [6] applied Lighthill–Whitham–Richards (LWR) shockwave theory to estimate queue length with the high-resolution traffic signal data. A detailed method of identifying break points was documented in their paper. The testing results showed that their method was able to estimate long queues with relatively high accuracy. However, their proposed model could not estimate the queue length under oversaturation. Using probe trajectory data for queue length estimation has been another hot topic for transportation researchers [7]. Cheng et al. [8] developed a cycle-by-cycle queue length estimation method with sampled vehicle trajectory data. Queue length was estimated based on the LWR shockwave theory. Later Hao et al. [9] developed a Bayesian Network (BN)-based method for cycle-by-cycle queue length estimation at signalized intersections. The travel times collected from mobile traffic sensors were the input data. The results showed that the BN-based method has a better performance than the method proposed in Cheng et al. [8]. Cai et al. [10] used fusing data from point and mobile sensors to estimate queue length at signalized intersections with the shockwave theory. This method assumed that at least one queue vehicle can be obtained in one cycle. Their model could not properly estimate the queue length when the arrival flow was unstable. The probe trajectory-based approaches suffer a major limitation: the sample rate can influence the accuracy of the queue length estimation [11].

Recent studies showed an increasing interest for queue length estimation using connected vehicle (CV) technology. Li et al. [12] used the probe trajectory and signal timing data extracted from the CV network to estimate real-time queue length with an event-based method. That paper used a lot of default values in the calculation (e.g., headway is 2.5 s, constant deceleration rate is 1.55 m/s^2). Those default values may not reflect the different drivers' driving behavior. Christofa et al. [13] developed two methods: gap-based method and shockwave-based method for queue spillback detection. The results showed that this approach can detect the occurrence of spillbacks for a range of penetration rates. Tiaprasert et al. [14] applied a discrete wavelet transform (DWT) for queue estimation using CV technology. One of the assumptions for the DWT is that the penetration ratio is known. For real situation, the detailed penetration ratio information may not be available. Yang and Menendez [15] proposed a convex optimization-based method for queue length estimation in a CV network. The validation showed that the estimation error went larger for oversaturated scenarios. The challenge in those methods using CV technology was the low penetration rate of CV on the roads, and the penetration rate of CV is expected to be low in the near future [16].

Though the queue length can be estimated—and the accuracy can be relatively high—those methods usually have their own assumptions, indicating that the methods may only work for some specific locations. Other than queue length estimation, researchers are also looking for approaches to directly detect the queue length. The image-based method can be an option for queue length detection [17]. Siyal and Fathy [18] applied the neural network to extract queue length from the image extracted from cameras. However, the computational load of this method was high since the neural network was applied. Cai et al. [19] used the texture difference and edge information to detect the vehicles and queue length from the videos. The practice showed that the inferences attached to vehicles and marks on the road can impact the accuracy of the queue length detection. Satzoda et al. [20] used the edges and dark features in the image for queue length detection. The evaluation showed that nearly 100% accuracy can be achieved in their testing database. The limitation of this method was that a lot of calibrations were required for driving detection zones in the images. The performance of the camera can also be greatly influenced by the light conditions. Xu et al. [21] developed a method for queue length detection by vehicle-to-RSU (V2R) communication. However, this approach was based on the assumption that all vehicles were installed with a communication system and a global positioning system (GPS). As a result, this method was also limited by the low penetration rate of CV on the roads.

The roadside LiDAR provides a solution for queue length detection regardless of whether vehicles can communicate with each other on the roads. The 360-degree LiDAR can scan all the objects in its detection range [22]. This means the penetration ratio of CV has no influence on the queue length detection using the roadside LiDAR. Furthermore, queue length can be detected using the LiDAR with or without the traffic signal information. Those advantages of LiDAR make the real-time queue length detection possible. This paper developed a systematic procedure for queue length detection with roadside LiDAR. The rest of the paper is structured as follows. Section 2 introduces the vehicle detection algorithm with roadside LiDAR data. The algorithm of queue length detection is documented in Section 3. Section 4 evaluates the proposed method using real-world collected LiDAR data. The last section summarizes the major contribution of this paper.

2. Materials and Preprocessing

The roadside LiDAR refers to the LiDAR deployed in a stationary location along the roadside. The roadside LiDAR (usually rotating LiDAR) has lower resolution and a lower price compared to the airborne LiDAR and on-board LiDAR (mobile LiDAR) considering the massive deployment in the near future. This paper used two types of LiDAR: VLP-16 and VLP-32c for data collection. For the detailed parameters of VLP-16 and VLP-32c, we refer the readers to [23].

But in theory, our proposed queue length detection method can work for any brand of rotating LiDAR after necessary calibration based on the different setting parameters of the LiDAR.

The roadside LiDAR can be installed permanently (on the top of a pedestrian signal) or temporarily (on a tripod) for data collection [24]. The recommended height for LiDAR installation is 2–3 m above the ground to avoid possible man-made destruction and to reduce occlusion issues considering the limited vertical field of view [25]. The scanning rate of the LiDAR is set as 10 Hz. The proposed vehicle detection procedure includes five major steps: background filtering [26], point clustering [27], object classification [28,29], lane identification [30,31] and object association [32]. Vehicle trajectories can be generated with the proposed method.

2.1. Background Filtering

For queue length detection, the objects of interest are the vehicles on the road. Background filtering is used to exclude the other irrelevant information (buildings, trees and ground points) and to keep the moving objects (vehicles, pedestrians and other road users) in the space at the same time. This paper applied a point density-based unsupervised algorithm named 3D-DSF developed by Wu et al. [25] for background filtering. The 3D-DSF first integrated the data collected in a time period (such as 5 min of data) into one space based on the XYZ coordinates of the LiDAR points. The whole space was then rasterized into small cubes with the same side length (0.1 m was used as the side length considering the accuracy and the computation load) [26]. The point density of the cubes representing the background should be higher than that of the cubes representing the moving objects after frame aggregation. By giving a pre-defined threshold of point density, the location of the cubes representing background can be then identified and stored in a 3D array. An automatic threshold identification method was well documented in the reference [26]. Any point located in the 3D array was then excluded from the space. Figure 1 shows an example of before-and-after background filtering. The previous studies [33–35] showed that 3D-DSF can exclude more than 95% of background points from the raw LiDAR data.

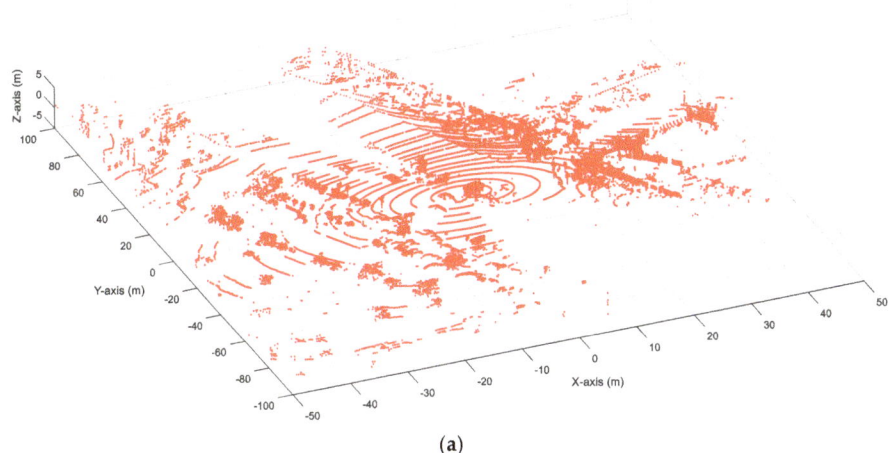

(a)

Figure 1. *Cont.*

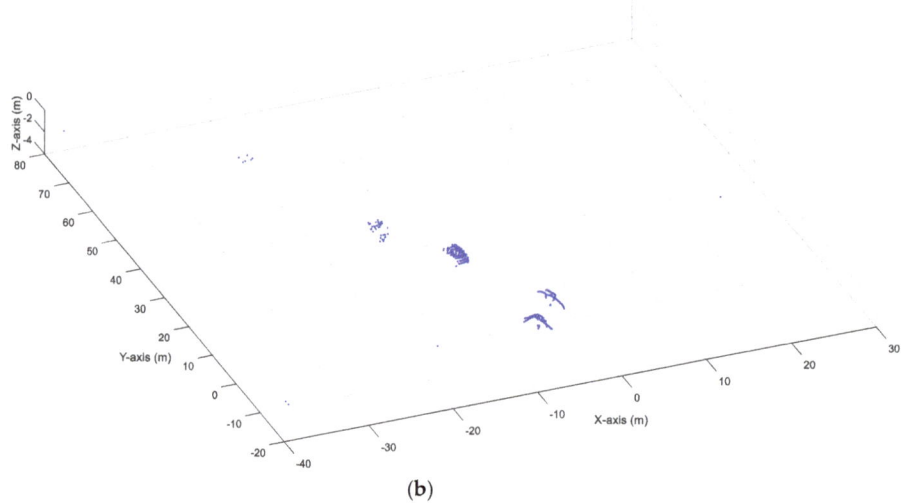

(b)

Figure 1. Before-and-after background filtering: (**a**) before background filtering; (**b**) after background filtering.

2.2. Point Clustering

The points in the LiDAR data are stored disorderly, indicating that the points representing the same object are not grouped together. Point clustering is used to find the points belonging to the same object and to provide the same ID to those points. Another function of object clustering is to exclude the noises left after background filtering since not all background points can be filtered with 3D-DSF. This paper applied a revised density-based spatial clustering of applications with noises (DBSCAN) for object clustering [36]. DBSCAN defined one cluster as a set of points with high density. Compared to the widely used K-means clustering, DBSCAN does not need to know the number of clusters in advance and can find any shape of the clusters in the data [27]. There are two initial parameters for DBSCAN: searching radius (eps) and minimum containing points (minPts). DBSCAN starts with a random unvisited point A and marks the points with a distance ≤ eps from point A as the neighbors of point A. The following criteria are applied.

a. If the number of the neighbors of point A ≥ minPts, then point A and its neighbors are marked as a cluster and point A is marked as a visited point. DBSCAN then uses the same method to process the points of other unvisited points in the same cluster to extend the range of the cluster.

b. If the number of the neighbors of point A < minPts, then point A will be marked as a noising point and a visited point.

Using the above-mentioned criteria, DBSCAN can process all those unvisited points. However, one major disadvantage of DBSCAN is that it could not cluster the points with uneven density effectively. For LiDAR data, the number of points representing the same object decreased with the increasing distance to the LiDAR, indicating the point density changes as the distance to the LiDAR increasing. To fix this issue, a revised DBSCAN with adaptive parameters are applied for point clustering. Different eps and minPts were applied for the points based on the distance from the LiDAR and the mechanical structure (field of view, angular resolution and the distance between two adjacent beams) of the LiDAR. The detailed calculation of eps and minPts was documented in our previous research [27]. The accuracy of the revised DBSCAN is 96.5% in average. Figure 2 shows an example of before-and-after point clustering. It is shown that the revised DBSCAN algorithm can successfully identify six objects and exclude the noise from the LiDAR data in Figure 2.

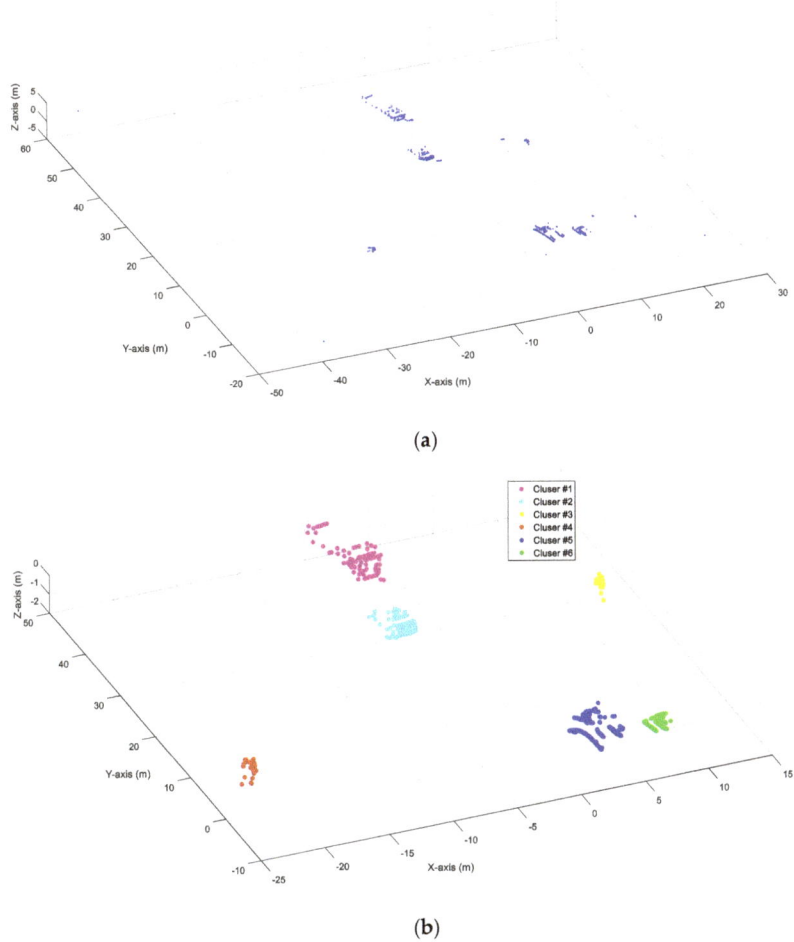

Figure 2. Before-and-after point clustering: (**a**) before point clustering; (**b**) after point clustering.

2.3. Object Classification

There may be different types of road users (vehicles, pedestrians and bicycles) on the road. It is then necessary to distinguish the vehicles from other types of road users for queue length detection. This paper aims to classify the objects into one of the four classes (passenger car/pickup, trucks/bus, pedestrians and bicycles). Six features (object length, object height, the difference between object height and object length, distance to LiDAR, number of points and object height profile) extracted from the point cloud were used to represent the difference between different classes. A random forest (RF) classifier was trained for object classification. The RF is a supervised algorithm that aggregates multiple decision trees. Our previous study [28] compared the performance of different methods (k-nearest neighbor, Naïve Bayes, RF and support vector machine) for roadside LiDAR classification. It was shown that RF can provide the best accuracy among those investigated methods. This paper used the public database [28] collected by the Center for Advanced Transportation Education and Research (CATER) in University of Nevada, Reno to train the RF classifier. The testing results showed that the RF can achieve an overall 95.3% accuracy for object classification. Figure 3 shows an example of the results of object classification. The RF classifier can correctly identify three passenger cars and five pedestrians in the point cloud.

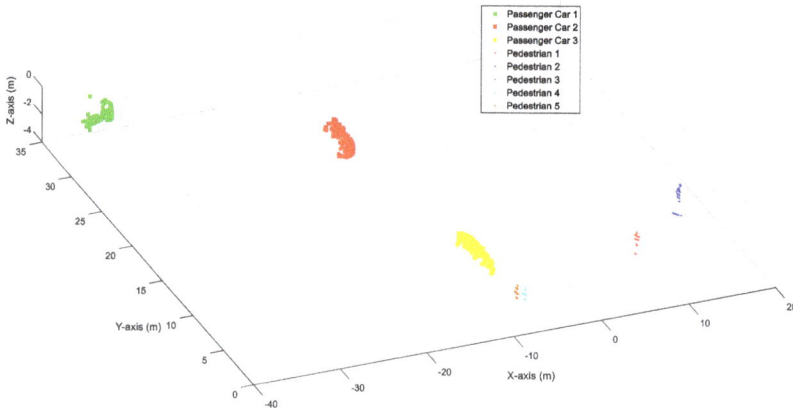

Figure 3. Vehicle classification.

2.4. Lane Identification

The lane information is important for lane-based queue length detection. It is assumed that the probability of the vehicles changing lanes near the intersection is low. This means the density of vehicle points on each lane is higher than that of vehicle points near the boundary area of the lane. We applied a revised grid-based clustering (RGBC) developed by the authors' team for lane identification [31]. The RGBC first integrates the vehicle points from multiple frames into one space. The whole space is segmented using a hierarchical segmentation to identify road areas and non-road areas. The whole space is divided into squares with a relatively larger side length, such as 10 m. This level is named as the first level. Each square will be identified as a road area and a non-road area based on the number of points in it. If there is no point in the square, this area will be identified as a non-road area. If there is at least one vehicle point in the square, this area will be considered as a road area. For the square considered as a road area, it will be further divided into a children level with a smaller side length. The same procedure will be used to judge whether each square represents the road area or not. Step by step, a children level will be provided for each parent level until a bottom layer with a pre-defined side length (such as 0.1 m) is generated. For the squares in the bottom layer, a pre-defined threshold of point density will be provided to distinguish the squares representing road area and the squares representing the non-road area. For the road area, DBSCAN was applied to cluster the points in the same lane. It should be mentioned that the lane boundary identified here may not be the exact boundary of the lane, but the boundary of vehicle points in the same lane. The previous study showed that more than 96% of vehicles can be assigned to the correct lanes. The errors were mostly caused by lane-change vehicles. Another task in the lane identification is to detect the location of the stop line. In this paper, the location of the stop line was manually identified by checking the point intensity in the LiDAR visualization software Veloview. The point intensity of the stop line is higher compared to other objects [37]. A linear regression line can then be generated by linking two points at the boundaries of the stop line [38].

2.5. Object Association

The speed of the vehicle is crucial for determining the end of the queue [39]. The speed can be calculated based on the distance traveled by the vehicle in a time period. Therefore, it is necessary to track the same vehicle continuously at different frames. This research applied a global nearest neighbor

(GNN) algorithm to link the point cloud representing the same vehicle at different frames. The GNN considers the vehicle in the current frame that has the minimum distance to the vehicle A in the last frame as vehicle A [40]. To calculate the speed, the algorithm used the point with the shortest distance to the LiDAR as a reference point for vehicle tracking. The speed (V) can then be calculated by:

$$V = F * \sqrt{(X_i - X_{i-1})^2 + (Y_i - Y_{i-1})^2 + (Z_i - Z_{i-1})^2} \quad (1)$$

where *XYZ* are the *XYZ* coordinates of the nearest point to the LiDAR, *i* is the frame ID and F is the rotating frequency of the LiDAR (unit: Hz). To evaluate the speed, we did a field test using a vehicle installed with a logger to extract the speed from the onboard diagnostics interface (OBD). The vehicle was also scanned by the roadside LiDAR and the speed was calculated. The OBD speed and the calculated speed were then compared. The testing results showed that about 98.8% of speed records calculated from GNN has a speed with a difference less than 2.4 km/h compared to the OBD speed [32].

3. Queue Length Detection

Inspired by the previous work [19,40], we assumed the following threshold to identify the end of a queue: if the speed of one vehicle is under 5 km/h, then this vehicle will be considered as in a queue; if the speed of one vehicle is equal to or higher than 5 km/h, this vehicle would not be considered as in a queue. The vehicle at the end of the queue (*VEQ*) is the key vehicle to determine the length of a queue. Since the *VEQ* may be relatively far away from the LiDAR if the road is congested, the length of the vehicle at the end of the queue (*LVEQ*) may not be fully detected (point density decreases with the increasing distance from the LiDAR) [41], as shown in Figure 4a. Therefore, it is necessary to estimate the LVEQ. We used a simple rule-based method for *LVEQ* estimation. We assumed that the average length of a passenger car is 6 m [42]. Therefore, if the detected vehicle length (*DVL*) is longer than 6 m, we use the detected length as the vehicle length. If the detected vehicle length is shorter than 6 m, we used 6 m as the length of the vehicle. The strategy of the simple rule-based method can be expressed as:

$$LVEQ = \begin{cases} DVL, & if\ DVL \geq 6\ m \\ 6\ m, & if\ DVL < 6\ m \end{cases} \quad (2)$$

Using Equation (2) can still have some errors in estimating the queue. But the influence of this error on the whole length of the queue should be very limited (We will present the result in the "Evaluation" part later).

If a truck or a larger vehicle is traveling in the lane close to the LiDAR, the vehicles on the other lanes may be blocked, which is called occlusion issue. The occlusion issue is a challenge to detect the end of the queue since the last one or several vehicles at the end of the queue may be blocked (invisible in the LiDAR). Figure 4b shows an example of occlusion issue. Vehicle E and part of vehicle D are invisible due to occlusion from vehicle F. Another challenge for detecting the end of the queue is the package loss issue. Package loss refers to the situation that some packages are lost due to the unstable connection between the LiDAR and the data storing device (usually a computer). As a result, there are a lot of sector-like areas which are invisible in the space. The package loss issue may also make the vehicle at the end of the queue invisible. An example of package loss issue is shown in Figure 4c. Vehicle E' and most part of vehicle F' are invisible since they are in the package loss area (sector-like area).

To fix those issues, the vehicle information in the past time (historical information) was used. The first task is to detect whether there is an occlusion issue or package loss issue. We assumed that drivers would slow down when they are approaching the queue, meaning the speed of the vehicle in the current frame should be less than or at least equal to the speed in the last frame. This assumption makes it possible to use the speed to estimate the location of the vehicle if the vehicle is invisible in the current frame. The following method was applied for *VEQ* estimation.

Assuming there are j vehicles traveling on the lane in frame i, the jth vehicle is the vehicle that farthest from the stop line. The speed of the jth vehicle is recorded as V. The jth vehicle in frame $i + 1$ can be total occluded, partial occluded or non-occluded. The following parts illustrated the VEQ identification method for three situations (total occluded, partial occluded and non-occluded).

In frame $i + 1$, if the jth vehicle is invisible (the ID of the jth vehicle in frame i could not be assigned to any vehicle in frame $i + 1$), then the location of the jth vehicle can first be assumed to be the end of the queue. The distance d between the jth vehicle and the $j - 1$th vehicle is directly copied from the distance between the $j - 1$th vehicle and the $j - 2$th vehicle. The speed of the jth vehicle can be then calculated as V'. If $V' \leq V$ and $V' \leq 5$ km/h, the jth vehicle is considered as the VEQ, as shown in Figure 4d. If $V' > V$ or $V' > 5$ km/h, the $j - 1$th vehicle is considered as the VEQ, as shown in Figure 4e. The algorithm can be illustrated as

$$L = \begin{cases} \sum_{1}^{j} l_i, & if\ V'_{(j+1)th} \leq V_{jth} \cap V'_{(j+1)th} \leq 5\ \text{km/h} \\ \sum_{1}^{j-1} l_i, & if\ V'_{(j+1)th} > V_{jth} \cup V'_{(j+1)th} > 5\ \text{km/h} \end{cases} \quad (3)$$

where $\sum_{1}^{j} l_i$ means the total length of the vehicles in frame i, V_{jth} is the vehicle of jth vehicle.

If the jth vehicle is visible in frame $i + 1$, the algorithm does not know whether the vehicle is partially occluded or non-occluded. The point of the jth vehicle that has the shortest distance to the LiDAR is selected as a key point. The key point is considered as the front corner of the jth vehicle. The length of the jth vehicle in frame i is used as the length of the jth vehicle in frame $i + 1$. If the front part of the jth vehicle is visible, then the key point can reflect the location of the jth vehicle correctly, as shown in Figure 4f. If the end part of the jth vehicle is visible, then the key point may be located at the middle of the length of the vehicle. As a result, there may be a distance error (Ed) between the estimated location and the actual location of the jth vehicle, as shown in Figure 4g. But Ed should be less than the length of the jth vehicle. The speed of the jth vehicle in frame $i + 1$ can then be calculated. The VEQ can be identified by checking the speeds of the vehicles in frame $i + 1$. The algorithm can be illustrated as

$$L = \begin{cases} \sum_{1}^{j} l_i, & if\ V'_{(j+1)th} \leq V_{jth} \cap V'_{(j+1)th} \leq 5\ \text{km/h} \\ \sum_{1}^{j-1} l_i, & if\ V'_{(j+1)th} > V_{jth} \cup V'_{(j+1)th} > 5\ \text{km/h} \end{cases} \quad (4)$$

where $\sum_{1}^{j} l_i$ means the total length of the vehicles in frame i, V_{jth} is the vehicle of jth vehicle.

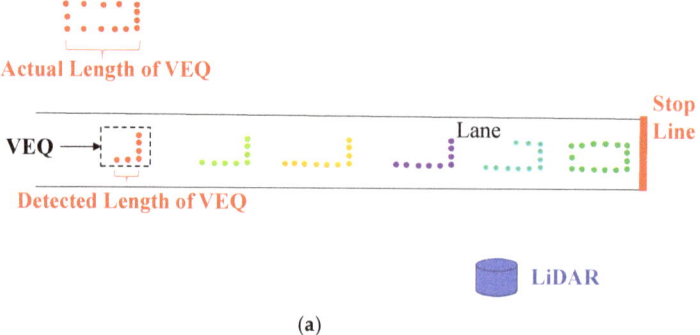

(a)

Figure 4. *Cont.*

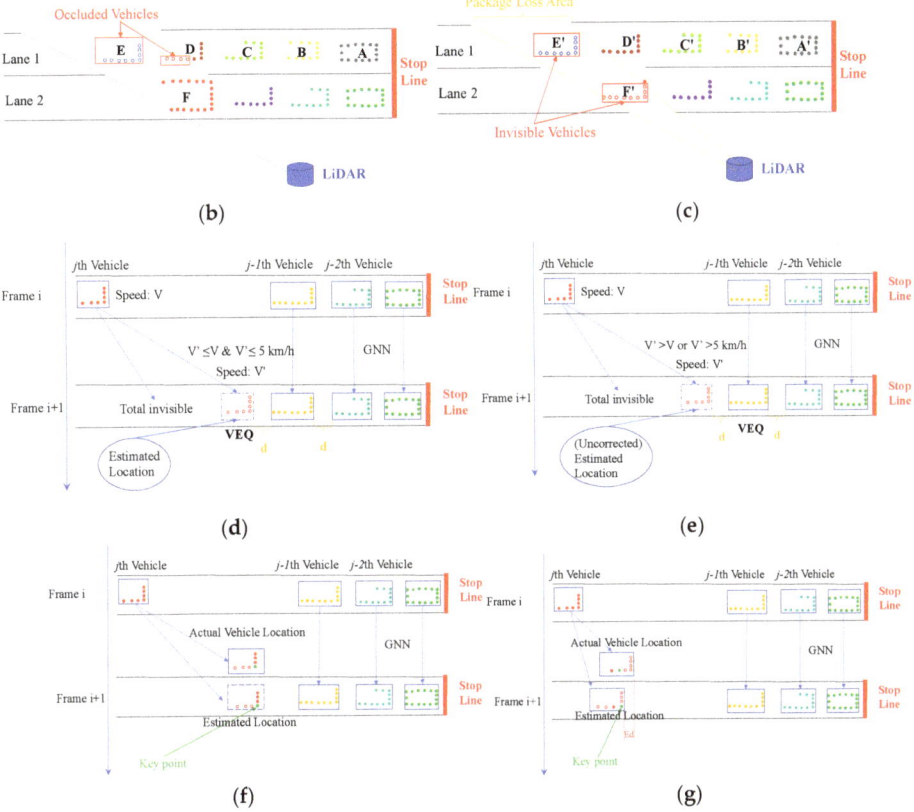

Figure 4. Queue length detection with occlusion and package loss issues: (**a**) occlusion issue, (**b**) occlusion issue, (**c**) package loss issue, (**d**) *VEQ* estimation: scenario 1, (**e**) *VEQ* estimation: scenario 2, (**f**) *VEQ* estimation: scenario 3, (**g**) *VEQ* estimation: scenario 4.

It should be mentioned that due to the different locations of the vehicles in the LiDAR data, the vehicles in the queue (not *VEQ*) may also be occluded or partially occluded due to the package loss or occlusion issue. Those occluded vehicles do not influence the queue length detection as long as the *VEQ* can be detected/predicted.

4. Evaluation

The performance of the LiDAR was evaluated with the ground-truth data extracted from the camera and the raw LiDAR visualized in the open-source visualization software—Veloview. Two trained graduate students were hired to manually extract the queue length from the camera installed at the selected sites. The number of vehicles in the queue was recorded by checking the camera and the LiDAR data in Veloview. The length of the queue was calculated by checking the location of the reference in camera and in Google Earth. The results extracted by the two graduate students were considered as the ground-truth data. It should be mentioned that the ground-truth data are not the data with 100% accuracy due to the distance calculation error in Google Earth and the inevitable human error.

Four sites with different road features in Reno, Nevada were selected for evaluation. The features of the three sites are documented in Table 1.

Table 1. Features of selected sites for evaluation.

Site	Speed Limit	Traffic Control	Areas for Queue Length Evaluation
I80 work zone	64.4 km/h (40 mph)	Left lane closed control	One westbound unclosed lane
Virginia St @ Artemesia Way	40.2 km/h (25 mph)	Signalized T-intersection	Two northbound through lanes
Baring Blvd at the front of the Reed High School	56.3 km/h (35 mph)	Yield sign for pedestrian crossing	Two westbound through lanes

Figure 5a shows the data collection location at the I80 work zone. Figure 5b shows the detected queue length and the measured queue length at the work zone in I80-W freeway. There are two lanes at the westbound and the left lane at the westbound was closed due to the pavement construction. It should be mentioned that there was not stop line near the work zone and the location of the LiDAR was not exactly located at the work zone since we did not get the permit to put LiDAR at the work zone. In other words, a zero value of queue length in Figure 5b does not really mean that there is no queue at the work zone, but no queue in the detection range of the LiDAR. The max queue length (among two lanes) in every minute was recorded by the proposed method and by checking the camera installed on top of the LiDAR. It is shown that from 13:05, the queue started to be formed. The camera can detect about a total of 169 m distance of the queue and the LiDAR can detect about a total of 300 m distance. The offsets between the queue lengths were manually extracted from the camera and the detected queue length from the LiDAR was small from 13:05 to 13:12. After 13:12, the queue length was difficult to be determined through the camera. Therefore, the green line in Figure 5b disappeared around 13:12. After 13:17, since the queue length was longer than the detection range of the LiDAR, the queue length could not be successfully detected. Therefore, the orange line disappeared around 13:17.

Figure 5c shows the data collection location at Virginia St @ Artemesia Way. The results of queue length detection are illustrated in Figure 5d. A total of six-minutes data were randomly selected for evaluation. The queue length for each lane was analyzed. It is shown that the detected queue length and the ground-truth queue length are close to each other for lane 2. However, there are some offsets between the detected queue length and the ground-truth queue length for lane 1 (as shown in the red rectangle in Figure 5d. Around 13:26, the detected queue length (28 m) was significantly higher than the ground-truth queue length (22 m). The longer detected queue length was caused by the definition of the end of the queue that the vehicle with the speed less than 5 km/h as the *VEQ*. But for ground-truth data, the queue length was identified by the graduate students based on their own judgment. Therefore, there were some offsets between the detected queue length and the ground-truth data. Figure 5e shows the data collection location at Baring Blvd. There is a pedestrian middle crossing at this site. Since there are not stop lines at this location, it is difficult to determine the start point of the queue. Therefore, we used the number of vehicles in the queue to represent the queue length. The results of the number of vehicles detection for each lane are illustrated in Figure 5f. It is shown that the detected number of vehicles and the ground-truth data matched very well though some errors existed in the detected results. There are two types of errors for detecting the number of vehicles. The first type of error can be seen at around 11:07 in Figure 5f. The detected number of vehicles in the queue was higher than the ground-truth number of vehicles in the queue in lane 2. By checking the camera data, it was found that the vehicle at the end of the queue was a commercial truck and it was chopped into two parts due to occlusion or package loss issue. As a result, the clustering algorithm clustered the truck as two vehicles. The second type of error can be seen at around 11:08 in Figure 5f. The detected number of vehicles in the queue was lower than the ground-truth number of vehicles in the queue in lane 1. The offsets were also caused by the different definitions of the end of the queue by the proposed algorithm and the graduate students.

Table 2 summarizes the distribution of the cumulative errors of the detected queue length and number of vehicles in the queue at the three sites. It is shown that 88.3% of calculated records had an error of less than 0.5 m in the queue length when compared to the ground-truth data and 96.2% of calculated records had an error of less than 3.0 m in the queue length when compared to the ground-truth data. As for the number of vehicles in the queue, 96.2% of the number of calculated vehicles in the queue was exactly matched to the human-counted records and 98.5% of calculated records had an offset within 1-vehicle count from the ground truth data. The maximum offset between the calculated records and the ground truth data were no more than 4 vehicles. The results indicated that the proposed method could achieve the relatively high accuracy for queue length detection.

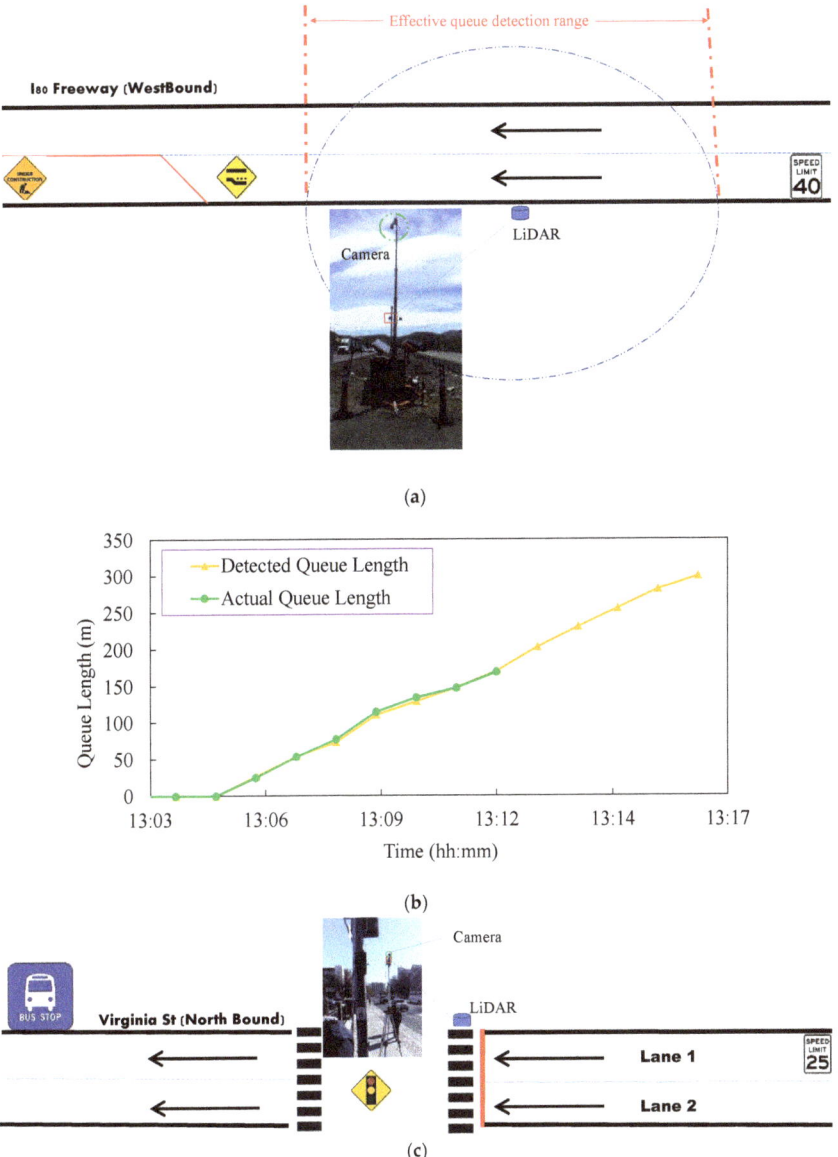

Figure 5. *Cont.*

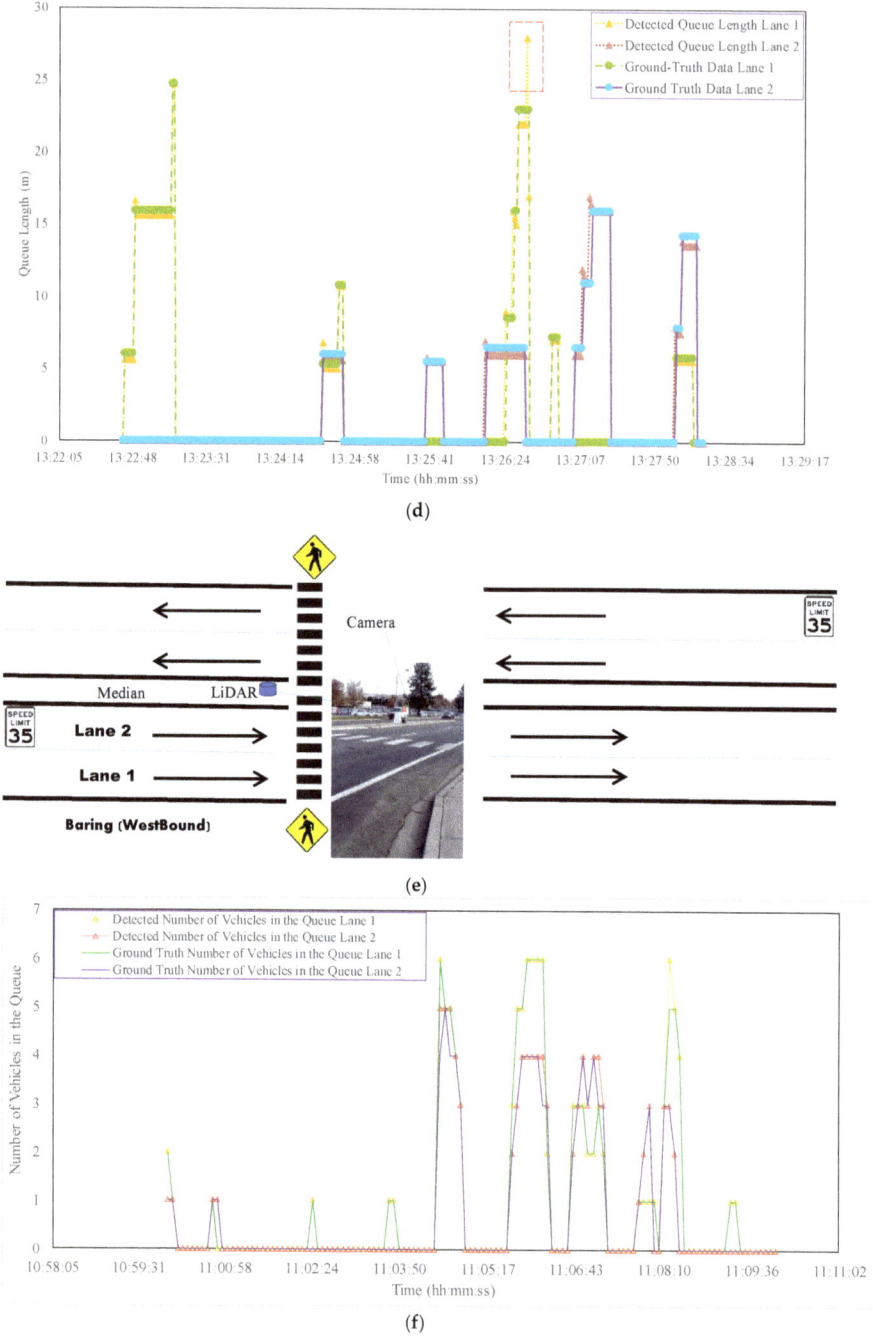

Figure 5. Evaluation of Queue Length Detection: (**a**) Site of I80 Westbound, (**b**) Results of Queue Length Detection at I80 Westbound, (**c**) Site of Virginia St @ Artemesia Way, (**d**) Results of Queue Length Detection at Virginia St @ Artemesia Way, (**e**) Site of Baring Blvd, (**f**) Results of Number of Vehicles in the Queue Detection at Baring Blvd.

Table 2. Cumulative errors of queue length detection.

Queue Length		Number of Vehicles in the Queue	
Error	Percentage (%)	Error	Percentage (%)
0–0.5 m	88.3	0 vehicle	96.2
0.5–1.0 m	89.8	1 vehicle	98.5
1.0–1.5 m	91.3	2 vehicles	99.1
1.5–2.0 m	94.5	3 vehicles	99.8
2.0–3.0 m	96.2	4 vehicles	100

5. Conclusions and Discussion

This paper presented a novel method for queue length detection using the roadside LiDAR data. Unlike the estimation methods used in most existing studies, the queue length can directly be detected with the proposed method. This proposed method can work for different road scenarios. The testing results showed that the proposed method can detect the queue length with high accuracy under different scenarios. The strategy of the proposed method is simple but effective in practice. With this proposed method, the accurate queue length can be provided in real-time for different applications. Therefore, the method is of great value in signal coordination especially in solving the initial queue estimation. Since the initial queue is unpredictable between cycle by cycle, this real-time measurement is proper to handle it. By capture the queue length, the proper offset can be calculated for each intersection along a corridor under each cycle to timely release the initial queue before the platoon arrives. This can be very helpful in adaptive signal control since the system is ready for real-time adjustment. Even for the traditional actuated-coordinated signals, the accurate trend of the initial queue change would help to decide the offset settings. On the other hand, it also directly supports the connected vehicle (which will avoid the queue blockage to the maximum extent) in the future. Vehicle to Vehicle or Vehicle to Infrastructure facilities can get the accurate queue information and in turn make adjustments from both vehicle approach and infrastructure approach with the least time loss in the process. The prospect of the method is bright. Queue length is also an important measure of effectiveness in the operational analysis. However, it is usually hard to estimate or measure; the problem has puzzled traffic engineers for a long time. The proposed method revolutionary decreases the measured time with the enhancement of the accuracy, which benefits a lot to operational analysis.

This paper did not compare the proposed queue length detection method with the existing queue length estimation methods since the data from the other sensors (such as loop detectors and signal timing) were not available. The future studies should also consider selecting one signalized intersection or a metered ramp to compare the proposed queue length detection method with other methods. For one LiDAR sensor, the longest detectable length of the queue is subject to the detection range of the LiDAR. If the queue length is out of the detection range, then another LiDAR to extend the detection range is needed. The ground-truth data were measured by two graduate students by checking the camera, Veloview and Google Earth. It is inevitable that errors exist in the ground-truth data in this paper. How to find a more accurate evaluation method for queue length detection is another research topic for future studies. Another limitation is that this paper used many assumptions (e.g., vehicle length is shorter than 6 m, vehicles with speed <5 km/h is considered in the queue), those assumptions can impact the accuracy of queue length detection. Future studies should consider to further improve the accuracy of queue length detection by reducing the assumptions of the proposed method. The performance of LiDAR can be reduced under adverse weather conditions, such as foggy, rainy, snowy weather. The proposed queue length detection highly relied on the accuracy of LiDAR detection. Therefore, future studies should also test the performance of the proposed queue length detection method under severe weather conditions.

Author Contributions: Conceptualization, J.W.; Y.Z. and H.X.; methodology, J.W. and X.S.; validation, J.W.; Y.T. and X.S.; formal analysis, J.W.; investigation, H.X.; resources, Y.Z.; data curation, Y.Z.; writing—original draft preparation, J.W.; writing—review and editing, X.S.; supervision, H.X.; funding acquisition, J.W. All authors have read and agreed to the published version of the manuscript.

Funding: This research was funded by the Qilu Young Scholar Program of Shandong University.

Acknowledgments: The authors thank Zong Tian and Hongchao Liu for their technical support in this research.

Conflicts of Interest: The authors declare no conflict of interest.

References

1. Yang, G.; Yue, R.; Tian, Z.; Xu, H. Modeling the Impacts of Traffic Flow Arrival Profiles on Ramp Metering Queues. *Transp. Res. Rec.* **2018**, *2672*, 85–92. [CrossRef]
2. Ban, X.J.; Hao, P.; Sun, Z. Real time queue length estimation for signalized intersections using travel times from mobile sensors. *Transp. Res. Part C Emerg. Technol.* **2011**, *19*, 1133–1156.
3. Sharma, A.; Bullock, D.M.; Bonneson, J.A. Input–output and hybrid techniques for real-time prediction of delay and maximum queue length at signalized intersections. *Transp. Res. Rec.* **2007**, *2035*, 69–80. [CrossRef]
4. Qian, G.; Lee, J.; Chung, E. Algorithm for queue estimation with loop detector of time occupancy in off-ramps on signalized motorways. *Transp. Res. Rec.* **2012**, *2278*, 50–56. [CrossRef]
5. Lee, J.; Jiang, R.; Chung, E. Traffic queue estimation for metered motorway on-ramps through use of loop detector time occupancies. *Transp. Res. Rec.* **2013**, *2396*, 45–53. [CrossRef]
6. Liu, H.X.; Wu, X.; Ma, W.; Hu, H. Real-time queue length estimation for congested signalized intersections. *Transp. Res. Part C Emerg. Technol.* **2009**, *17*, 412–427. [CrossRef]
7. Neumann, T. Efficient queue length detection at traffic signals using probe vehicle data and data fusion. In Proceedings of the ITS 16th World Congress, Stockholm, Sweden, 21–25 September 2009.
8. Cheng, Y.; Qin, X.; Jin, J.; Ran, B.; Anderson, J. Cycle-by-cycle queue length estimation for signalized intersections using sampled trajectory data. *Transp. Res. Rec.* **2011**, *2257*, 87–94. [CrossRef]
9. Hao, P.; Ban, X.J.; Guo, D.; Ji, Q. Cycle-by-cycle intersection queue length distribution estimation using sample travel times. *Transp. Res. Part B Methodol.* **2014**, *68*, 185–204. [CrossRef]
10. Cai, Q.; Wang, Z.; Zheng, L.; Wu, B.; Wang, Y. Shock wave approach for estimating queue length at signalized intersections by fusing data from point and mobile sensors. *Transp. Res. Rec.* **2014**, *2422*, 79–87. [CrossRef]
11. Sun, Z.; Ban, X.J. Vehicle trajectory reconstruction for signalized intersections using mobile traffic sensors. *Transp. Res. Part C Emerg. Technol.* **2013**, *36*, 268–283. [CrossRef]
12. Li, J.Q.; Zhou, K.; Shladover, S.E.; Skabardonis, A. Estimating queue length under connected vehicle technology: Using probe vehicle, loop detector, and fused data. *Transp. Res. Rec.* **2013**, *2356*, 17–22. [CrossRef]
13. Christofa, E.; Argote, J.; Skabardonis, A. Arterial queue spillback detection and signal control based on connected vehicle technology. *Transp. Res. Rec.* **2013**, *2366*, 61–70. [CrossRef]
14. Tiaprasert, K.; Zhang, Y.; Wang, X.B.; Zeng, X. Queue length estimation using connected vehicle technology for adaptive signal control. *IEEE Trans. Intell. Transp. Syst.* **2015**, *16*, 2129–2140. [CrossRef]
15. Yang, K.; Menendez, M. Queue estimation in a connected vehicle environment: A convex approach. *IEEE Trans. Intell. Transp. Syst.* **2018**, *20*, 2480–2496. [CrossRef]
16. Zhang, Z.; Zheng, J.; Xu, H.; Wang, X. Vehicle Detection and Tracking in Complex Traffic Circumstances with Roadside LiDAR. *Transp. Res. Rec.* **2019**, *2673*, 62–71. [CrossRef]
17. Zanin, M.; Messelodi, S.; Modena, C.M. An efficient vehicle queue detection system based on image processing. In Proceedings of the 12th International Conference on Image Analysis and Processing, Mantova, Italy, 17–19 September 2003; pp. 232–237.
18. Siyal, M.Y.; Fathy, M. A neural-vision based approach to measure traffic queue parameters in real-time. *Pattern Recognit. Lett.* **1999**, *20*, 761–770. [CrossRef]
19. Cai, Y.; Zhang, W.; Wang, H. Measurement of vehicle queue length based on video processing in intelligent traffic signal control system. In Proceedings of the International Conference on Measuring Technology and Mechatronics Automation, Changsha, China, 13–14 March 2010; Volume 2, pp. 615–618.
20. Satzoda, R.K.; Suchitra, S.; Srikanthan, T.; Chia, J.Y. Vision-based vehicle queue detection at traffic junctions. In Proceedings of the 7th IEEE Conference on Industrial Electronics and Applications (ICIEA), Singapore, 18–20 July 2012; pp. 90–95.

21. Xu, Y.; Wu, Y.; Xu, J.; Ni, D.; Wu, G.; Sun, L. A queue-length-based detection scheme for urban traffic congestion by VANETs. In Proceedings of the IEEE Seventh International Conference on Networking, Architecture, and Storage, Xiamen, China, 28–30 June 2012; pp. 252–259.
22. Lv, B.; Xu, H.; Wu, J.; Tian, Y.; Zhang, Y.; Zheng, Y.; Yuan, C.; Tian, S. LiDAR-Enhanced Connected Infrastructures Sensing and Broadcasting High-Resolution Traffic Information Serving Smart Cities. *IEEE Access* **2019**, *7*, 79895–79907. [CrossRef]
23. Wu, J.; Tian, Y.; Xu, H.; Yue, R.; Wang, A.; Song, X. Automatic ground points filtering of roadside LiDAR data using a channel-based filtering algorithm. *Opt. Laser Technol.* **2019**, *115*, 374–383. [CrossRef]
24. Wu, J.; Xu, H.; Yue, R.; Tian, Z.; Tian, Y.; Tian, Y. An automatic skateboarder detection method with roadside LiDAR data. *J. Transp. Saf. Secur.* **2019**, 1–20. [CrossRef]
25. Wu, J.; Xu, H.; Zheng, J. Automatic background filtering and lane identification with roadside LiDAR data. In Proceedings of the 2017 IEEE 20th International Conference on Intelligent Transportation Systems (ITSC), Yokohama, Japan, 16–19 October 2017; pp. 1–6.
26. Wu, J.; Xu, H.; Sun, Y.; Zheng, J.; Yue, R. Automatic background filtering method for roadside LiDAR data. *Transp. Res. Rec.* **2018**, *2672*, 106–114. [CrossRef]
27. Zhao, J.; Xu, H.; Liu, H.; Wu, J.; Zheng, Y.; Wu, D. Detection and tracking of pedestrians and vehicles using roadside LiDAR sensors. *Transp. Res. Part C Emerg. Technol.* **2019**, *100*, 68–87. [CrossRef]
28. Wu, J.; Xu, H.; Zheng, Y.; Zhang, Y.; Lv, B.; Tian, Z. Automatic Vehicle Classification using Roadside LiDAR Data. *Transp. Res. Rec.* **2019**, *2673*, 153–164. [CrossRef]
29. Chen, J.; Xu, H.; Wu, J.; Yue, R.; Yuan, C.; Wang, L. Deer Crossing Road Detection with Roadside LiDAR Sensor. *IEEE Access* **2019**, *7*, 65944–65954. [CrossRef]
30. Wu, J.; Xu, H.; Zhao, J. Automatic lane identification using the roadside LiDAR sensors. *IEEE Intell. Transp. Syst. Mag.* **2020**, *12*, 25–34. [CrossRef]
31. Cui, Y.; Xu, H.; Wu, J.; Sun, Y.; Zhao, J. Automatic Vehicle Tracking with Roadside LiDAR Data for the Connected-Vehicles System. *IEEE Intell. Syst.* **2019**, *34*, 44–51. [CrossRef]
32. Wu, J. An automatic procedure for vehicle tracking with a roadside LiDAR sensor. *Inst. Transp. Eng. ITE J.* **2018**, *88*, 32–37.
33. Lv, B.; Xu, H.; Wu, J.; Tian, Y.; Yuan, C. Raster-based Background Filtering for Roadside LiDAR Data. *IEEE Access* **2019**, *7*, 2169–3536. [CrossRef]
34. Lv, B.; Xu, H.; Wu, J.; Tian, Y.; Tian, S.; Feng, S. Revolution and rotation-based method for roadside LiDAR data integration. *Opt. Laser Technol.* **2019**, *119*, 105571. [CrossRef]
35. Zheng, J.; Xu, B.; Wang, X.; Fan, X.; Xu, H.; Sun, G. A portable roadside vehicle detection system based on multi-sensing fusion. *Int. J. Sens. Netw.* **2019**, *29*, 38–47. [CrossRef]
36. Ester, M.; Kriegel, H.P.; Sander, J.; Xu, X. A density-based algorithm for discovering clusters in large spatial databases with noise. *KDD* **1996**, *96*, 226–231.
37. Yang, B.; Fang, L.; Li, Q.; Li, J. Automated extraction of road markings from mobile LiDAR point clouds. *Photogramm. Eng. Remote Sens.* **2012**, *78*, 331–338. [CrossRef]
38. Cheng, M.; Zhang, H.; Wang, C.; Li, J. Extraction and classification of road markings using mobile laser scanning point clouds. *IEEE J. Sel. Top. Appl. Earth Obs. Remote Sens.* **2016**, *10*, 1182–1196. [CrossRef]
39. Yu, L.; Dan-pu, Z.; Xian-qing, T.; Zhi-li, L. The queue length estimation for congested signalized intersections based on shockwave theory. In Proceedings of the International Conference on Remote Sensing, Environment and Transportation Engineering, Nanjing, China, 26–28 July 2013.
40. Wu, A.; Qi, L.; Yang, X. Mechanism analysis and optimization of signalized intersection coordinated control under oversaturated status. *Procedia Soc. Behav. Sci.* **2013**, *96*, 1433–1442. [CrossRef]
41. Yue, R.; Xu, H.; Wu, J.; Sun, R.; Yuan, C. Data Registration with Ground Points for Roadside LiDAR Sensors. *Remote Sens.* **2019**, *11*, 1354. [CrossRef]
42. Cheung, S.Y.; Coleri, S.; Dundar, B.; Ganesh, S.; Tan, C.W.; Varaiya, P. Traffic measurement and vehicle classification with single magnetic sensor. *Transp. Res. Rec.* **2005**, *1917*, 173–181. [CrossRef]

© 2020 by the authors. Licensee MDPI, Basel, Switzerland. This article is an open access article distributed under the terms and conditions of the Creative Commons Attribution (CC BY) license (http://creativecommons.org/licenses/by/4.0/).

Article

Vehicle Detection under Adverse Weather from Roadside LiDAR Data

Jianqing Wu [1], Hao Xu [2], Yuan Tian [2], Rendong Pi [1] and Rui Yue [2,*]

1. School of Qilu Transportation, Shandong University, Jinan 250061, China; jianqingwusdu@sdu.edu.cn (J.W.); pirendong@mail.sdu.edu.cn (R.P.)
2. Department of Civil and Environmental Engineering, University of Nevada, Reno, NV 89557, USA; haox@unr.edu (H.X.); yuantian@nevada.unr.edu (Y.T.)
* Correspondence: ryue@nevada.unr.edu; Tel.: +17-7-5870-7637

Received: 22 May 2020; Accepted: 14 June 2020; Published: 17 June 2020

Abstract: Roadside light detection and ranging (LiDAR) is an emerging traffic data collection device and has recently been deployed in different transportation areas. The current data processing algorithms for roadside LiDAR are usually developed assuming normal weather conditions. Adverse weather conditions, such as windy and snowy conditions, could be challenges for data processing. This paper examines the performance of the state-of-the-art data processing algorithms developed for roadside LiDAR under adverse weather and then composed an improved background filtering and object clustering method in order to process the roadside LiDAR data, which was proven to perform better under windy and snowy weather. The testing results showed that the accuracy of the background filtering and point clustering was greatly improved compared to the state-of-the-art methods. With this new approach, vehicles can be identified with relatively high accuracy under windy and snowy weather.

Keywords: vehicle detection; adverse weather; roadside LiDAR; data processing

1. Introduction

Adverse weather can negatively influence transportation performance in two aspects: decreasing the operational efficiency and increasing the crash risk. Fortunately, as connected vehicle (CV) technology becomes more realistic, the overall operational efficiency and traffic safety can greatly benefit from CV technology, especially under adverse weather conditions. However, effectively employing CV technology on the road requires accurate traffic data. The quality of these data could also be influenced by adverse weather, which confuses the judgment of the CV network and causes the loss of operational efficiency and crashes. Therefore, investigating how to improve the accuracy of traffic data under adverse weather is significantly important for current CV technology. Light detection and ranging (LiDAR), an emerging sensor for intelligent transportation systems, has the potential of providing traffic data under good weather conditions [1]. The new 360-degree LiDAR can detect all road users and surrounding environments in a 360-degree horizontal field of view (FOV). Compared to traditional sensors, such as cameras, loop detectors, and radar, LiDAR can work day and night and has higher accuracy for object detection [2]. Airborne and on-board LiDAR (mobile LiDAR) are the traditional installation methods for object detection and remote sensing [3]. Recently, the roadside LiDAR has been a new deployment method for transportation applications. The LiDAR can be installed on a tripod for short-term data collection or on roadside infrastructures (such as a wire pole) for long-term data collection [4,5]. The roadside LiDAR sensor is able to scan the surfaces of all road vehicles (including both connected vehicles and unconnected vehicles) within the detection range by generating 3D point clouds, which provides a perfect solution for filling the data gap of the transition period from unconnected vehicles to connected vehicles [6]. Here, connected vehicles refer to those vehicles that can be engaged in the connected vehicle environment. The high-resolution trajectories of all road users can then be extracted from the roadside LiDAR and can provide valuable information such as driver behavior analysis, fuel consumption, near-crash identification, and prediction [7–10].

A significant number of studies have been conducted to extract useful traffic information from roadside LiDAR data. The roadside LiDAR data processing procedure typically includes four steps: background filtering, object clustering, object classification, and object tracking [11]. This paper focuses on the first two parts: background filtering and object clustering. The background in roadside LiDAR data usually includes stationary objects such as buildings and the ground surface, and dynamic objects such as waving trees, grasses, and bushes. When referring to stationary objects, the location of the same LiDAR point at different frames is not strictly fixed due to the slight shaking of the LiDAR laser beams [5], which results in difficulties for background filtering. The original method for filtering the background was to search the frames without road users within the detection range [12,13]. However, it may be difficult to select the correct number of frames without any road users at high-volume traffic road segments or intersections. Zhang et al. [14] developed a point association (PA)-based method for background filtering. A frame without any road users was manually selected as a reference frame. Then, a predefined distance threshold was assigned to the background points in the reference frame. Any point with a distance to the roadside LiDAR shorter than the threshold was identified as a background point. However, the threshold needed to be selected based on the users' experience, which limited the actual application of the PA-based method. Wu et al. [15] developed a point density-based method named 3D density statistic filtering (3D-DSF) for background filtering. The 3D-DSF method does not need to manually select the suitable frames. In their method, the whole detection range is divided into amounts of small cubes, and the point density of each cube in each frame is calculated. Then, by frame aggregation, the sum of the point density over all frames of each cube can be found. A predefined threshold is used to distinguish background cubes from non-background cubes. More details about the 3D-DSF are referred to in [16]. The assumption of this study was that the sum of the point density of the background cube will be much larger than that of the cube with road users. However, a limitation of the 3D-DSF is that it is unable to exclude the background points effectively under congested intersections. Lv et al. [17] developed a raster-based (RA) method using the change in point density as a feature for background filtering. Any cube with a change in point density larger than two in two adjacent frames was considered as background. The testing results showed that the raster-based method could exclude more than 98% of the background points in the three investigated sites. However, all the above-mentioned methods were performed under normal weather. The performance of those background filtering methods under harsh environments, such as strong wind and snow, was not evaluated.

Point clustering means to cluster the points belonging to one object into one group. Zhang et al. [18] used the Euclidean clustering extraction (ECE) algorithm for point clustering. ECE uses two parameters, the cluster size (S) and the tolerance (d), to search the points belonging to one object. Since there are no standard methods for parameter selection, heuristic testing is required to determine the optimal value for different datasets. Wu [5] applied the density-based spatial clustering of applications with noise (DBSCAN) for clustering. The advantage of DBSCAN is that it does not need to know the number of objects in advance. DBSCAN uses epsilon and the minimum number of points to determine whether a point belongs to a group or not. Wu [5] (Wu, 2018) suggested using 1.2 m as epsilon and 10 as the minimum number of points for the input of DBSCAN. Later, Zhao et al. [19] found that the fixed parameters of DBSCAN could not group the points correctly when the object was far away from the LiDAR. The principal reason was that the density of the same object changed with a different distance to the roadside LiDAR. Zhao et al. [19] developed a revised DBSCAN for object clustering based on the distribution feature of the LiDAR point within the space. However, the DBSCAN related algorithms are computationally expensive since they require an extensive search of all points in the point cloud. A previous study [20] also found that the method proposed by Zhao et al. [19] could not cluster the points correctly under snowy weather.

In fact, a large amount of research has been done to process LiDAR data under severe weather conditions [21–29]. Wojtanowski et al. [22] found that LiDAR is susceptible to adverse weather conditions. Charron et al. [23] developed a dynamic 3D outlier detection method to remove snow noise from the onboard LiDAR data. The testing results showed that the proposed method could achieve more than 90% precision. Jokela et al. [24] found that LiDAR sensors' performance decreased with the increasing density of fog and the distance between the target and the LiDAR. The visible range for object detection in the LiDAR relied on the different types of LiDAR. Kutila et al. [25] evaluated the performance of automotive LiDAR in fog and rain. It was found that fog can be a challenge for object detection using the LiDAR at a 905 nm wavelength due to light being scattered by fog particles and a 1550 nm wavelength was recommended to be used in the LiDAR in order to reduce the impact of fog particles. Bijelic et al. [26,27] compared the performance of four different state-of-the-art LiDAR systems. The results showed that all the LiDAR systems decreased in fog and that changing the internal parameters in the LiDAR could improve their functions under adverse weather.

The above-mentioned studies have shown that adverse weather can reduce the resolution of the roadside LiDAR data qualitatively. It is still necessary to quantitatively analyze the influence of different adverse conditions on the roadside LiDAR and to develop new methods that can accommodate background filtering and point clustering for adverse weather conditions.

2. Background Filtering

One advantage of roadside LiDAR is that past information (historical frames) can be used to process the current data [30,31]. With this feature, the accuracy of data processing can be greatly improved. In fact, the previously mentioned methods, such as 3D-DSF, RA, and PA, all used historical information to enhance the accuracy of the background filtering. However, for temporary data collection, the wind may influence the resolution of the LiDAR data, especially at windy spots. As a result, non-background points can be misrecognized as background points and background points can be misrecognized as non-background points. For background filtering, 3D-DSF is still the most widely used method for roadside LiDAR data processing [32–35]. Here, we examined the performance of 3D-DSF under snowy and windy weather conditions. One road segment along the I-80 freeway in Reno was selected as the testing site. The site's location is shown in Figure 1.

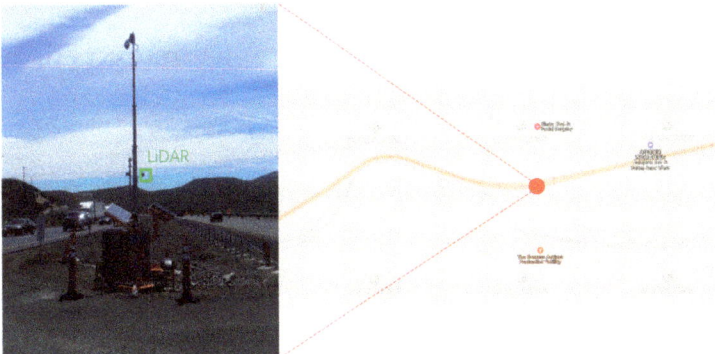

Figure 1. Testing site.

Figure 2 shows an example of 3D-DSF under windy and non-windy weather conditions. Figure 2a,b shows that under normal (non-windy) weather, 3D-DSF can exclude most background points and leave the non-background points in the space. In Figure 2b, we can clearly see where the cluster points are, as they are highlighted in green. Previous research has shown that vehicles can easily be identified after data are applied with 3D-DSF. However, under windy weather, 3D-DSF could not effectively exclude the ground surface, as shown in Figure 2c,d. In Figure 2d, although the background points are partially eliminated, the non-background points and background points are still unseparated after applying 3D-DSF. The extraction results are significantly different from Figure 2b. The wind may cause a relatively large offset between the ground points at different frames, indicating that past information may not provide a good reference for background filtering. Under windy weather, the point density of the cubes containing some ground points may not meet the predefined threshold. As a result, the ground points may be identified as non-background points.

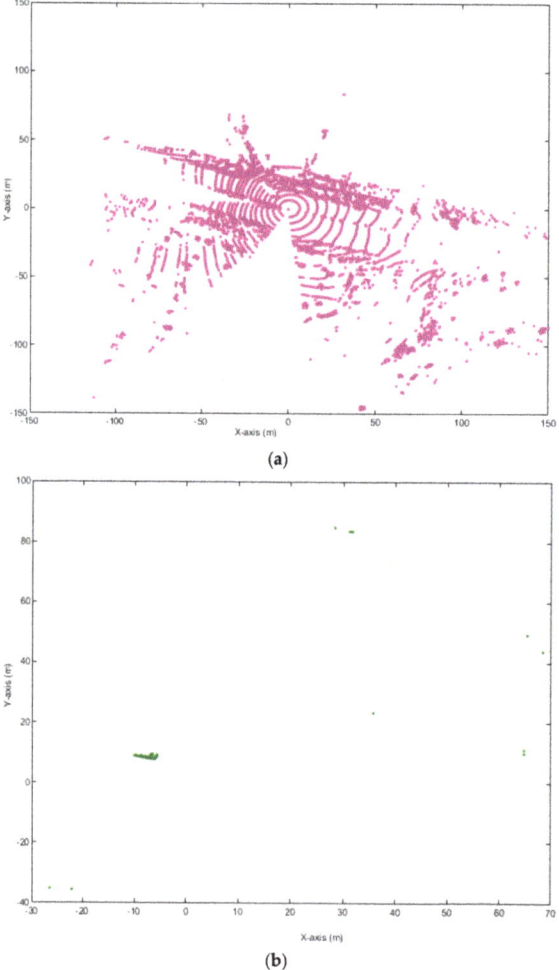

Figure 2. *Cont.*

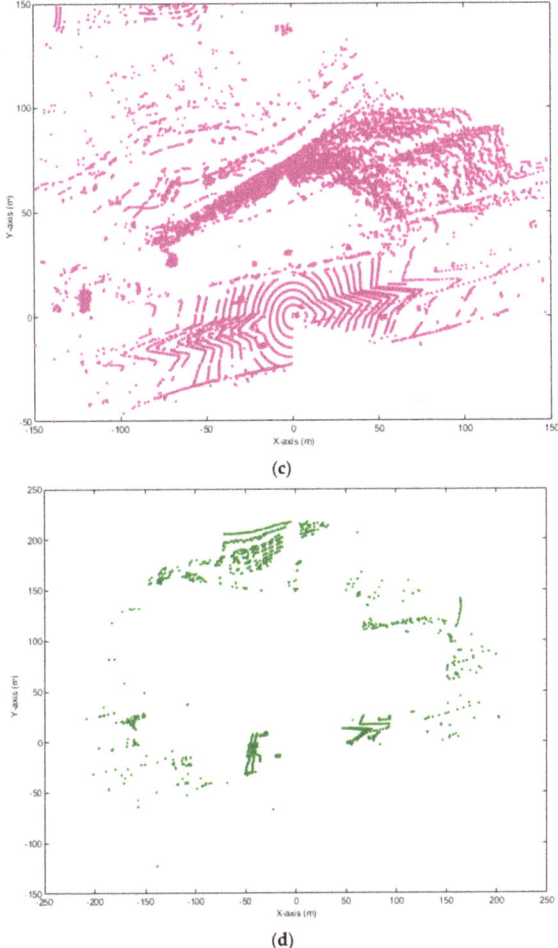

Figure 2. Performance of 3D density statistic filtering (3D-DSF) under windy and non-windy weather: (**a**) No wind before applying 3D-DSF, (**b**) No wind after applying 3D-DSF, (**c**) Strong wind before applying 3D-DSF, (**d**) Strong wind after applying 3D-DSF.

The errors of background filtering under windy weather usually occur on the ground surface, because the ground surface on the road is usually smooth, and the distance between two ground circles is larger than other objects [36]. As a result, a small disturbance in the position of the LiDAR may lead to a larger offset in the location of ground surfaces. The offset in the ground surface may then cause a reduced point density in the cubes representing the ground surface, and it may increase the point density in the nearby non-background cubes. Therefore, the emphasis is on improving the accuracy of background filtering under windy weather in order to find a method to exclude the ground points effectively. This paper develops a ground surface-enhanced density statistic filtering method (GS-DSF) for background filtering. The details of the GS-DSF are documented as follows.

The idea of ground surface exclusion is inspired by the ground surface exclusion used for on-board LiDAR serving autonomous vehicles [36]. The rotating LiDAR generates different circles for ground points with different distances from the LiDAR. When there is an object in the space, the slope created by the object points between two adjacent frames significantly differs from the slope created by the ground points, as shown in Figure 3.

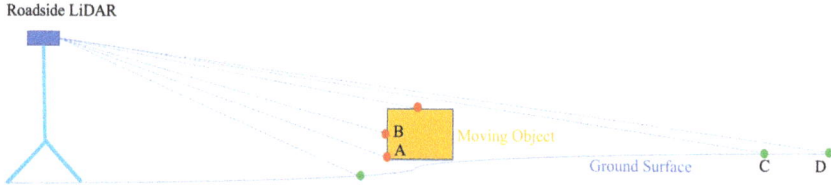

Figure 3. Slope difference created by moving object and ground surface.

It is shown that when a moving object appears, the slope created by the points in the moving object in two adjacent frames is much steeper than the slope created by the points in the ground surface. Equation (1) further illustrates the example in Figure 3.

$$(\sin(\alpha) = \frac{Sqrt((X_A-X_B)^2+(Y_A-Y_B)^2+(Z_A-Z_B)^2)}{Z_A-Z_B}) \gg (\sin(\beta) = \frac{Sqrt((X_C-X_D)^2+(Y_C-Y_D)^2+(Z_C-Z_D)^2)}{Z_C-Z_D}) \quad (1)$$

where sin (α) and sin (β) represent the slopes of the moving object and the ground surface, A and B represent two points in the moving object, and C and D represent two points on the ground surface. X, Y, and Z are the XYZ coordinates (location in space) of the point. The previous study [37] found that α was usually less than 30 degrees and β was usually close to 90 degrees. In this research, we used 45 degrees as a threshold to distinguish background points and non-background points, which is named the slope-based method [37]. Since the computational load of directly applying the slope-based method on the raw LiDAR data was heavy, this paper firstly applies density statistic filtering (DSF) on the raw LiDAR data and then uses the slope-based method to exclude the ground points after DSF. The GS-DSF used here is an updated version of the traditional 3D-DSF. As mentioned before, a limitation of 3D-DSF is that the background points could not be effectively excluded under windy weather. The GS-DSF used here fixes this issue with the following updates.

The first improvement made by the GS-DSF used here is to randomly pick up the frames instead of using continuous frames. For each selected frame, the frame identity (ID) is stored (a larger ID means the frame is picked up later). The random selection can reduce the probability of picking up the frames with moving objects captured in the space. The second update of the GS-DSF which is used here is that the neighbor information is applied for background filtering. The updated GS-DSF picks up point A with the frame with the smallest ID (initial frame). Then, the neighbor of point A in other frames (except the initial frame) within a predefined distance (D) can be obtained. D is determined by the horizontal and vertical resolution. Assuming there are N randomly selected frames and n number of neighbors of point A, then the following criteria can be applied:

$$\begin{cases} A \text{ is a background point, } if\ n = N \\ A \text{ requires further investigation, } if\ n < N \end{cases} \quad (2)$$

If n = N, this means that point A appears in each frame in the investigated frames, indicating A is a background point. If n < N, there are two possible reasons. The first possible reason is that point A is a background point if it is blocked by the moving object in some frames. The second possible reason is that point A is a non-background point. When a moving object shows up, a vector-like blocked area is created, as shown in Figure 4.

Both Figure 4a,b have an occlusion area named the "system occlusion area". This area was produced by the background points (such as wire pole) blocking the LiDAR. This area is invisible. As for Figure 4b, there is an occlusion area created by the moving vehicle. This occlusion area does not exist in Figure 4a. It can be clearly shown that for the occluded area, the slope between the two adjacent frames should be less than the slope created by the moving object (the same trend between α and β in Figure 3).

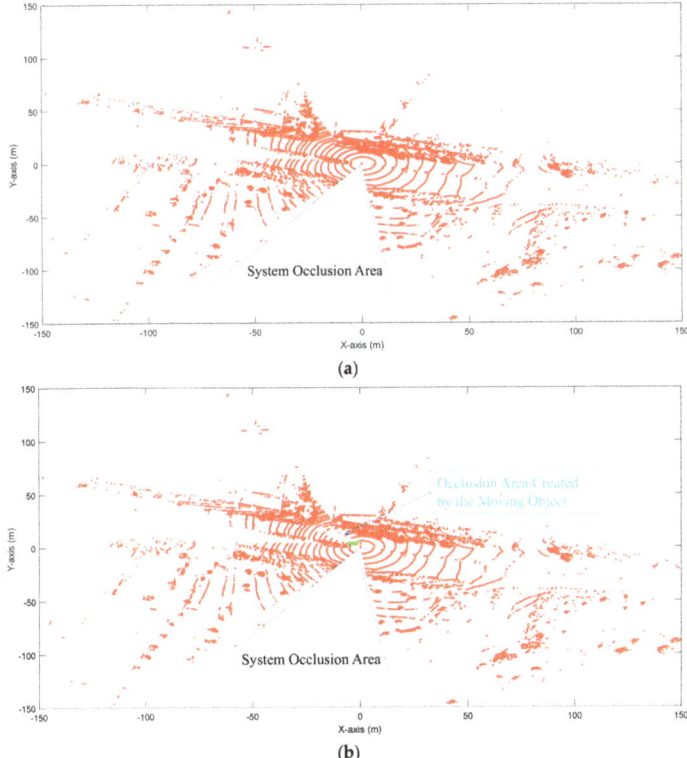

Figure 4. Occlusion issue: (**a**) Non-occlusion area created by moving objects, (**b**) Occlusion area created by moving objects.

If n < N, it means that point A did not show up in some frames. Assuming point A did not show up in frame i, then all the points that did not show up in frame i were extracted. The slope between the two adjacent frames can then be calculated. If the average slope was shorter than 45 degrees, those points were identified as background. Otherwise, they were identified as non-background points. Figure 5 shows the results of background filtering using GS-DSF and 3D-DSF under windy weather.

It is shown that the performance of GS-DSF is better than 3D-DSF under windy weather in both free-flow and congested situations. The 3D-DSF left a lot of ground points after background filtering. When the traffic was congested, the 3D-DSF misidentified the truck which had stopped on the road as a background point. As for GS-DSF, it could exclude the background points and correctly identify the vehicle which had temporarily stopped on the road as a non-background point. To quantitatively evaluate the performance of GS-DSF, 20 frames were randomly selected under windy weather in free-flow situations and another 20 frames were randomly selected under windy weather in congested situations. Table 1 shows an example of the performance of GS-DSF and 3D-DSF (one frame in a free-flow situation and one frame in a congested situation).

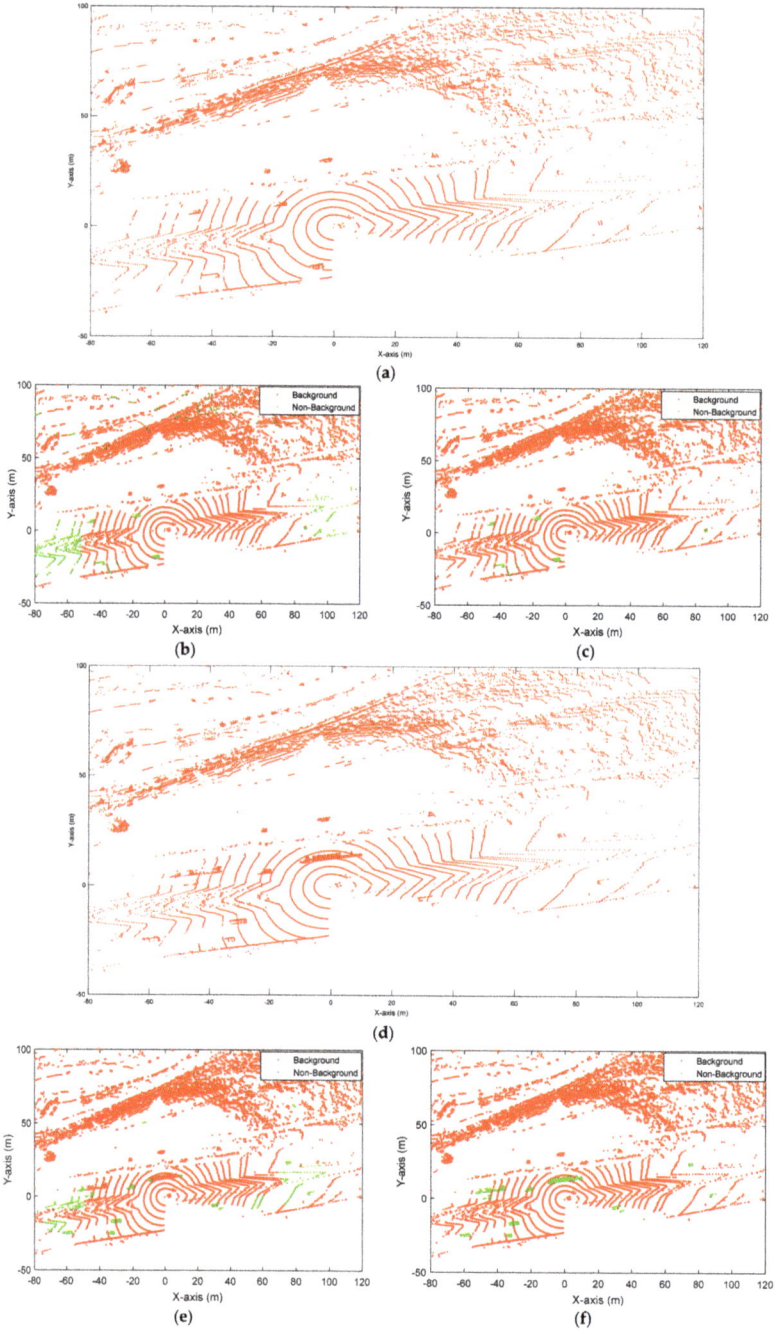

Figure 5. Performance of 3D-DSF and ground surface-enhanced density statistic filtering method (GS-DSF) under windy weather: (**a**) Free-flow: Raw light detection and ranging (LiDAR) data before background filtering, (**b**) Free-flow: 3D-DSF, (**c**) Free-flow: GS-DSF, (**d**) Congested traffic: Raw LiDAR data before background filtering, (**e**) Congested traffic: 3D-DSF, (**f**) Congested traffic: GS-DSF.

Table 1. Quantitative Evaluation of ground surface-enhanced density statistic filtering (GS-DSF) and 3D density statistic filtering (3D-DSF).

	Background Points (BP)	Vehicle Points (VP)	Background Points after Filtering (BPF)		Vehicles Points after Filtering (VPF)		Type 1 Error	Type 2 Error
Free-Flow	598,512	9873	GS-DSF	59	GS-DSF	9789	0.0098%	0.8508%
			3D-DSF	3615	3D-DSF	9802	0.6040%	0.7191%
Congested Situation	599,982	20,172	GS-DSF	71	GS-DSF	20,150	0.0118%	0.1091%
			3D-DSF	3429	3D-DSF	2578	0.5715%	87.2199%

The Type 1 error in Table 1 indicates the acceptance of background points as non-background points and the Type 2 error indicates the acceptance of non-background points as background points. These two types of errors can be represented as:

$$\begin{cases} Type\ 1\ error = \frac{BPF}{BP} \times 100\% \\ Type\ 2\ error = \frac{VP-VPF}{VP} \times 100\% \end{cases} \quad (3)$$

It is clearly shown that both Type 1 and Type 2 errors remain low for GS-DSF under free-flow and congested situations. The two types of errors for 3D-DSF are much higher compared to GS-DSF. The Type 2 error even reached 87.2% under congested situations for 3D-DSF, indicating that a large proportion of vehicle points were misidentified as background points and were excluded from the database. The average Type 1 error and Type 2 error of GS-DSF are 0.013% and 0.642% for free-flow situations and congested situations, respectively. The average Type 1 error and Type 2 error of 3D-DSF are 0.633% and 50.614% for free-flow situations and congested situations, respectively.

Figure 6 shows an example of GS-DSF background filtering under rainy and snowy weather.

It is shown that water drops (not under heavy rain) are invisible in the LiDAR sensors. The LiDAR points behind the water drops were blocked, leading to discontinuous ground circles and an incomplete vehicle shape, as the vehicle shape overlapped with the ground circles, as shown in Figure 6a. Under rainy weather, GS-DSF can successfully distinguish background points and non-background points, and the extracted vehicle shape is shown in Figure 6b in green. When the weather is snowy, a lot of snowflakes showed up in the LiDAR data (small dots in Figure 6c). Due to the free fall of the snowflakes, the positions of the snowflakes change in different frames. As a result, GS-DSF could not exclude the snowflakes effectively during the background filtering step (sparse dots in the center), as shown in Figure 6d. Therefore, snowflake exclusion needs to be performed in the following steps.

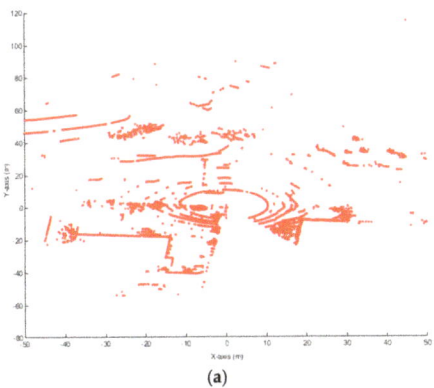

(a)

Figure 6. *Cont.*

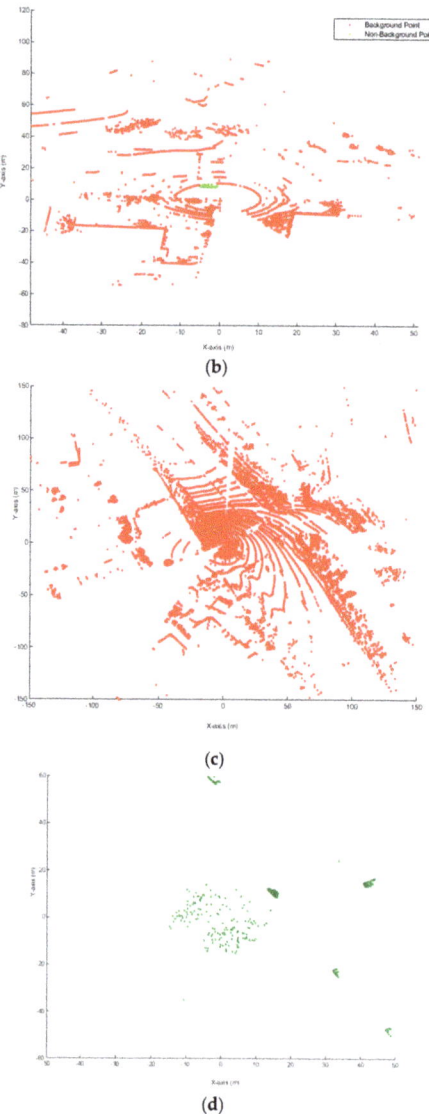

Figure 6. GS-DSF under rainy and snowy weather: (**a**) Rainy weather: Before GS-DSF, (**b**) Rainy weather: After GS-DSF, (**c**) Snowy weather: Before GS-DSF, (**d**) Snowy weather: After GS-DSF.

3. Point Clustering

The purpose of point clustering is to cluster the points belonging to one object into the same group. As for the roadside LiDAR data, several researchers have applied the DBSCAN-related algorithms for point clustering [32,33]. Since DBSCAN purely uses the distribution of point density as the threshold for clustering, when there are snowflakes in the space and if the snowflakes are around the object, it is possible that the snowflakes can be degree-clustered as the points object. If the mis-clustered snowflake is the point close to the roadside LiDAR (corner point), then the calculation of the speed and location of the object is inaccurate [38]. The other widely used k-means method requires an initial estimate of the number of clusters in the dataset [39]. Other researchers have used height information to cluster the LiDAR points in a space [40], but the random locations of the snowflakes can lead to false clustering results using the height-based method. Another limitation of the existing method is the heavy computational load, caused by the traversal search. Therefore, these existing methods could not meet the point clustering task under windy weather. This paper develops a fast and efficient method for point clustering. Instead of searching the point directly, this paper uses a voxelization-based method to process the data. The core of the voxelization-based method is to convert the LiDAR point into a volumetric space. The whole space is firstly divided into small cubes. Each cube can be identified as "an occupied cube" or "a non-occupied cube". The key challenge here is how to find a reasonable side length for the cube and how to find a threshold to distinguish the occupied cube and the non-occupied cube.

The point distribution feature of the snowflakes was firstly analyzed. About 10 h of LiDAR data under heavy snow weather were collected. A total of 200 frames were randomly selected for investigation. The maximum distance of the snowflakes among the 200 frames is shown in Figure 7.

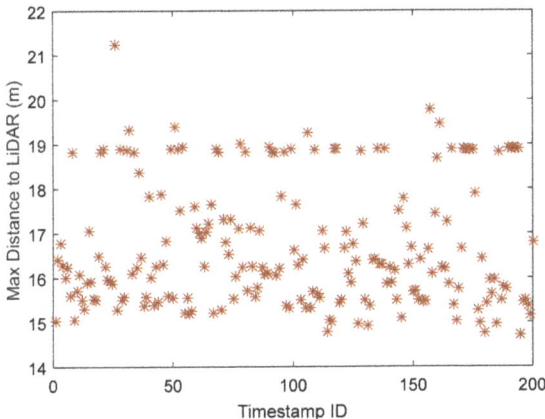

Figure 7. Maximum distance distribution of the snowflakes to the LiDAR.

It is shown that the maximum distance of the snowflake is less than 22 m in all frames. When the distance is longer than 22 m, the reflection of the snowflake is too weak to be detected by the LiDAR. This feature indicates that the influence range of the snowflakes on the data is limited to 22 m from the LiDAR. The reason for this phenomenon is that the snowflakes can scatter the laser and reduce the intensity of the reflection.

The LiDAR measures the reflectivity of an object with 256-bit resolution, independent of laser power and distance over a range from 1 m to 100 m. Commercially available reflectivity standards and retro-reflectors are used for the absolute calibration of the reflectivity.

- Diffuse reflectors report values from 0–100 for the range of reflectivity from 0% to 100%.
- Retro-reflectors report values from 101 to 255 with 255 being the reported reflectivity for an ideal retro-reflector and 101–254 being the reported reflectivity for partially obstructed or imperfect retro-reflectors.

The distribution of intensity of the snowflakes and the vehicles is shown in Figure 8.

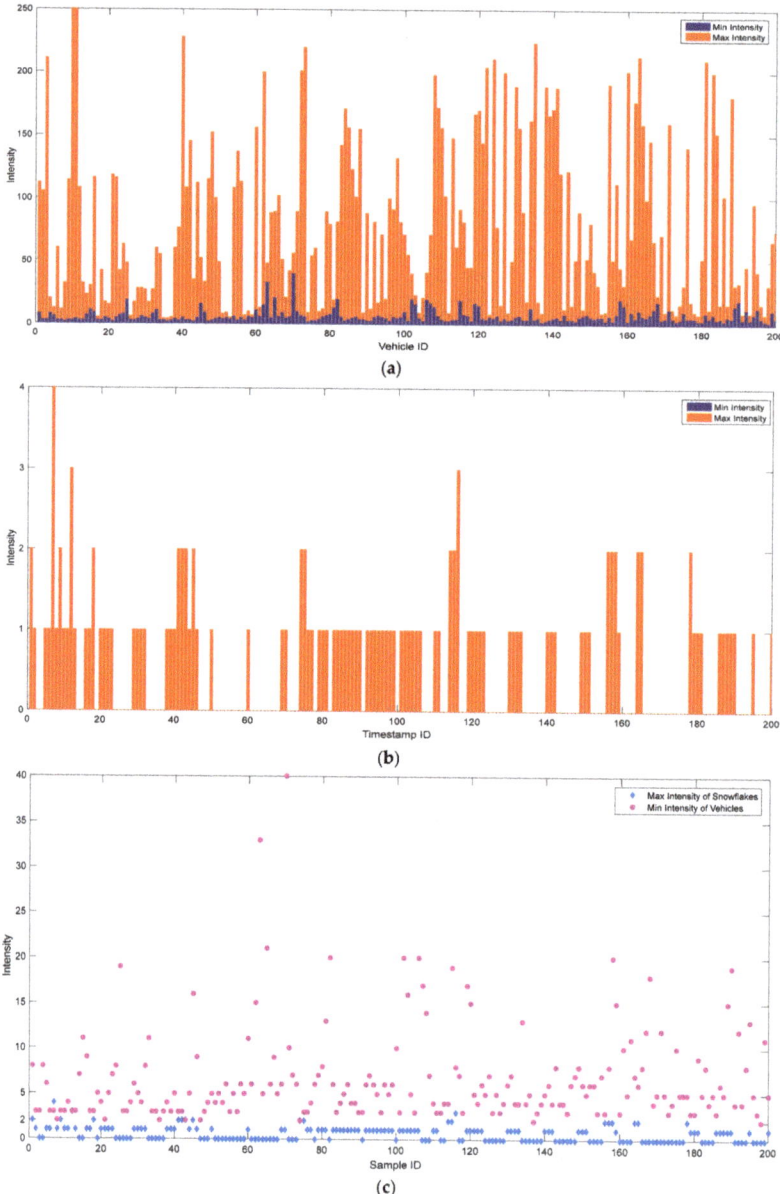

Figure 8. Intensity features of vehicles and snowflakes: (**a**) Intensity of vehicles, (**b**) Intensity of snowflakes, (**c**) Comparison of maximum intensity of snowflakes and minimum intensity of vehicles.

It can be seen that the maximum intensity of the vehicles varied in a larger range compared to that of the snowflakes. The absolute value of the maximum intensity of the vehicles is also larger than that of the snowflakes. Therefore, to better distinguish vehicles and snowflakes, we used the minimum intensity of vehicles. Then, the comparison of the maximum intensity of the snowflakes and the minimum intensity of the vehicles showed that the maximum intensity of most snowflakes was less than the minimum intensity of the vehicles, which suggested that the two indexes could help distinguish vehicles and snowflakes. By analyzing 100 randomly selected frames, it was also found that 98.5% of snowflakes had a maximum intensity of less than two and 96% of vehicles had a minimum intensity larger than two. The minimum intensity of the snowflakes was zero, indicating that the LiDAR did not receive the signal that it sent out. As for the snowflakes, the minimum intensity was zero and the maximum intensity was two (for 98.5%), but for the vehicles, the minimum intensity was usually more than two. Therefore, the value of two was selected as a threshold to distinguish the snowflakes and vehicles. The points with a minimum intensity higher than two were considered as non-snowflakes and the points with a maximum intensity less than two were considered as snowflakes and were removed from the space. For the points with an intensity equal to two, they were left in the space and clustered based on the revised DBSCAN algorithm proposed by Zhao et al. [19]. Figure 9 shows the point clustering with the proposed method and the revised DBSCAN algorithm developed in [19]. A cluster refers to points that can be categorized into one group. It can be seen that there were no obvious differences in Clusters 1–3 using the two methods. The influence of the snowflakes only occurred within 20 m of the LiDAR [20]. Therefore, only Cluster 4 was different under the two methods. For Cluster 4, the revised DBSCAN algorithm mis-clustered a lot of snowflakes around the vehicle as vehicle points while the proposed algorithm successfully excludes snowflakes and keeps the vehicle points in the space.

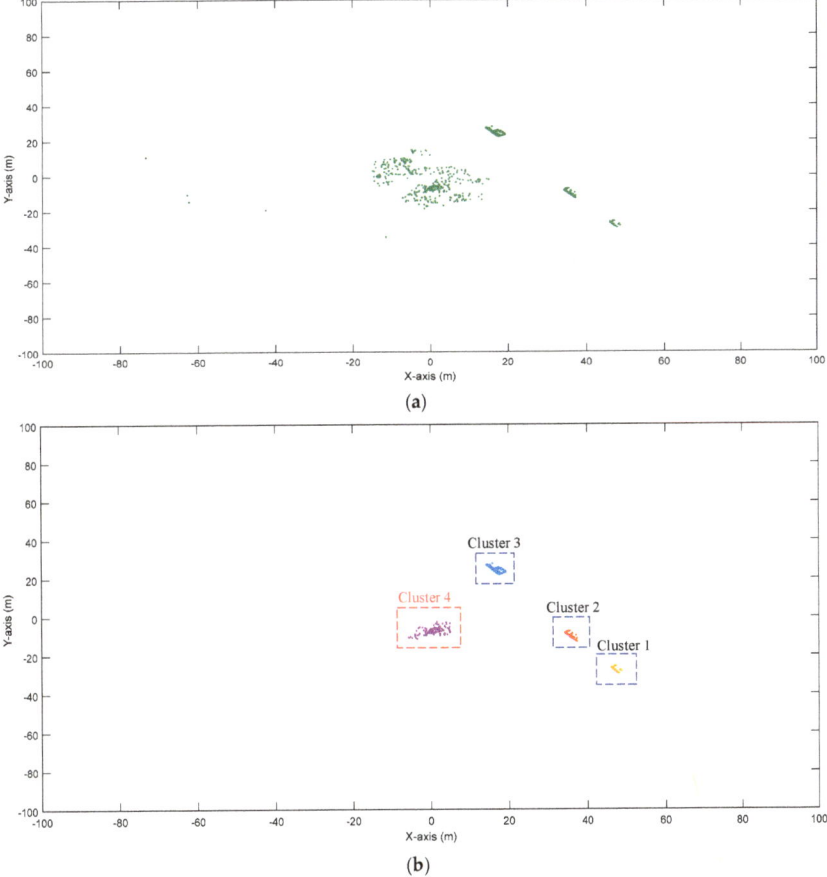

Figure 9. *Cont.*

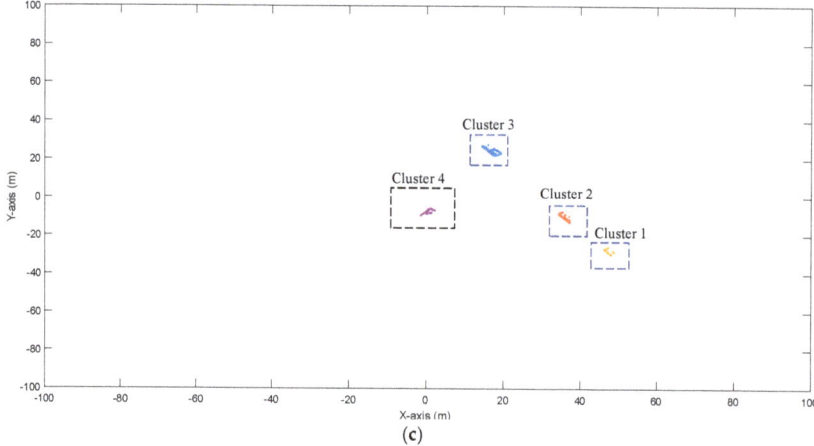

(c)

Figure 9. Point clustering: (**a**) Before point clustering, (**b**) Revised density-based spatial clustering of applications with noise (DBSCAN), (**c**) Proposed method.

To further evaluate the performance of the proposed method, the proposed method and the methods developed in [5] and [19] were used to process the same LiDAR databases collected in windy weather and snowy weather (125 and 651 data for each scenario, respectively). Table 2 summarizes the results of the three methods. Though there were still some errors in counting the vehicle volume under both snowy and windy weather using the proposed method, the accuracy was greatly improved compared to the methods in [5] and [19]. An overall accuracy of more than 90% can be achieved with the proposed method. The evaluation shows that the performance of the proposed method is superior compared to the state-of-the-art methods.

Table 2. Performance evaluation.

	Snowy Weather			Windy Weather		
Actual Number of Vehicles	125			651		
Methods	Proposed Method	Method in [5]	Method in [19]	Proposed Method	Method in [5]	Method in [19]
Detected number of vehicles	135	190	145	689	781	725
Error (%)	8.0	52.0	16.0	5.8	19.9	11.4

4. Conclusions and Discussion

This paper evaluates the performance of the state-of-the-art methods of background filtering and point clustering for roadside LiDAR data under windy and snowy weather. The results showed that the existing background filtering and point clustering methods could not process the roadside LiDAR data effectively. This paper develops a ground surface-enhanced point density statistics filtering method to exclude the background points under windy weather. The intensity information was used to improve the accuracy of the revised DBSCAN algorithm developed by Zhao et al. [19]. The testing results showed that the proposed methods can exclude the background points and cluster the vehicle points into one group effectively under windy and snowy weather.

There are already some algorithms developed for autonomous vehicles, such as those in [23]. However, those algorithms serving for autonomous vehicles could not be directly applied to the connected vehicles since the working environment and region of interest are different. There are still some limitations that can be improved in the future. Foggy weather can also significantly decrease the quality of the LiDAR data. However, LiDAR data under foggy weather was not available for this research. Future studies should evaluate the performance of the proposed methods using the LiDAR data under foggy and smoggy weather. This paper manually selects two as the intensity value to identify the snowflakes, but a more advanced method to automatically select the threshold is still needed.

Author Contributions: Conceptualization, J.W. and H.X.; methodology, J.W.; validation, J.W., Y.T. and R.P.; formal analysis, J.W. and Y.T.; investigation, H.X.; resources, R.Y.; data curation, R.P.; writing—original draft preparation, J.W. and R.Y.; writing—review and editing, R.Y.; supervision, H.X.; funding acquisition, J.W. All authors have read and agreed to the published version of the manuscript.

Funding: This research was funded by Qilu Young Scholar Program of Shandong University.

Acknowledgments: The authors thank Zong Tian and Hongchao Liu for their technical support in this research.

Conflicts of Interest: The authors declare no conflict of interest.

References

1. Chang, J.C.; Findley, D.J.; Cunningham, C.M.; Tsai, M.K. Considerations for Effective Lidar Deployment by Transportation Agencies. *Transp. Res. Record* **2014**, 1–8. [CrossRef]
2. Thornton, D.A.; Rechnill, K.; Coffman, B. Automated parking surveys from a LIDAR equipped vehicle. *Transp. Res. Part C-Emerg. Technol.* **2014**, *39*, 23–35. [CrossRef]
3. Williams, K.; Olsen, M.J.; Roe, G.V.; Glennie, C. Synthesis of Transportation Applications of Mobile LIDAR. *Remote Sens.* **2013**, *5*, 4652–4692. [CrossRef]
4. Lv, B.; Xu, H.; Wu, J.Q.; Tian, Y.; Tian, S.; Feng, S.Y. Revolution and rotation-based method for roadside LiDAR data integration. *Opt. Laser Technol.* **2019**, *119*. [CrossRef]
5. Wu, J.Q. An Automatic Procedure for Vehicle Tracking with a Roadside LiDAR Sensor. *ITE J.-Inst. Transp. Eng.* **2018**, *88*, 32–37.
6. Chen, J.R.; Xu, H.; Wu, J.Q.; Yue, R.; Yuan, C.W.; Wang, L. Deer Crossing Road Detection With Roadside LiDAR Sensor. *IEEE Access* **2019**, *7*, 65944–65954. [CrossRef]
7. Lv, B.; Xu, H.; Wu, J.Q.; Tian, Y.; Zhang, Y.S.; Zheng, Y.C.; Yuan, C.W.; Tian, S. LiDAR-Enhanced Connected Infrastructures Sensing and Broadcasting High-Resolution Traffic Information Serving Smart Cities. *IEEE Access* **2019**, *7*, 79895–79907. [CrossRef]
8. Wu, J.Q.; Xu, H.; Zheng, Y.C.; Tian, Z. A novel method of vehicle-pedestrian near-crash identification with roadside LiDAR data. *Accid. Anal. Prev.* **2018**, *121*, 238–249. [CrossRef]
9. Yue, R.; Xu, H.; Wu, J.Q.; Sun, R.J.; Yuan, C.W. Data Registration with Ground Points for Roadside LiDAR Sensors. *Remote Sens.* **2019**, *11*, 1354. [CrossRef]
10. Zhao, J.X.; Xu, H.; Wu, J.Q.; Zheng, Y.C.; Liu, H.C. Trajectory tracking and prediction of pedestrian's crossing intention using roadside LiDAR. *IET Intell. Transp. Syst.* **2019**, *13*, 789–795. [CrossRef]
11. Sun, Y.; Xu, H.; Wu, J.Q.; Zheng, J.Y.; Dietrich, K.M. 3-D Data Processing to Extract Vehicle Trajectories from Roadside LiDAR Data. *Transp. Res. Record* **2018**, *2672*, 14–22. [CrossRef]
12. Tarko, A. Application of the Lomax distribution to estimate the conditional probability of crash. In Proceedings of the 18th International Conference Road Safety on Five Continents (RS5C 2018), Jeju Island, Korea, 16–18 May 2018.
13. Lee, H.; Coifman, B. Side-Fire Lidar-Based Vehicle Classification. *Transp. Res. Record* **2012**, 173–183. [CrossRef]
14. Zhang, Z.Y.; Zheng, J.Y.; Wang, X.; Fan, X.L. Background Filtering and Vehicle Detection with Roadside Lidar Based on Point Association. In Proceedings of the 37th Chinese Control Conference (CCC), Wuhan, China, 25–27 July 2018; pp. 7938–7943.
15. Wu, J.Y.; Xu, H.; Zheng, J.Y.; IEEE. Automatic Background Filtering and Lane Identification with Roadside LiDAR Data. In Proceedings of the 20th IEEE International Conference on Intelligent Transportation Systems (ITSC), Yokohama, Japan, 16–19 October 2017.
16. Wu, J.Q.; Xu, H.; Sun, Y.; Zheng, J.Y.; Yue, R. Automatic Background Filtering Method for Roadside LiDAR Data. *Transp. Res. Record* **2018**, *2672*, 106–114. [CrossRef]

17. Lv, B.; Xu, H.; Wu, J.Q.; Tian, Y.; Yuan, C.W. Raster-Based Background Filtering for Roadside LiDAR Data. *IEEE Access* **2019**, *7*, 76779–76788. [CrossRef]
18. Zhang, J.; Xiao, W.; Coifman, B.; Mills, J. Image-based Vehicle Tracking From Roadside Lidar Data. In Proceedings of the ISPRS Geospatial Week, Enschede, The Netherlands, 10–14 June 2019.
19. Zhao, J.X.; Xu, H.; Liu, H.C.; Wu, J.Q.; Zheng, Y.C.; Wu, D.Y. Detection and tracking of pedestrians and vehicles using roadside LiDAR sensors. *Transp. Res. Part C-Emerg. Technol.* **2019**, *100*, 68–87. [CrossRef]
20. Wu, J.; Xu, H.; Zheng, J.; Zhao, J. Automatic vehicle detection with roadside LiDAR data under rainy and snowy conditions. *IEEE Intell. Transp. Syst. Mag.* **2020**. [CrossRef]
21. Zheng, J.Y.; Xu, B.; Wang, X.; Fan, X.L.; Xu, H.; Sun, G. A portable roadside vehicle detection system based on multi-sensing fusion. *Int. J. Sens. Netw.* **2019**, *29*, 38–47. [CrossRef]
22. Wojtanowski, J.; Zygmunt, M.; Kaszczuk, M.; Mierczyk, Z.; Muzal, M. Comparison of 905 nm and 1550 nm semiconductor laser rangefinders' performance deterioration due to adverse environmental conditions. *Opto-Electron. Rev.* **2014**, *22*, 183–190. [CrossRef]
23. Charron, N.; Phillips, S.; Waslander, S.L. De-noising of Lidar point clouds corrupted by snowfall. In Proceedings of the 2018 15th Conference on Computer and Robot Vision (CRV), Toronto, Canada, 9–11 May 2018; pp. 254–261.
24. Jokela, M.; Kutila, M.; Pyykonen, P. Testing and Validation of Automotive Point-Cloud Sensors in Adverse Weather Conditions. *Appl. Sci. -Basel* **2019**, *9*, 2341. [CrossRef]
25. Kutila, M.; Pyykonen, P.; Holzhuter, H.; Colomb, M.; Duthon, P.; IEEE. Automotive LiDAR performance verification in fog and rain. In Proceedings of the 21st IEEE International Conference on Intelligent Transportation Systems (ITSC), Maui, HI, USA, 4–7 November 2006; pp. 1695–1701.
26. Bijelic, M.; Mannan, F.; Gruber, T.; Ritter, W.; Dietmayer, K.; Heide, F. Seeing through fog without seeing fog: Deep sensor fusion in the absence of labeled training data. *arXiv* **2019**, arXiv:1902.08913.
27. Bijelic, M.; Gruber, T.; Ritter, W. A benchmark for lidar sensors in fog: Is detection breaking down? In Proceedings of the 2018 IEEE Intelligent Vehicles Symposium (IV), Changshu, China, 26–30 June 2018; IEEE: Piscataway, NJ, USA; pp. 760–767.
28. Heinzler, R.; Schindler, P.; Seekircher, J.; Ritter, W.; Stork, W. Weather Influence and Classification with Automotive Lidar Sensors. In Proceedings of the 2019 IEEE Intelligent Vehicles Symposium (IV 2019), Paris, France, 9–12 June 2019; pp. 1527–1534.
29. Phillips, T.G.; Guenther, N.; McAree, P.R. When the Dust Settles: The Four Behaviors of LiDAR in the Presence of Fine Airborne Particulates. *J. Field Robot.* **2017**, *34*, 985–1009. [CrossRef]
30. Nezafat, R.V.; Sahin, O.; Cetin, M. Transfer Learning Using Deep Neural Networks for Classification of Truck Body Types Based on Side-Fire Lidar Data. *J. Big Data Anal. Transp.* **2019**, *1*, 71–82. [CrossRef]
31. Wu, J.Q.; Xu, H.; Zhao, J.X. Automatic Lane Identification Using the Roadside LiDAR Sensors. *IEEE Intell. Transp. Syst. Mag.* **2020**, *12*, 25–34. [CrossRef]
32. Zhang, Z.Y.; Zheng, J.Y.; Xu, H.; Wang, X. Vehicle Detection and Tracking in Complex Traffic Circumstances with Roadside LiDAR. *Transp. Res. Record* **2019**, *2673*, 62–71. [CrossRef]
33. Wu, J.Q.; Xu, H.; Zheng, Y.C.; Zhang, Y.S.; Lv, B.; Tian, Z. Automatic Vehicle Classification using Roadside LiDAR Data. *Transp. Res. Record* **2019**, *2673*, 153–164. [CrossRef]
34. Wu, J.; Xu, H.; Yue, R.; Tian, Z.; Tian, Y.; Tian, Y. An automatic skateboarder detection method with roadside LiDAR data. *J. Transp. Saf. Secur.* **2019**, 1–20. [CrossRef]
35. Wu, J.Q.; Xu, H.; Lv, B.; Yue, R.; Li, Y. Automatic Ground Points Identification Method for Roadside LiDAR Data. *Transp. Res. Record* **2019**, *2673*, 140–152. [CrossRef]
36. Choi, Y.W.; Jang, Y.W.; Lee, H.J.; Cho, G.S. Three-Dimensional LiDAR Data Classifying to Extract Road Point in Urban Area. *IEEE Geosci. Remote Sens. Lett.* **2008**, *5*, 725–729. [CrossRef]
37. Wu, J.Q.; Tian, Y.; Xu, H.; Yue, R.; Wang, A.B.; Song, X.G. Automatic ground points filtering of roadside LiDAR data using a channel-based filtering algorithm. *Opt. Laser Technol.* **2019**, *115*, 374–383. [CrossRef]
38. Shan, J.; Sampath, A. Building extraction from LiDAR point clouds based on clustering techniques. In *Topographic Laser Ranging and Scanning: Principles and Processing*; Toth, C.K., Shan, J., Eds.; CRC Press: Boca Raton, FL, USA, 2008; pp. 423–446.

39. Lee, H.; Slatton, K.C.; Roth, B.E.; Cropper, W.P. Adaptive clustering of airborne LiDAR data to segment individual tree crowns in managed pine forests. *Int. J. Remote Sens.* **2010**, *31*, 117–139. [CrossRef]
40. Yu, Y.T.; Li, J.; Guan, H.Y.; Wang, C.; Yu, J. Semiautomated Extraction of Street Light Poles from Mobile LiDAR Point-Clouds. *IEEE Trans. Geosci. Remote Sens.* **2015**, *53*, 1374–1386. [CrossRef]

 © 2020 by the authors. Licensee MDPI, Basel, Switzerland. This article is an open access article distributed under the terms and conditions of the Creative Commons Attribution (CC BY) license (http://creativecommons.org/licenses/by/4.0/).

Article

Improving Road Traffic Forecasting Using Air Pollution and Atmospheric Data: Experiments Based on LSTM Recurrent Neural Networks

Faraz Malik Awan *, Roberto Minerva and and Noel Crespi

Telecom SudParis, Institut Polytechnique de Paris, CNRS UMR5697, 91000 Evry, France; roberto.minerva@telecom-sudparis.eu (R.M.); noel.crespi@mines-telecom.fr (N.C.)
* Correspondence: faraz_malik.awan@telecom-sudparis.eu; Tel.: +33-7533-8866-7

Received: 27 May 2020; Accepted: 1 July 2020; Published: 4 July 2020

Abstract: Traffic flow forecasting is one of the most important use cases related to smart cities. In addition to assisting traffic management authorities, traffic forecasting can help drivers to choose the best path to their destinations. Accurate traffic forecasting is a basic requirement for traffic management. We propose a traffic forecasting approach that utilizes air pollution and atmospheric parameters. Air pollution levels are often associated with traffic intensity, and much work is already available in which air pollution has been predicted using road traffic. However, to the best of our knowledge, an attempt to improve forecasting road traffic using air pollution and atmospheric parameters is not yet available in the literature. In our preliminary experiments, we found out the relation between traffic intensity, air pollution, and atmospheric parameters. Therefore, we believe that addition of air pollutants and atmospheric parameters can improve the traffic forecasting. Our method uses air pollution gases, including CO, NO, NO_2, NO_x, and O_3. We chose these gases because they are associated with road traffic. Some atmospheric parameters, including pressure, temperature, wind direction, and wind speed have also been considered, as these parameters can play an important role in the dispersion of the above-mentioned gases. Data related to traffic flow, air pollution, and the atmosphere were collected from the open data portal of Madrid, Spain. The long short-term memory (LSTM) recurrent neural network (RNN) was used in this paper to perform traffic forecasting.

Keywords: air pollution; atmospheric data; deep learning; IoT; LSTM; machine learning; RNN; Sensors; smart cities; traffic flow; traffic forecasting

---

1. Introduction

1.1. Motivation

Vehicular traffic management is a major issue in cities and metropolitanareas [1]. Traffic has a relevant impact on different aspects of daily life, from time spent in traffic jams to higher level of pollution produced, from gas and resources consumption to infrastructural investments and maintenance of road and transportation systems [2]. Traffic management and optimization are essential parts in every smart city platform. Smart mobility is one of the most important services of smart city platform. It has a direct impact on the quality of life of citizens and on the ability of the city to support the exchange of people and goods within the urban environment. Traffic regulation and orchestration are key components. With a city's large number of vehicles, problems related to traffic are critical for the effective functioning of the city and the health of its citizens. Traffic congestion is a major problem, especially when it is associated with an increasing number of vehicles in use (e.g., in cities with inadequate public transportation). It leads to environmental, social, and economic issues [3].

The timely prediction of traffic flow can be helpful to avoid congestion, as drivers can choose the most comfortable and less congested path to reach their destination, or modify their time schedule for their journey in order to compensate for the expected time of arrival caused by the traffic. Road traffic forecasting is defined as the estimation or prediction of the traffic flow in the (near) future. Another aspect of traffic levels in cities is car and truck generated air pollution. Many cities suffer from air pollution. Increasing traffic emissions is one of the major contributors to urban air pollution [4]. According to the World Health Organization (WHO) [5], a large portion of air pollution is contributed by the transport sector. These two phenomena are linked, and many cities are tackling this problem by deploying sensors for measuring traffic intensity and air quality. Air pollution generated by traffic depends on several factors, ranging from the types of vehicles (gasoline, diesel, electric), to the level of congestion and the time spent in traffic jams, the atmospheric or geographical characteristics of the environment, and many more.

A large networks of sensors have already been deployed in several cities (e.g., Madrid, Santander, and Barcelona in Spain, Singapore, Seoul, Copenhagen). Data generated by these sensors are very useful for forecasting. For example, around 4000 traffic intensity sensors are deployed in Madrid, Spain (Figure 1) [6]. These sensors provide information about the number of vehicles passing per hour (actually every 15 minutes). Similarly, there are 24 stations measuring air pollution (Figure 2) and 26 stations collecting atmospheric data such as local temperature, pressure, wind speed, and wind direction (Figure 3). Madrid's data, then, offer the possibility to further analyze the correlations between traffic intensity, levels of pollution, and meteorological condition. Figures 1–3 show that traffic intensity sensors are greater in number as compared to air pollution sensors. Air pollution sensor data are not so granular as the traffic intensity ones. Therefore, in our experiments, we chose traffic sensors in close proximity (upto 500 m) (Figure 4c) to air pollution sensors and, vice versa, we selected air pollution sensor stations close to big roads or crossroads. Air pollutants such a CO, NO, NO_2, NO_x, and O_3 are associated with road traffic [7–9]. The combination of large quantities of curated data with machine/deep learning models can provide useful insights for the correlation of traffic with air pollution. Many studies demonstrate how data about traffic flow can be used to predict air pollution. For example, Batterman et al. [10] used a dispersion model, called the Research Line Source (R-LINE) model, and emission inventory to predict the air pollutants $PM2.5$ and NO_x. Ly et al. [11] predicted the concentration of NO_2 and CO by using multisensor devices data and weather data, including temperature, relative humidity, and absolute humidity. In this work, they used the data of an Italian city (unnamed city) between March 2004 and February 2005. Similarly, Lana et al. [12] used a Random Forest regression model to predict the air pollution level with respect to road traffic utilizing open data from Madrid for the year 2015. Russo et al. [13] used atmospheric data, including temperature, wind direction, wind intensity, along with other air pollutants, including NO_2, NO, and CO as input variables to neural network to forecast the concentration of $PM10$. However, in their experiments, they did not take traffic intensity into account. Brunello et al. [14] investigated temporal information management to assess the relationships between air pollutants, including NO_2, NO_x, and $PM2.5$, and road traffic. In all of these studies, thanks to the direct link between road traffic and air pollutants, road traffic was used to predict air pollution. Air pollution and traffic intensity data are collected as time-series of values and are generally made available for analysis and study. However, to the best of our knowledge, there has not yet been an attempt to use air pollution to improve the traffic forecasting. Traffic intensity is a major contributor to air pollution. The presence of certain pollutants in the air is most likely determined (or largely contributed) by vehicle traffic. Being able to correlate the actual level of these pollutants, on a timely basis for an area close to an air pollution station, to the expected level of traffic in the same area can be of help in better predicting the traffic intensity. Hypothetically, if the only source of pollution was car traffic, a strong correlation between the air pollution level and the intensity of traffic could be drawn. Cities and urban conglomerates are complex systems and there are other major contributors to air pollutions (home heating, factories and transformation implants, and others). Besides this, also meteorological condition can influence the air quality, e.g., strong winds can spread and disseminate pollutants in large areas making it more difficult to find strong correlations between traffic, air pollution and other contributors. In spite of the complexity of these causal relations, Madrid offers an impressive wealth of data for approaching and further study the correlation between traffic intensity and air pollution. The analysis considers the current level of pollution in a specific area at a specific time interval "t" as an evidence of presence of traffic. This evidence is also reinforced by the ability to know the traffic intensity levels before the time "t". Using these data could lead to a better prediction of the traffic intensity. Generally speaking, the approach of considering air pollution data as a means to predict traffic intensity can be undertaken in two ways: to use air pollution data together with traffic intensity data to improve the prediction of traffic intensity, or to use the air pollution data and numerical models to infer the expected traffic intensity. This paper evaluates the first option, while the second one is left for further study.

Cities are systems that attract peoples, goods and activities and their impact is not limited to the city limits, but extend to cities, towns, and villages in the surrounding area. According to a World Economic Forum report [15], people prefer living, staying, studying, and growing up in cities. In fact, big cities exert a strong attraction effect and have a considerable impact on very large areas. The traffic and pollution issues involved may therefore be better analyzed if the extended areas are considered. Sometimes, air quality measurements are also assessed in decentralized areas. Thanks to the availability of several open datasets, it is possible to investigate the correlation between air pollution and traffic intensity that may have contributed to the level of pollution in large monitored areas. This information will in turn offer the possibility to focus on air quality analysis and to correlate it to the expected traffic intensity. This paper investigates this possibility, starting from a highly-sensed and populated area (Madrid and its surrounding area). In Madrid's data portal, datasets related to air pollution and atmospheric data are available timely each hour. On the other hand, data for traffic flow is updated every 15 minutes. Historic data of traffic flow, air pollution, and atmospheric variables for each month is made available at the end of the month. One expected outcome of this work is to validate (or reject) the usage of current air pollution measurements and levels combined with atmospheric data to improve the prediction of the traffic intensity levels.

Traffic intensity is the major cause of the pollution problem. So not surprising, measuring or using the resultant levels of pollution generated can be a means to understand how many vehicles may be present. Pant et al. [16] performed an analysis to characterize the traffic-related PM emissions in a tunnel environment. For this purpose, they chose 545 meters long, one of the major tunnels in Birmingham, called A38 Queensway Tunnel. Around 25,000 vehicles travel through this tunnel daily. They deployed the PM sensors at the distance of 1.5 m on emergency layby. A similar experiment can be done with different number of vehicles to observe the volume of the pollution produced. A set of vehicles operating for a specific period of time in the same area will produce a very similar quantity of pollutants (imagine 100 cars in a closed environment, they will produce the same amount of pollutants when operating for the same period of time). Measuring the levels of pollutants over time may create a dataset usable to predict level of pollution as well as from the pollution levels to determine how many cars were contributing. Hypothetically, measuring the level of pollution at a certain instant may allow to determine how many cars were operating. In the real-world, things are more dynamic, for instance:

- the concentration of pollutants is greater close to big roads [17] (this is also why we tried to consider traffic intensity sensors close to the pollution sensors).
- the set of vehicles may be dynamic in composition (more diesel, more electric, and so on) during the days.
- the pollution level generated can be impacted by the meteorological condition.

However, the traffic in a city shows patterns and in spite of the dynamic of the composition/aggregation of vehicles producing pollutants, there are patterns also in how people use the cars (e.g., similar number of commuters in peak hours of traffic). These patterns are also well-known by users, they, in fact, expect to have different traffic condition during the day and the week (with large differences between working days and week-ends). Over a long period of time, these patterns repeat and the levels of pollution can be considered as signatures of traffic intensity. The hypothesis to verify is if the levels of pollution may correspond on the average to certain levels of traffic and if these measurements of pollution can be used to improve the traffic predictions. Having time-series of the pollution signatures together with time-series of traffic intensity will allow to better predict the traffic intensity.

The objective is also to determine if such an approach is practical and if it can give useful and improved results over an analysis that considers only the traffic intensity time-series. Determining the relations between levels of pollution and traffic intensity may lead to important consequences such as: to better control the air quality in more parts of the city and still maintain the desired levels of monitoring of vehicular traffic situation; the reduction of the number of traffic sensors, which can lead to reduced maintenance costs that could go in favor of a more capillary environment management infrastructure; moving from specific sensing and monitoring to general-purpose sensing for large urban environments [18]; the integration and exploitation of other forms of environmental control (e.g., satellite data).

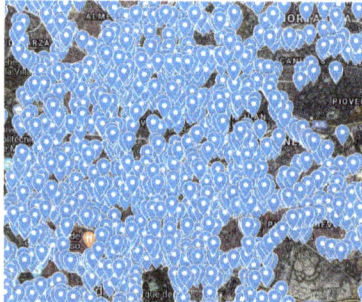

Figure 1. Traffic intensity sensors in Madrid.

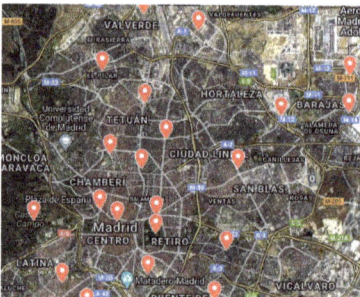

Figure 2. Air pollution sensors in Madrid.

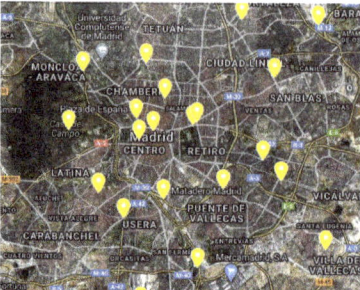

Figure 3. Weather stations in Madrid.

LSTM recurrent neural network is very popular for dealing with time-series data [19]. In the case at hand, the relationship between traffic intensity and pollution levels are aligned (see Section 3.1 and Figures 5 and 6), other time the relationship is blurred by other factors (e.g., meteorological factor). Neural Network can be fruitfully used to capture the evident and the more hidden patterns. For instance, in a week period different patterns (working days versus week-end may show different courses). An adequate period of time for a repeated number of time (e.g., a weekly observation for a duration of a year of data) may disclose relevant correlations. Therefore, we adopted a long short-term memory (LSTM) recurrent neural network (RNN)-based approach which uses air pollutants, including CO, NO, NO_2, NO_x, and O_3, along with some atmospheric variables including pressure, temperature, wind direction, and wind speed to improve road traffic forecasting in Madrid, Spain. The experiments presented in this paper are based on one year of data collected from Madrid's open data source. Complete details about the dataset are provided in Section 4.

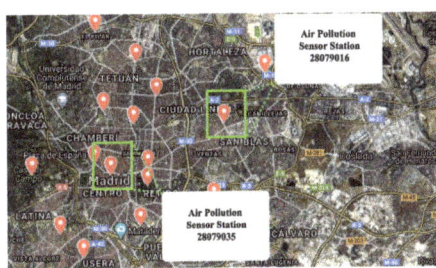

(a) Two air pollution sensor stations, considered for experiments.

(b) Traffic intensity sensors used for one air pollution sensor 28079016.

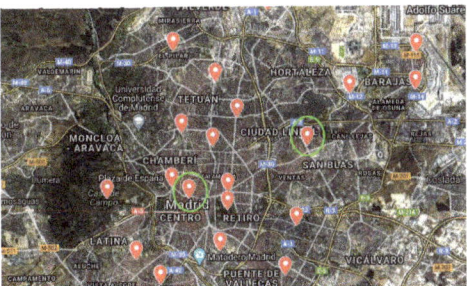

(c) Ariel view of Madrid's map showing the areas considered within 500m radius of both air pollution sensor stations.

Figure 4. Considered air pollution sensor stations, traffic intensity sensors, and areas in Madrid.

1.2. Contribution

With this paper, we have made the following contributions:

- We provide a detailed statistical analysis based on the relationship between air pollutants, atmospheric variables, and road traffic;
- To the best of our knowledge, this is the first attempt to use air pollutants in combination with atmospheric variables to improve traffic forecasting in a smart city;
- Our approach uses a well-known LSTM RNN for time-series traffic data forecasting; and
- We provide some proof of the validity of our approach and avenues for future work.

1.3. Organization

The paper is organized as follows. Section 2 offers a summary of the related work, and Section 3 explains the methodology. The dataset information and performance evaluation are provided in Section 4, and Section 5 concludes the paper and indicates promising directions for future work.

2. Related Work

In this section, we summarize the existing work on traffic forecasting available in the literature. Ji et al. [20] used a deep learning, LSTM RNN-based model exploiting long-term evolution (LTE) access data as an input to their model for the prediction of real-time speed of the traffic. Similarly, Wei et al. [21] proposed an AutoEncoder and LSTM-based method to predict traffic flow. They collected data from the Caltrans Performance Measurement System (PeMS) and considered only three features: (1) traffic flow, (2) occupancy, and (3) speed. Li et al. [22], in their paper, provide an overview of the machine learning approaches for short-term traffic forecasting. Ketabi et al. [23] provide a comparative analysis of multiple variant recurrent neural network and conventional methods for traffic density prediction. They used 40 day data, generated by 58 cameras in London, of the time slot between 9:30 AM and 6:30 PM. Their work considered two features: time and traffic density. Zhu et al. [24] used GPS information data to develop a traffic flow prediction model. Based on data clustering using historic GPS data, their artificial neural network-based prediction model utilized a weighted optimal path algorithm to predict short-term traffic flow. This prediction, based only on the departure time, was then used as input to an A-Dijkstra algorithm to find an optimal path.

Hou et al. [25] proposed a hybrid model that combines an autoregressive integrated moving average (ARIMA) algorithm and a wavelet neural network algorithm for short-term traffic prediction. Their experiment is based on a case study of the Wenhuadong/Tongyi intersection in Weihai City, and only considers weekdays. They collected data over three workdays, using the data from first two days for training and 3rd day's data for testing. Time and traffic flow were the only two features considered. Similarly, Tang et al. [26] proposed a hybrid model, comprising denoising schemes and support-vector machines for traffic flow prediction. To conduct their experiments, they collected data from three traffic flow loop detectors deployed on a highway in Minneapolis, MN (USA). They considered five denoising methods (Empirical Mode Decomposition, Ensemble Empirical Mode Decomposition, Moving Average, Butterworth filter, and Wavelet) for performance evaluation purposes. Their data contained three features: volume, speed, and occupancy. Wang et al. [27] presented an integrated method, combining Group method of data handling (GMDH) and seasonal autoregressive integrated moving average (SARIMA), for traffic flow prediction in the Nanming district of Guiyang, Guizhou province, China. They collected data for five working days; data from the 1st four days were used for training while the last day's data were used for testing. They used residue series as features and labels, respectively to train the model. Rajabzadeh et al. [28] proposed an hybrid approach for short-term road traffic prediction. Based on stochastic differential equations, their approach ultimately improves the short-term prediction. They divided their approach into two steps: (1) a Hull-White model implementation to obtain a prediction model from previous days and (2) the implementation of an extended Vasicek model in order to model a difference between predictions and observations. Two datasets were used: one from a highway in Tehran, and the other an open dataset of PeMS time and traffic volume as inputs. Goudarzi et al. [29] proposed an approach based on self-organizing vehicular network to predict traffic flow. They used a probabilistic generative neural network technique, called deep belief neural networks, to predict traffic flow. Data generated by road side units (RSUs) were used for experiments, with traffic volume and time as inputs. Abadi et al. [30] used traffic flow series that indicate the trends in traffic flow; wavelet decomposition provided basis series and deviation series from the traffic flow data. In addition, local weighted partial least squares and Kalman filtering were used to predict the basis series. One day's data (8:00 AM to 8:00 PM) from the website of the ministry of communication of Taiwan were used for their experiments. Zhang et al. [31] used atmospheric data (average wind speed, temperature, ice fog, freezing fog, smoke) as input to gated recurrent neural network to predict the traffic flow. Rey del Castillo [6] presented an analysis on Madrid's traffic. In this work, short-term indicators of traffic evolution have been produced. Similarly, Lagunas [32] used different machine learning algorithms, including K-means, K-nearest neighbors, and Decision Tree, combined with traffic data, weather data, and data related to events in Madrid to predict the traffic congestion in an area.

The majority of the above-mentioned works used traffic intensity and time in order to forecast traffic. However, we believe that some other parameters like atmospheric conditions can effect the traffic flow which have not been considered in above-mentioned works. Tsirigotis et al. [33] considered only rainfall, along with traffic volume and speed to forecast the traffic. Similarly, Xu et al. [34] considered temperature and humidity, along with taxi trajectory data to forecast traffic flow. They took travel time, pick-up & drop time, and distance into account to forecast traffic flow. Only one month's data (01 January 2015 to 31 January 2015) were considered. We believe, traffic pattern can vary in different days and months. For example, we might observe different traffic pattern during weekends. Similarly, according to a case study in Copenhagen, Denmark, 80% journeys are made on foot in city center and 14% are made by bicycle in summer [35]. On the other hand, traffic forecasting based-on taxi trajectory might have other flaws too. For example, road lines leading to airports might have heavy traffic flow as compared to other lines in surrounding areas. Traffic forecasting for surrounding areas, based on taxi traveling in the lines with heavy traffic flow might result an inaccurate forecasting. In this paper, we are introducing the use of air pollutants and atmospheric parameters (pressure, temperature, wind direction, and wind speed) to forecast traffic. These are the two motivations for using atmospheric parameters: they influence the level of air pollutants in the air, and they also can influence the human behavior. For example, Badii et al. [36] used weather conditions, including temperature, humidity, and rainfall to predict the availability of parking spots inside parking garages, given the fact that depending on the weather condition, people's choice of parking may vary. For example, in thunderstorm, people will prefer indoor parking. Similarly, on different occasions, people may prefer to use public transport which may affect the occupancy of parking lots.

3. Methodology

In this section, we describe the methodology for forecasting traffic flow using traffic intensity values. A first step was to use traffic intensity data combined with air pollution and atmospheric data in order to forecast the traffic. We correlate traffic intensity data to air pollution and atmospheric variables (as we also want to study the relationship between traffic and pollution). As described earlier, air pollutants are often linked to the road traffic levels. Using that link, we propose to use air pollutants and atmospheric variables to forecast the traffic flow. In the second step, we used only time-stamped traffic intensity data, excluding air pollutants and atmospheric data, to forecast the traffic flow. The results produced from step one and step two were then compared to observe how air pollution and atmospheric data, combined with traffic intensity data, could be used to forecast traffic flow. Our experiments were organized into two categories: (1) statistical analysis and (2) traffic forecasting using LSTM RNN. For our experiments, we used open data, collected by the city of Madrid, Spain [37]. The first category of experiments was instrumental for analyzing the quality of available data and to identify macroscopic properties of the data sets.

3.1. Statistical Analysis

As the initial step, we chose one of the air pollution measuring stations and selected two traffic flow sensors at different distances (Figure 7). We collected hourly data from 01 January 2019 to 31 December 2019. This data contained the number of vehicles per hour that passed the sensors, and the air pollutants (CO, NO, NO_2, NO_x, and O_3) levels. Subsequently, we used the accumulated data in order to have an initial view on the possible correlations and to determine a set of parameters that could have an impact on the correlation. We plotted the data on graphs in order to observe the traffic flow patterns with respect to air pollution, as shown in Figure 5. Figure 5 represent the hourly graph of traffic flow measures of one of the selected traffic flow sensors with respect to air pollutants CO, NO, NO$_2$, and NO$_x$. These graphs represent the values of each hour of each day of the year 2019. The graphs in green represent the traffic intensity while the corresponding graphs in red represent the air pollutant levels. In these graphs, blue dotted lines divide the graphs into four time intervals. During the first 2 intervals, all the measured air pollutants follow the traffic flow trend, with few exceptions. In the first interval, the pollutant levels decrease when the traffic is decreasing. Similarly, during the second interval, the pollutant levels increase when the traffic is increasing. A similar pattern can be seen during the fourth interval. However, during the 3rd interval, the pollutants do not seem to be following the traffic flow pattern. To investigate this phenomenon, we studied air pollution dispersion aspects and considered wind speed as one of the factors in air pollution dispersion [38].

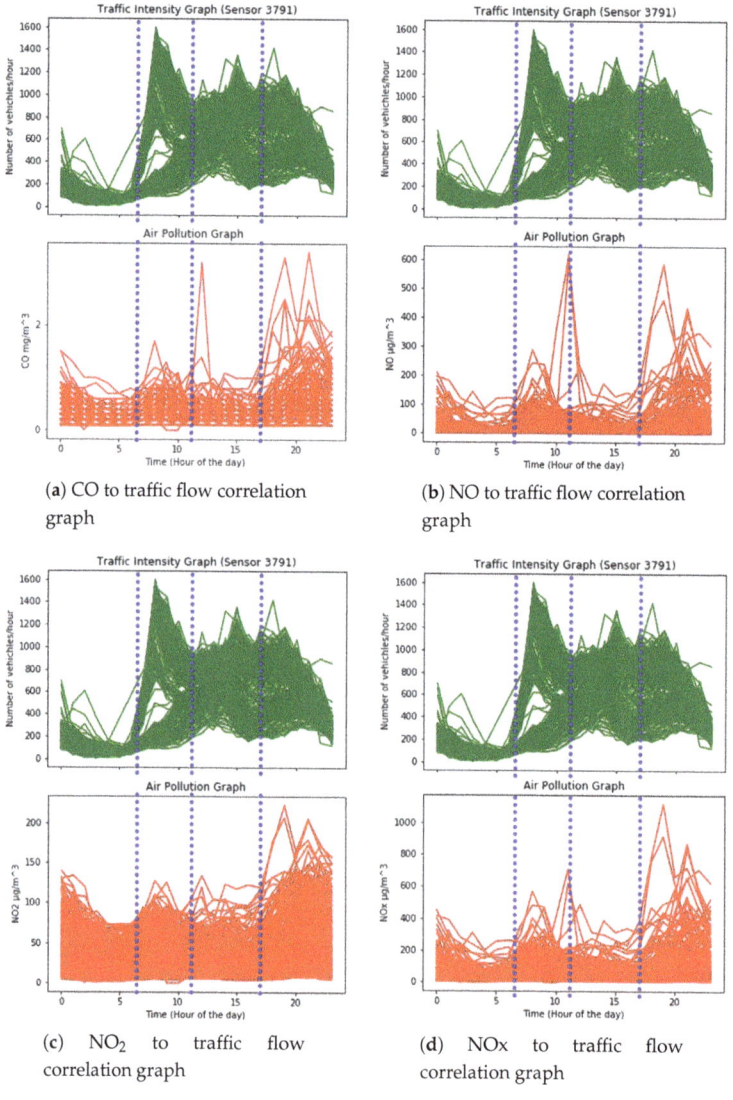

Figure 5. Correlation graphs of traffic flow and air pollutants with respect to each hour of the day.

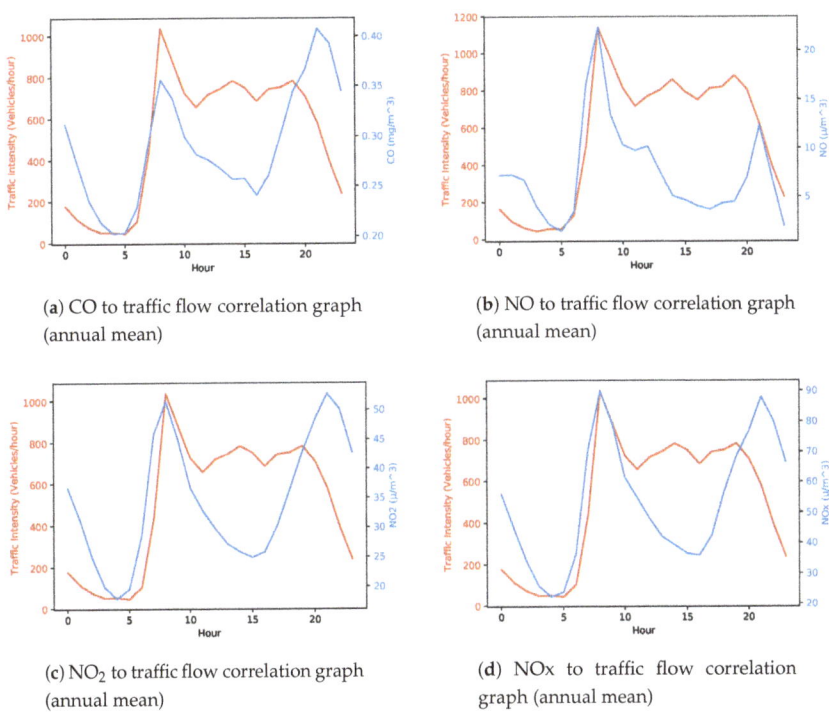

(a) CO to traffic flow correlation graph (annual mean)

(b) NO to traffic flow correlation graph (annual mean)

(c) NO_2 to traffic flow correlation graph (annual mean)

(d) NOx to traffic flow correlation graph (annual mean)

Figure 6. Correlation graphs of traffic flow and air pollutants with respect to each hour of the day (annual mean).

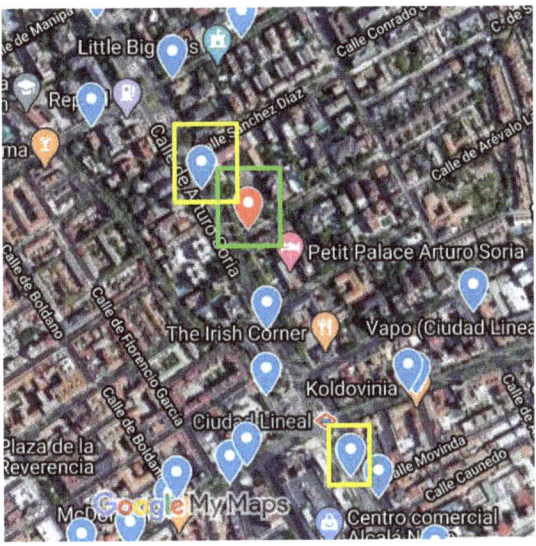

Figure 7. Considered air pollution station (highlighted by the green rectangle) and traffic flow sensors (highlighted by the yellow rectangles).

Hence, as a further verification, we plotted a graph representing the average annual wind speed for each hour (Figure 8), which reveals that wind speed is constantly increasing during the time interval when air pollution does not follow the traffic flow pattern. Given the air pollution dispersion values and the available data, we consider that wind speed is one of the factors that influence air pollution dispersion. As mentioned above, we noticed from statistical analysis that there are similarities in the growth of traffic and the growth of pollution during the morning, and there is a shift in the growth of traffic and the growth of pollution during the evening. In the mid of the day, the correlation is more difficult to capture. This is why we used RNN in order to determine some correlations beyond the statistical ones. The same algorithm using only traffic intensity data and using traffic intensity + meteorological + pollution data show different levels of precision in favor of the analysis that considers more contextual information (a comparative analysis is provided in the Section 4.2). Figure 5 presents the correlation between air pollutants and traffic intensity with respect to each hour of each day of the year. However, in order to provide more insights related to correlation, we have plotted an annual mean graphs for all the considered air pollutants (Figure 6). Phase shift can be seen in Figure 6 too, however, phase shift in Figure 6 is different than that of in Figure 5 because of average annual values.

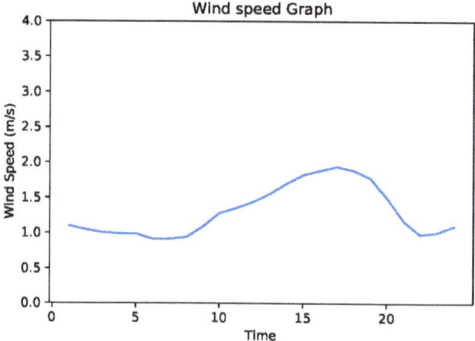

Figure 8. Average annual wind speed.

3.2. Linear Interpolation

Missing values from the data is another major issue when dealing with time-series data. Even though the available open data of the city of Madrid is well maintained, minor glitches in sensors are almost inevitable. Sensors may go offline because of technical issues, or there is a possibility that received data could not be stored on a server. While conducting our initial data analysis, we observed that some of the traffic flow sensors had missing values for some timestamps. Though these missed values were not numerous, it was necessary to fill the gap because we were dealing with time-series data. In order to deal with this issue, we used a well-known method, linear interpolation. Linear interpolation is a popular technique to fill the missing values in a dataset [39]. This technique seeks to identify timestamps that are similar to those that are missing their values, and fills each missing value with an average value [40]. Linear interpolation states that there is a constant gradient in the rate of change between one sample point and the next point. Considering this assumption, if the amplitude of the ith point is x_i and the amplitude of the $i+1^{th}$ point is x_{i+1}, then keeping the constant gradient, the jth point between x_i and x_{i+1} can be calculated as follows [41]:

$$\frac{x_{i+1} - x_i}{(i+1) - i} = \frac{x_j - x_i}{j - i} \tag{1}$$

or

$$x_j = (j - i)(x_j - x_i) + x_i \tag{2}$$

3.3. Traffic Forecasting Using LSTM Recurrent Neural Network

When dealing with time-series data or spatial temporal reasoning, the LSTM RNN is considered one of the best options. As shown in Figure 9, unlike traditional neural networks, the LSTM RNN has memory units instead of neurons. With traditional fully connected neural networks, there is a full connection between the neurons of two adjacent layers. However, there is no connection between the neurons within the same layer. This lack of connection in traditional neural networks could create problems, and may likely cause total failure in terms of spatial temporal reasoning [42]. In RNNs, a hidden unit (memory unit) receives the feedback. This feedback goes from previous state to the current state. We used $timestamp, day\_of\_the\_week, CO, NO, NO_2, NO_x, O_3, pressure, temperature, wind\_direction, wind\_speed,$ and $traffic\_flow$ as the features for our RNN. If we denote the input for the model as $x = (x_1, x_2, x_3, ..., x_T)$ and the output as $y = (y_1, y_2, y_3, ..., y_T)$, with the T in x and y is the prediction time, the traffic flow prediction at time t can be calculated iteratively using the following equations [43]:

$$i_t = \sigma(W_{ix}x_t + W_{im}m_{t-1} + W_{ic}c_{t-1} + b_i) \tag{3}$$

$$f_t = \sigma(W_{fx}x_t + W_{fm}m_{t-1} + W_{fc}c_{t-1} + b_f) \tag{4}$$

$$c_t = f_t \odot c_{t-1} + i_t \odot g(W_{cx}x_t + W_{cm}m_{t-1} + b_c) \tag{5}$$

$$o_t = \sigma(W_{ox}x_t + W_{om}m_{t-1} + W_{oc}c_t + b_o) \tag{6}$$

$$m_t = o_t \odot h(c_t) \tag{7}$$

$$y_t = W_{ym}m_t + b_y \tag{8}$$

In the above equations, $\sigma()$ represents the sigmoid function, which is defined as:

$$\sigma(x) = \frac{1}{1 + e^{-x}} \tag{9}$$

and the $\odot$ in Equations (3)–(8) represents the dot product (also known as scalar product). A memory block, shown in Figure 10, has an input gate, an output gate, and a forget gate. The output of the input gate is represented as i_t, that of the output gate as o_t, and the output of the forget gate as f_t, where c_t and m_t represent the cell and memory activation vectors, respectively. Similarly, W and b represent the weight and the bias matrix which are used to establish connections between input layer, memory block, and output layer. $g(x)$ and $h(x)$ are centered logistic sigmoid functions.

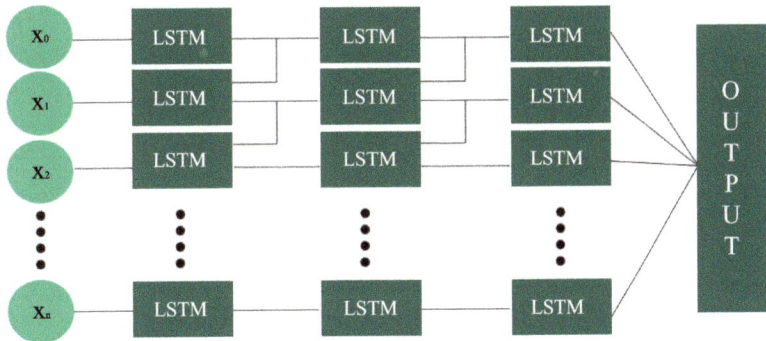

Figure 9. LSTM Recurrent Neural Network Architecture.

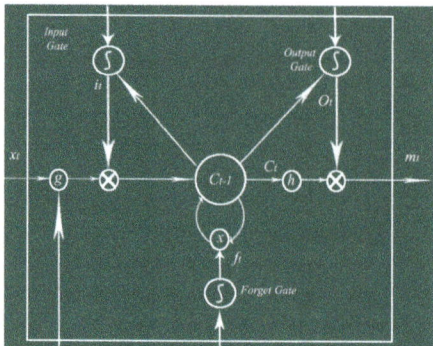

Figure 10. Architecture of a LSTM Memory Unit in Hidden Layers.

3.4. Data Normalization

Data normalization is one of the most important steps in data pre-processing. It guarantees the quality of the data before we use as the input to machine/deep learning models [44]. Data normalization is required when features have different ranges of values. For example, in our dataset, the traffic intensity values range approximately between 0 and 1500 while the value ranges for CO and NO_2 are 0–3.4 and 0–616, respectively. This difference of scale may lead to the poor performance of a machine/deep learning model. Data normalization helps to deal with data that contains values that have different scales. Moreover, it also helps to reduce the training time. Different kind of data normalization techniques are available, including min-max, median normalization, and Z-score decimal scaling. In this paper, we used the most popular normalization technique, min-max normalization [45].

Min-Max Normalization

Min-max normalization maps data into pre-defined ranges i.e., [0,1] or [−1,1]. The values of each attribute in the data are defined according to their minimum and maximum value. If we denote the attribute in the data by "*Atr*", its value by "*a_val*", its normalized value as "*a_norm*", and pre-defined range as [*lower_lim*, *higher_lim*], then following equation [44] can be used to calculate normalized values between the range [*lower_lim*, *higher_lim*]:

$$a\_norm = lower\_lim + \frac{(higher\_lim - lower\_lim) \times (a\_valu - min(Atr))}{max(Atr) - min(Atr)} \qquad (10)$$

3.5. Hyperparameter

We used the following configuration of a LSTM RNN to forecast traffic flow using Madrid's open data:

- 3 LSTM layers;
- **Dropout:** To keep our model from going into overfitting, we applied dropout [46] at each LSTM layer with a value of 0.7;
- **Early Stopping:** To stop the training before the model approaches overfitting, we used early stopping [47] with the patience value of 5;
- **Look Back Steps:** In order to do prediction at time t, "look back" shows how many previous time steps need to be considered. We set the "look back" steps value at 168, which represents the total number of hours in a week. We chose 168 hours (one week) as "look back" period. The plan is to capture the evolution of the air pollutants over a period in which different, but recursive patterns may occur, e.g., working day traffic vs. Week-end traffic. We wanted to grasp the differences between working days and week-end. In addition, in such a period, the pollutants have time to consolidate (some pollutant can float for hours or more). Moreover, this time period could result a

better forecasting. Traffic intensity shows different patterns between weekdays and weekends. Pollution "signatures" refer to longer and more complex situations. A week within a particular month (e.g., December before Christmas time) can be characterized by higher volume of traffic and hence pollution. Different months can have very different levels of traffic and pollution. The choice of considering one week is due to the possibility to grasp these variations, while still maintaining a short period for observation and data capture. With respect to pollution, a longer period of time (e.g., a month) would allow a more specific characterization of the traffic in that specific month and the related pollution signature could be used in order to help the prediction. A shorter period of time (one day, two days) is not able to capture these variations in traffic intensity and pollution measurements. However, the choice of one week is a starting point and, for further work, a better tuning of the time could be envisaged.

4. Dataset and Performance Evaluation

This section describes the dataset and its features, and evaluates the performance achieved by LSTM RNN for traffic flow forecasting using air pollution and atmospheric data. Open data from Madrid, Spain [37] collected and normalized for 1 year of observations. A large set of data related to traffic intensity was collected in the first step. This dataset also contained weather and pollution-related features. We conducted experiments using the data from two air pollution sensor stations (Figure 4a) to forecast traffic flow. These stations measure CO, NO, NO_2, NO_x and O_3 values in the air. In addition, we used timestamp, traffic intensity, and atmospheric data, including temperature, pressure, wind speed, and wind direction from nearby weather stations. For a comparison, in the second step, we only used traffic intensity and timestamp values (with no air pollutant or atmospheric parameters) to forecast the traffic flow, and compared the results to see the effect of considering air pollutant and atmospheric data.

We chose 25 traffic flow sensors in a 500 m radius of the two air pollution sensor stations (Figure 4b). Traffic flow data is available after every 15 minutes, however, other data, including CO, NO, NO_2, NO_x, O_3, *Pressure*, *Temperature*, *Wind Speed*, and *Wind Direction* are updated hourly.

As the air pollutant data and atmospheric data are available hourly, therefore, we collected the hourly traffic data to keep it coherent with air pollution and atmospheric data. Table 1 represents the details of the features used to train the model. As our data were organized hourly (from 01 January 2019 to 31 December 2019), we had 8760 records in total; 67% of our data were used for training and 33% were used for testing. In order to extract the traffic flow insights for the roads where sensors are deployed, Table 2 represents the statistics of 25 traffic flow sensors within the chosen distance from the associated air pollution sensor station, and the minimum, maximum, and average traffic flow in the year 2019. Out of 25 sensors, 9 were faulty and gave either null value or garbage values. For those sensors, the minimum, maximum, and average flow values are represented as "NA" in Table 2.

Table 1. Features used for training the model.

Feature	Value/Unit
Month	1–12
Day	1–28/29/30/31
Weekday	1–7
Hour	0–23
CO	mg/m^3
NO	$\mu g/m^3$
NO_2	$\mu g/m^3$
NO_x	$\mu g/m^3$
O_3	$\mu g/m^3$
Pressure	mb
Temperature	°C
Wind Direction	Angle
Wind Speed	m/s
Traffic Flow	Vehicles/h

Table 2. Traffic flow sensors' statistics.

Air Pollution Sensor Station	Traffic Flow Sensor	Distance from Air Pollution Sensor Station	Minimum Flow (Annual)	Maximum Flow (Annual)	Average Flow (Annual)
28079016	6037	240 m	0	384	112.344
	3791	79 m	4	1601	493.693
	3775	294 m	17	1166	468.615
	5938	205 m	0	220	32.011
	5939	125 m	5	1980	522.943
	10124	242 m	NA	NA	NA
	6058	214 m	NA	NA	NA
	3594	296 m	NA	NA	NA
	5922	366 m	NA	NA	NA
	10128	500 m	4	1413	437.701
	10125	455 m	NA	NA	NA
	5941	303 m	0	1324	135.017
	5923	426 m	5	1334	437.864
	5994	483 m	0	480	135.389
	5940	369 m	NA	NA	NA
	5942	336 m	0	1523	534.091
	5944	349 m	0	182	72.176
	5921	374 m	23	1214	481.669
	3776	425 m	17	1208	476.911
	5937	484 m	0	313	86.216
28079035	3731	26 m	NA	NA	NA
	4303	39 m	0	181	52.188
	3730	133 m	NA	NA	NA
	4301	137 m	NA	NA	NA
	10387	196 m	40	1260	608.482

4.1. Evaluation Metrics

In order to evaluate the results of the experiments, we defined some metrics to be used for the evaluation of our model. We used two of the most-used evaluation metrics *Mean Absolute Error (MAE)* and *Means Squared Error (MSE)*. Their mathematical representations are [48,49]:

$$MAE = \frac{1}{N} \sum_{i=1}^{N} |y_i^{predicted} - y_i^{observed}| \quad (11)$$

$$MSE = \frac{1}{N} \sum_{i=1}^{N} (y^{predicted} - y^{observed})^2 \quad (12)$$

MAE is not sensitive to outliers. It does not deal well with big errors. It is very useful for continuous variable data. MSE is very useful when the dataset contains outliers. At the beginning of the analysis, we wanted to be sure to grasp insights from very different data and patterns (traffic intensity and air pollutants). For this reason, we decided to check our results using both MSE and MAE. However, in our case, we found out that MAE alone could be used to evaluate the whole performance. Therefore, in future work, for additional experiments, we will use MAE for the evaluation. We used the training loss and the validation loss in the learning curve in order to be sure that our model was not overfitting.

4.2. Results

This section provides the MAE and MSE scores of the LSTM RNN model for each of the operational traffic flow sensors (excluding faulty sensors). As explained in the previous section, 25 traffic intensity sensors were considered, and out of those 25, 9 sensors were faulty and so were eliminated from the dataset during the experiments. Hence, Table 3 presents the MAE and MSE scores of 16 traffic flow sensors. We performed an hourly forecast. In order to do that, we determined the traffic intensity at time t by considering traffic intensity data, air pollution data, and atmospheric data from $[0, t-1]$ and, air pollution data and atmospheric data from time t.

The maximum MAE produced by the LSTM RNN for the traffic sensors within the radius of 500 m of air pollution sensor "28079016" was 0.214 while the minimum MAE was 0.061. Similarly, the maximum MSE was 0.60 and the minimum MSE was 0.009. In order to evaluate our LSTM RNN model further, we conducted the same experiments for air pollution sensor station "28079035" and 5 traffic flow sensors within its 500 m radius. Out of those 5 traffic flow sensors, 3 were faulty. Hence, Table 3 presents the values of 2 of the operational traffic flow sensors (4303 and 10387) around the station "28079035". The LSTM RNN produced values 0.105 MAE and 0.017 MSE for traffic flow sensor "4303", and 0.136 MAE and 0.029 MSE for traffic flow sensor "10387".

Table 3. Mean absolute error (MAE) and mean squared error (MSE) for two considered traffic flow forecasting for considered traffic flow sensors.

Air Pollution Sensor Station	Traffic Flow Sensor	MAE	MSE
28079016	6037	0.183	0.045
	3791	0.206	0.056
	3775	0.206	0.054
	5938	0.073	0.009
	5939	0.166	0.035
	10128	0.203	0.053
	5941	0.061	0.005
	5923	0.188	0.046
	5994	0.173	0.047
	5942	0.214	0.060
	5944	0.208	0.056
	5921	0.200	0.051
	3776	0.193	0.051
	5937	0.160	0.030
28079035	4303	0.105	0.017
	10387	0.136	0.029

In order to observe the effect of introducing air pollutants and atmospheric parameters, we randomly selected five traffic intensity sensors and performed forecasting, considering only timestamped traffic intensity values. Figures 11 and 12 represent the comparative analysis of the mean absolute error and the mean squared error, respectively, with and without using air pollutants and atmospheric parameters as input features. It is clear that air pollutants and atmospheric parameters improve the MAE and the MSE. Our LSTM recurrent neural network-based approach performed better for all of the five considered traffic intensity sensors when air pollutants and atmospheric parameters were used along with the timestamped traffic intensity values.

Sensors **2020**, *20*, 3749

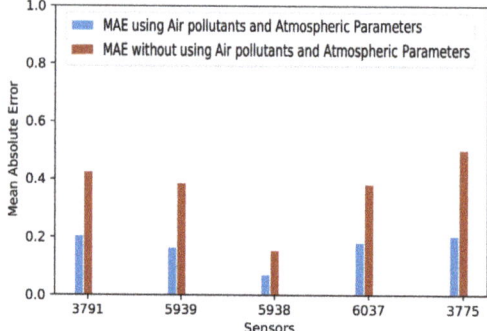

Figure 11. MAE with and without using air pollutants and atmospheric parameters.

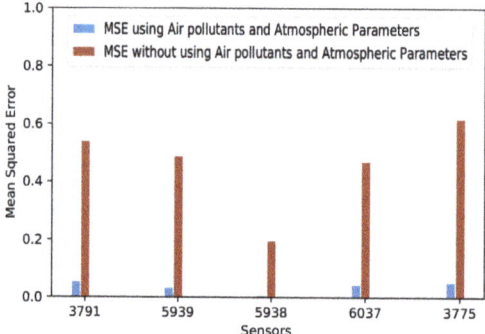

Figure 12. MSE with and without using air pollutants and atmospheric parameters.

4.3. Further Evaluation

To further evaluate the LSTM RNN model, we determined if our model was overfitting or not. One of the most-widely used methods for verifying overfitting [50,51] is to plot learning curves. A learning curve plots a model's training loss and validation loss. These curves give information about overfitting and underfitting:

- **Overfitting** represents the ability of the model to learn too much during the training process, so that when unseen data are provided for prediction, it shows poor performance. Overfitting can be diagnosed by plotting learning curves. If the training loss is decreasing but validation loss starts increasing after a specific point, this shows that a model is overfitting [51].
- **Underfitting** represents the inability of the model to learn from training data. If a learning curve shows either of the following two behaviors, the model is underfitting:

    – Validation loss is very high and training loss is flat regardless of training time.
    – Training loss is continuously decreasing without being stable until the training is complete.

Given above definitions, we plotted learning curves to observe the behavior of our model. Figure 13 shows that the learning curve of our model is not following any of the above-mentioned definitions of overfitting and underfitting. Training loss is decreasing and after a specific point it becomes stable. Similarly, validation loss becomes stable and remains close to the training loss. Both of these observations show that our model is a good fit.

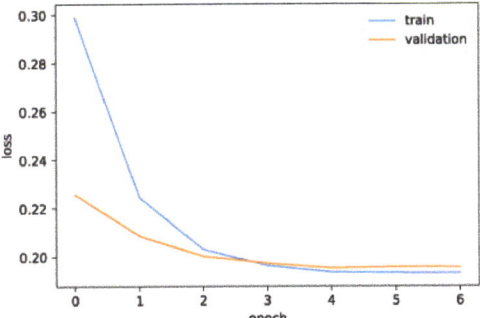

Figure 13. Learning curve representing training and validation losses of the LSTM RNN model for traffic flow forecasting.

4.4. Threat to Validity

The model utilized with the currently available data in Madrid. The penetration of electric vehicles may be a factor impacting the generation of pollution in major cities. This could have also a long term impact on our forecasts. However, the substitution of older vehicles with hybrid or electric ones will be relatively quick but not immediate. This delay will give the model some time to adapt and learn the new patterns. Given the ongoing concerns about air pollution, the use of electric vehicles is increasing around the world. For example, the national electric mobility mission plan is anticipating the sale of around 7 million electric vehicles yearly from 2020 onwards [52]. While it will take a long time to completely eliminate conventional vehicles, the elimination of conventional fuel vehicles could be a threat to our approach's validity, as it is partially dependent upon vehicular pollution emission.

5. Conclusions

Traffic forecasting is one of the most important tasks for big cities. Accurate traffic flow forecasting can help drivers to better plan their trips. To provide accurate traffic flow forecasting, this work, first combined air pollutants and atmospheric data with traffic intensity data to forecast traffic flow in Madrid, Spain. In the second step, only timestamped traffic intensity data were used to forecast traffic flow, and then those results were compared with the results from the experiments at step one. The comparison was carried out to observe the effect of adding air pollutants and atmospheric data to forecast the traffic flow. We used a long short-term memory recurrent neural network (LSTM RNN) to perform traffic flow forecasting, with time-series traffic flow, air pollution, and atmospheric data collected from the open datasets of Madrid, Spain. Air pollutants (CO, NO, NO_2, NO_X, and O_3), which are associated with road traffic, were considered as the input features, along with atmospheric variables (wind speed, wind direction, temperature and pressure), because in air pollution dispersion models, these features influence the dispersion of air pollution. Together these features helped the model to better forecast the traffic flow. Experimental results show that addition of air pollutant and atmospheric information with timestamp improved the performance.

Future Work

In future work, we plan to extend our experiments to assess the effects of seasons, e.g., summer and winter. Traffic patterns are likely to be different in August in Europe, as many people leave cities and go on vacations. Moreover, we want to identify the percentage of air pollution contributed by road traffic and heating/cooling systems in homes, offices, and factories. In addition, we are planning to take air pollution dispersion models like Ausplume and Calpuss into account to better understand the behavior of air pollution. The correlation between air pollution and traffic intensity may differ in different areas of the city. Density of the infrastructure can have an impact on the correlation. In this paper, we only considered two areas in Madrid. However in the future, we plan to take multiple areas and their infrastructure into account to observe the correlation between traffic flow and air pollutants. As a goal, we want to understand if it is possible to analyze the 'signatures'/traces of pollution in order to derive and predict information for correlated phenomena. At the same time, satellite pollution measurements will be taken into consideration in order to understand if they can be used together with ground values to better identify the correlations. In this paper, we considered one of the popular neural network models, i.e., LSTM recurrent neural network. However, some studies [53] show that traditional machine learning models can sometimes perform better than deep learning techniques. In addition to traditional machine learning models, statistical models have also been found to perform better than machine learning models [54]. Hence, it is an open research question to choose the better machine/deep learning model combined with air pollution and atmospheric data.

In addition, we want to investigate how to optimize the fusion of different sources of information to improve the prediction for relevant phenomena in the cities. The deployment and maintenance of a large sensor network for traffic and air quality monitoring is a large investment that requires careful planning in order to be effective and practical. There are a few cities (Madrid is one), that have similar deployment and provide open access to data [37,55,56]. Many other cities cannot afford such an investment. This means that monitoring may be very active in certain areas while areas nearby are not similarly controlled. We will work on pollution data analysis to verify if it is possible to adequately monitor pollution and to derive and predict phenomena related/associated to it. Another aspect that will be further studied is the possibility offered by the fusion of data in reducing the number of sensors in a city without lowering the information quality, which will ultimately lead to a reduction in cost. For instance, in Madrid, some traffic sensors could be eliminated in favor of more air control sensors if a strong relationship can be verified between traffic and pollution levels.

Author Contributions: Conceptualization: F.M.A., R.M., and N.C. Data curation: F.M.A. Formal analysis: F.M.A., R.M. and N.C. Methodology: F.M.A. Writing-original draft: F.M.A., R.M., and N.C. Writing-review & editing: R.M., and N.C. All authors have read and agreed to the published version of the manuscript.

Funding: This research received no external funding.

Conflicts of Interest: The authors declare no conflict of interest.

References

1. Schmidt, J.M.; Tendwa, O.; Bruwer, M.M. Traffic impact of the its time event. In Proceedings of the 37th Annual Southern African Transport Conference, Pretoria, South Africa, 9–12 July 2018; pp. 704–716.
2. Kuang, Y.; Yen, B.T.; Suprun, E.; Sahin, O. A soft traffic management approach for achieving environmentally sustainable and economically viable outcomes: An Australian case study. *J. Environ. Manag.* **2019**, *237*, 379–386. [CrossRef] [PubMed]
3. Bogaerts, T.; Masegosa, A.D.; Angarita-Zapata, J.S.; Onieva, E.; Hellinckx, P. A graph CNN-LSTM neural network for short and long-term traffic forecasting based on trajectory data. *Transp. Res. Part C Emerg. Technol.* **2020**, *112*, 62–77. [CrossRef]
4. Lazić, L.; Urošević, M.A.; Mijić, Z.; Vuković, G; Ilić, L. Traffic contribution to air pollution in urban street canyons: Integrated application of the OSPM, moss biomonitoring and spectral analysis. *Atmos. Environ.* **2016**, *141*, 347–360 [CrossRef]
5. World Health Organization. Air Pollution. Available online: https://www.euro.who.int/en/health-topics/environment-and-health/Transport-and-health/data-and-statistics/air-pollution-and-climate-change2 (accessed on 27 March 2020).
6. Analyzing Traffic Flows in Madrid City. Available online: https://ec.europa.eu/eurostat/cros/system/files/s06p2-analizing-traffic-flows-in-madrid-city.pdf (accessed on 23 June 2020).
7. Maciag, P.S.; Kasabov, N.; Kryszkiewicz, M.; Bembenik, R. Air pollution prediction with clustering-based ensemble of evolving spiking neural networks and a case study for London area. *Environ. Mod. Soft.* **2019**, *118*, 262–280. [CrossRef]

8. Rosenlund, M.; Forastiere, F.; Stafoggia, M.; Porta, D.; Perucci, M.; Ranzi, A.; Nussio, F.; Perucci, C.A. Comparison of regression models with land-use and emissions data to predict the spatial distribution of traffic-related air pollution in Rome. *J. Expo. Sci. Environ. Epidem.* **2008**, *18*, 192–199. [CrossRef]
9. Crouse, D.L.; Goldberg, M.S.; Ross, N.A. A prediction-based approach to modelling temporal and spatial variability of traffic-related air pollution in Montreal, Canada. *Atmos. Environ.* **2009**, *43*, 5075–5084. [CrossRef]
10. Batterman, S.; Ganguly, R.; Harbin, P. High resolution spatial and temporal mapping of traffic-related air pollutants. *Int. J. Environ. Res. Public Health* **2015**, *12*, 3646–3666. [CrossRef]
11. Ly, H.B.; Le, L.M.; Phi, L.V.; Phan, V.H.; Tran, V.Q.; Pham, B.T.; Le, T.T.; Derrible, S. Development of an AI model to measure traffic air pollution from multisensor and weather data. *Sensors* **2019**, *19*, 4941. [CrossRef]
12. Laña, I.; Del Ser, J.; Padró, A.; Vélez, M.; Casanova-Mateo, C. The role of local urban traffic and meteorological conditions in air pollution: A data-based case study in Madrid, Spain. *Atmos. Environ.* **2016**, *145*, 424–438. [CrossRef]
13. Russo, A.; Lind, P.G.; Raischel, F.; Trigo, R.; Mendes, M. Neural network forecast of daily pollution concentration using optimal meteorological data at synoptic and local scales. *Atmos. Pollut. Res.* **2015**, *6*, 540–549. [CrossRef]
14. Brunello, A.; Kamińska, J.; Marzano, E.; Montanari, A.; Sciavicco, G.; Turek, T. Assessing the Role of Temporal Information in Modelling Short-Term Air Pollution Effects Based on Traffic and Meteorological Conditions: A Case Study in Wrocław. In Proceedings of the European Conference on Advances in Databases and Information Systems, Bled, Slovenia, 8–11 September 2019; pp. 463–474.
15. World Economic Forum, This Is Why People Live, Work, and Stay in a Growing City. Available online: https://www.weforum.org/agenda/2018/10/this-is-why-people-live-work-stay-leave-in-growing-city/ (accessed on 27 March 2020)
16. Pant, P.; Shi, Z.; Pope, F.D.; Harrison, R.M.; Characterization of traffic-related particulate matter emissions in a road tunnel in Birmingham, UK: Trace metals and organic molecular markers. *Aerosol. Air. Qual. Res.* **2016**, *17*, 117–130. [CrossRef]
17. Zhang, X.; Craft, E; Zhang, K.; Characterizing spatial variability of air pollution from vehicle traffic around the Houston Ship Channel area. *Atmos. Environ.* **2017**, *161*, 167–175. [CrossRef]
18. Laput, G.; Zhang, Y.; Harrison, C. Synthetic sensors: Towards general-purpose sensing. In Proceedings of the 1st CHI Conference on Human Factors in Computing Systems, Colorado, CO, USA, 6–11 May 2017; pp. 3986–3999
19. Guo, T.; Xu, Z.; Yao, X.; Chen, H.; Aberer, K.; Funaya, K. Robust online time-series prediction with recurrent neural networks. In Proceedings of the IEEE International Conference on Data Science and Advanced Analytics, Montreal, Canada, 17–19 October 2016; pp. 816–825.
20. Ji, B.; Hong, E.J.; Deep-learning-based real-time road traffic prediction using long-term evolution access data. *Sensors* **2019**, *19*, 5327. [CrossRef] [PubMed]
21. Wei, W.; Wu, H.; Ma, H. An autoencoder and LSTM-based traffic flow prediction method. *Sensors* **2019**, *19*, 2946. [CrossRef]
22. Li, Y.; Shahabi, C. A brief overview of machine learning methods for short-term traffic forecasting and future directions. *Sigspatial Spec.* **2018**, *10*, 3–9. [CrossRef]
23. Ketabi, R.; Al-Qathrady, M.; Alipour, B.; Helmy, A. Vehicular Traffic Density Forecasting through the Eyes of Traffic Cameras; a Spatio-Temporal Machine Learning Study. In Proceedings of the 9th ACM Symposium on Design and Analysis of Intelligent Vehicular Networks and Applications, Miami, MIA, USA, 25–29 November 2019; pp. 81–88.
24. Zhu, D.; Du, H.; Sun, Y.; Cao, N. Research on path planning model based on short-term traffic flow prediction in intelligent transportation system. *Sensors* **2018**, *18*, 4275. [CrossRef]
25. Hou, Q.; Leng, J.; Ma, G.; Liu, W.; Cheng, Y. An adaptive hybrid model for short-term urban traffic flow prediction. *Phys. A Stat. Mech. Appl.* **2019**, *527*, 121065. [CrossRef]
26. Tang, J.; Chen, X.; Hu, Z.; Zong, F.; Han, C.; Li, L.; Traffic flow prediction based on combination of support vector machine and data denoising schemes. *Phys. A Stat. Mech. Appl.* **2019**, *534*, 120642. [CrossRef]
27. Wang, W.; Zhang, H.; Li, T.; Guo, J.; Huang, W.; Wei, Y.; Cao, J. An interpretable model for short term traffic flow prediction. *Math. Comp. Simul.* **2019**, *171*, 264–278. [CrossRef]
28. Rajbzadeh, Y.; Rezaie, A.H.; Amindavar, H. Short-term traffic flow prediction using time-varying Vasicek model. *Transp. Res. Part C Emerg. Technol.* **2017**, *74*, 168–181. [CrossRef]

29. Goudarzi, S.; Kama, M.N.; Anisi, M.H.; Soleymani, S.A.; Doctor, F. Self-organizing traffic flow prediction with an optimized deep belief network for internet of vehicles. *Sensors* **2018**, *18*, 3459. [CrossRef] [PubMed]
30. Abadi, A.; Rajabioun, T.; Ioannou, P.A. Traffic flow prediction for road transportation networks with limited traffic data. *IEEE Trans. Intell. Transp. Syst.* **2014**, *16*, 653–662. [CrossRef]
31. Zhang, D.; Kabuka, M.R. Combining weather condition data to predict traffic flow: A GRU-based deep learning approach. *IET Intell. Transp. Syst.* **2018**, *12*, 578–585. [CrossRef]
32. Analyzing Traffic Flows in Madrid City. Available online: https://eprints.ucm.es/49461/1/TFM-201809-4.0%20-%20Pina%20Lagunas%20-%20Sergio.pdf (accessed on 23 June 2020).
33. Tsirigotis, L.; Vlahogianni, E.I.; Karlaftis, M.G. Does information on weather affect the performance of short-term traffic forecasting models?. *Int. J. Intell. Transp. Syst. Res.* **2012**, *10*, 1–10. [CrossRef]
34. Xu, X.; Su, B.; Zhao, X.; Xu, Z.; Sheng, Q.Z. Effective traffic flow forecasting using taxi and weather data. In Proceedings of the International Conference on Advanced Data Mining and Applications, Gold Coast, Australia, 12–15 December 2016; pp. 507–519.
35. European Commission Directorate-General for the Environment. Available online: https://ec.europa.eu/environment/pubs/pdf/streets-people.pdf (accessed on 7 May 2020)
36. Badii, C.; Nesi, P.; Paoli, I. Predicting available parking slots on critical and regular services by exploiting a range of open data. *IEEE Access* **2018**, *6*, 44059–44071. [CrossRef]
37. Open data portal of the Madrid City Council. Availble online: https://datos.madrid.es/portal/site/egob (accessed on 2 February 2020)
38. Baldauf, R.; Watkins, N.; Heist, D.; Bailey, C.; Rowley, P.; Shores, R. Near-road air quality monitoring: Factors affecting network design and interpretation of data. *Air Qual. Atmos. Health* **2009**, *2*, 1–9. [CrossRef]
39. Che, Z.; Purushotham, S.; Cho, K.; Sontag, D.; Liu, Y. Recurrent neural networks for multivariate time-series with missing values. *Sci. Rep.* **2018**, *8*, 6085. [CrossRef]
40. Li, L.; Zhang, J.; Wang, Y.; Ran, B. Missing value imputation for traffic-related time-series data based on a multi-view learning method. *IEEE Trans. Intell. Transp. Syst.* **2018**, *20*, 2933–2943. [CrossRef]
41. Usman, K.; Ramdhani, M. Comparison of Classical Interpolation Methods and Compressive Sensing for Missing Data Reconstruction. In Proceedings of the IEEE International Conference on Signals and Systems, Bandung, Indonesia, 16–18 July 2019; pp. 29–33.
42. Zhao, Z.; Chen, W.; Wu, X.; Chen, P.C.; Liu, J. LSTM network: A deep learning approach for short-term traffic forecast. *IET Intell. Transp. Syst.* **2017**, *11*, 68–75. [CrossRef]
43. Ma, X.; Tao, Z.; Wang, Y.; Yu, H.; Wang, Y. Long short-term memory neural network for traffic speed prediction using remote microwave sensor data. *Transp. Res. Part C Emerg. Technol.* **2015**, *54*, 187–197. [CrossRef]
44. Nayak, S.C.; Misra, B.B.; Behera, H.S. Impact of data normalization on stock index forecasting. *Int. J. Comput. Inf. Syst. Ind. Manag. Appl.* **2014**, *6*, 357–369.
45. Gajera, V.; Gupta, R.; Jana, P.K. An effective multi-objective task scheduling algorithm using min-max normalization in cloud computing. In Proceedings of the 2nd International Conference on Applied and Theoretical Computing and Communication Technology, Bengaluru, India, 21–23 July 2016; pp. 812–816.
46. Srivastava, N.; Hinton, G.; Krizhevsky, A.; Sutskever, I.; Salakhutdinov, R. Dropout: A simple way to prevent neural networks from overfitting. *J. Mach. Learn. Res.* **2014**, *15*, 1929–1958.
47. Prechelt, L. *In Neural Network: Tricks of the Trade*; Springer: Heiderlberg, Germany, 1998.
48. Zhang, L.; Liu, Q.; Yang, W.; Wei, N.; Dong, D. An improved k-nearest neighbor model for short-term traffic flow prediction. *Procedia-Soc. Behav. Sci.* **2013**, *96*, 653–662. [CrossRef]
49. Li, L.; Su, X.; Wang, Y.; Lin, Y.; Li, Z.; Li, Y. Robust causal dependence mining in big data network and its application to traffic flow predictions. *Transp. Res. Part C Emerg. Technol.* **2015**, *58*, 292–307. [CrossRef]
50. Perlich, C.; Provost, F.; Simonoff, J.S. Tree induction vs. logistic regression: A learning-curve analysis. *J. Mach. Learn. Res.* **2003**, *4*, 211–255.
51. Perlich C. *Encyclopedia of Machine Learning*; Springer: Boston, MA, USA, 2011.
52. Nimesh, V.; Sharma, D.; Reddy, V.M.; Goswami, A.K. Implication viability assessment of shift to electric vehicles for present power generation scenario of India. *Energy* **2020**, *195*, 116976. [CrossRef]
53. Awan, F.M.; Saleem, Y.; Minerva, R.; Crespi, N. A Comparative Analysis of Machine/Deep Learning Models for Parking Space Availability Prediction. *Sensors* **2020**, *20*, 322. [CrossRef]

54. Makridakis, S., Spiliotis, E.; Assimakopoulos, V. Statistical and Machine Learning forecasting methods: Concerns and ways forward. *PLoS ONE* **2018**, *13*, e0194889. [CrossRef]
55. Open Data Portal of the Barcelona City. Available online: https://opendata-ajuntament.barcelona.cat/data/es/dataset (accessed on 25 March 2020).
56. Open data portal of the Turin City. Available online: https://www.torinocitylab.it/en/assetto/open-data (accessed on 25 March 2020)

© 2020 by the authors. Licensee MDPI, Basel, Switzerland. This article is an open access article distributed under the terms and conditions of the Creative Commons Attribution (CC BY) license (http://creativecommons.org/licenses/by/4.0/).

Article

Xbee-Based WSN Architecture for Monitoring of Banana Ripening Process Using Knowledge-Level Artificial Intelligent Technique

Saud Altaf [1], Shafiq Ahmad [2,*], Mazen Zaindin [3] and Muhammad Waseem Soomro [4]

1. University Institute of Information Technology, Pir Mehr Ali Shah Arid Agriculture University Rawalpindi, Rawalpindi 48312, Pakistan; saud@uaar.edu.pk
2. Department of Industrial Engineering, King Saud University, Riyadh 11451, Saudi Arabia
3. Department of Statistics and Operations Research, King Saud University, Riyadh 11451, Saudi Arabia; zaindin@ksu.edu.sa
4. Manukau Institute of Technology, Auckland 2023, New Zealand; mwaseem@manukau.ac.nz
* Correspondence: ashafiq@ksu.edu.sa

Received: 21 June 2020; Accepted: 17 July 2020; Published: 20 July 2020

Abstract: Real-time monitoring of fruit ripeness in storage and during logistics allows traders to minimize the chances of financial losses and maximize the quality of the fruit during storage through accurate prediction of the present condition of fruits. In Pakistan, banana production faces different difficulties from production, post-harvest management, and trade marketing due to atmosphere and mismanagement in storage containers. In recent research development, Wireless Sensor Networks (WSNs) are progressively under investigation in the field of fruit ripening due to their remote monitoring capability. Focused on fruit ripening monitoring, this paper demonstrates an Xbee-based wireless sensor nodes network. The role of the network architecture of the Xbee sensor node and sink end-node is discussed in detail regarding their ability to monitor the condition of all the required diagnosis parameters and stages of banana ripening. Furthermore, different features are extracted using the gas sensor, which is based on diverse values. These features are utilized for training in the Artificial Neural Network (ANN) through the Back Propagation (BP) algorithm for further data validation. The experimental results demonstrate that the projected WSN architecture can identify the banana condition in the storage area. The proposed Neural Network (NN) architectural design works well with selecting the feature data sets. It seems that the experimental and simulation outcomes and accuracy in banana ripening condition monitoring in the given feature vectors is attained and acceptable, through the classification performance, to make a better decision for effective monitoring of current fruit condition.

Keywords: wireless sensor network; fruit condition monitoring; artificial neural network; ethylene gas; banana ripening

1. Introduction

Fresh produce, especially fruits and vegetables, is considered an important part of our day to day diet because it is a major source of vitamins, minerals, organic acids, dietary fibers, and also antioxidants. According to the food guide pyramid, a balanced diet should include at least 2–4 servings of fruit every day [1]. The consumption of fruits and vegetables has increased recently with greater consumer awareness about the health benefits of fresh produce over processed foods. Fruits and vegetables are highly perishable commodities, so proper post-harvest handling is required to avoid unwanted losses and to retain the freshness and quality. During long-distance transportation and distribution, the risk of post-harvest losses may increase, and therefore, proper care and handling have been emphasized in recent years for post-harvest commodities [2]. There are several causes of post-harvest losses, including increased respiration rate, hormone production (i.e., ethylene), physiological disorders, general senescence, and compositional and morphological changes. However, the excess of ethylene (plant growth hormone) production is mainly liable for higher post-harvest losses, particularly for climacteric fruits. For this research, bananas have been chosen as the model for several reasons.

Banana is a major harvesting fruit yield in Pakistan and is grown in the large area of the province Sindh with an approximate production of 155 K tons in the farming season because of the favorable climatic and soil conditions for its successful farming. Major farming areas are Badin, Tando Allahyar, Naushero Feroz, Hyderabad, Nawabshah, Sangar, Thatta, and Tando Muhammad Khan, and farming has been extended to some other northern areas of the province of Sindh. These areas of production amount to 87–90% of total production in Pakistan [1].

Normally, fruit cold storage units are built near the cultivation field for easy transfer of fruits for storing and transportation. Therefore, it is necessary to improve the management capability through remote and automatic monitoring procedures. Fruit cold storage is usually constructed in large square meter areas and different fruit types are stored according to the season [3]. After finishing one season, storage reusability sometimes requires the sensor's locations to be changed, and traditional wired connectivity will cost a great deal of time [4]. To acquire and process the monitoring data, a Wireless Sensor Network (WSN) has various advantages such as low cost, wide coverage, self-organization, flexible deployment, and low power consumption and can effectively be used in home automation, the military and several civil fields [5]. However, little research has been reported in applications that are related to fruit condition monitoring and cold storage [3,4].

Methylecyloprpene (MCP), an ethylene antagonist compound, has been of keen interest to post-harvest biologists for the past few years. However, the commercialization of MCP is still limited to apples, pears, tomatoes, melons, and flowers [5]. Thus, researchers are attempting to provide more data on the potential application of MCP for other plant commodities. MCP application for delaying the ripening of bananas has also been studied widely by researches, but inconsistent responses received by researchers for its effects are limiting the commercialization of MCP application for bananas [6]. Hence, further research to study the effects of MCP on bananas using different exposure techniques would be useful for establishing its commercial application.

Bananas are the model for this study due to a combination of scientific and agricultural reasons. They have a distinctive climacteric form for ethylene production and exhalation rate and exhibit ripening by a change in color, flavor, aroma, texture, and other physiological characteristics [7]. Thus, it is very easy to observe the ripening and quality-associated changes during the study. Nutritionally, fresh bananas are a good source of carbohydrates, protein, and fibers with ultimately a good amount of calories and low fat content. They contain approximately 35% carbohydrates, 6–7% fiber, 1–2% protein, and also contain essential features such as phosphorus, vitamin A, potassium, magnesium, iron, calcium, B6, and C [8,9].

Ethylene can greatly affect the value of harvested fruit produce. It can be advantageous or deleterious depending on the product, its ripening stage, and its desired use [10]. Ethylene production is greatly affected by the storage temperature of produce, and ethylene production is generally reduced at low temperatures. However, a lower temperature can cause chilling injury in chilling sensitive produce like banana and can enhance ethylene production. Excess ethylene gas produced during stress-like situations including a senescent breakdown of fruit, chilling-related disorder, and ethylene-induced disorders can cause superficial scald (e.g., in apples), browning (e.g., internal flesh browning of avocados, pineapple), undesirable chemical changes, softening of tissue, and many other negative effects in produce [11]. Fruits are highly perishable commodities; from the moment they are picked. They need proper management of ethylene in post-harvest treatment to maintain their quality, maximum freshness, and shelf life from the field to cold storage and the consumer. To slow down the ripening process of fresh produce, we need to inhibit or slow down the action of ethylene gas. Thus, there will be slow ripening due to less available ethylene [12]. A ZigBee-based monitoring system was demonstrated in [8], to capture feature data (pressure, humidity, sunlight, and temperature) from a remote location for present fruit conditions in containers. Different sensor nodes and Xbee motes are used for transmitting, storing, and analyzing data at the base station. Recently, ethylene antagonist agents have been used for blocking all effects of ethylene gas at the receptor level to provide significant effects for monitoring the ripening process and related other chances [9]. 1-MCP is a well-known ethylene antagonist that suppresses ethylene action by blocking ethylene receptor sites [13]. The alternate of 1-MCP for the ethylene receptor is about ten times better than that of ethylene [10]. There are many papers on the proficiency of 1-MCP ethylene antagonist on constraining the effects of ethylene on the green life of bananas, and 1-MCP concentration mixtures, temperature, and duration of treatment have been under investigation [14,15]. There is no reported, commercially available technique that can be used for handling banana production with 1-MCP. A common technique used to treat fresh produce (generally for all types of produce) with 1-MCP is by exposing fresh produce for several hours to a fixed 1-MCP concentration in a controlled room [16]. For bananas, generally, the same procedure is being used by researchers to treat them at the green stage, before any exogenous ethylene application, which is found to be effective to extend the green life (mature, but unripe stage) of banana. However, there are limited research studies showing its effects on yellow life (at and after partially ripened stage) of banana. An efficient technique to decrease the ethylene-induced ripening of bananas by cooling to 14 °C and using Modified Atmosphere Packaging (MAP) processes has shown auspicious results [15] using WSN-based architecture for remote quality monitoring. However, bananas have to be repacked after the ethylene action into a polymeric film in which the appropriate modified atmosphere will be established. Due to the wide variation in respiration rates of fruits and the different permeability of packaging, MAP is not a feasible independent technique for commercial application [14]. Treatment with 1-MCP seems to be a more convenient method since repacking would not be required. Hence, there is a need for an alternative technique that can provide continuous exposure of 1-MCP to bananas to further delay ripening even after the partially ripened stage. A novel technology known as Controlled Release Packaging (CRP) is being utilized for the delivery of antioxidants and antimicrobials, which can be further extended for the delivery of an ethylene antagonist from the active packaging layer to delay the ripening of bananas. Before establishing the CRP system, study of the physiological responses of partially ripe bananas to planned release (controlled exposure) of 1-MCP and testing its effects on bananas in the packaging system is required.

The contribution of this paper is to achieve improvements in management capability through remote and automatic monitoring. A practical architecture of a WSN-based banana ripening monitoring system is proposed and tested with multiple ANN classification architectures for efficient decision making, and sensor data validation.

The next section discusses the banana ripening process and shows the conceptual illustration of the CRP system. The following section shows the tiered architecture and analyzes the technical requirements (hardware and software) including the role of sensor nodes in monitoring. The following parts of the paper present the ANN tested architecture for data validation and demonstrate the experimental results of the network performance from the sink nodes and a satisfactory diagnosis percentage through classification performance to make a better decision for better monitoring of the present banana condition.

2. Banana Ripening Process

The ripening process brings a sequence of biochemical modifications that are responsible for the pigment formation, change of color, unpredictable smell, starch breakdown, abscission, and finally textural changes of banana [11]. During the stages of ripening, the peel color of banana changes from green to yellow and then a brownish color, as shown in Figure 1. The peel color of banana is the most used indicator to observe the quality by the consumer to decide the actual and consumption quality. During ripening, the firmness of banana decreases, which can also be used as a quality indicator. The tempering of banana mainly instigated by the enzyme activities in the cell wall involves polygalacturonase (PG), Pectate Lyase (PL), Pectin Methyl Esterase (PME), and cellulose, and activities of these enzymes are mainly ethylene dependent [17].

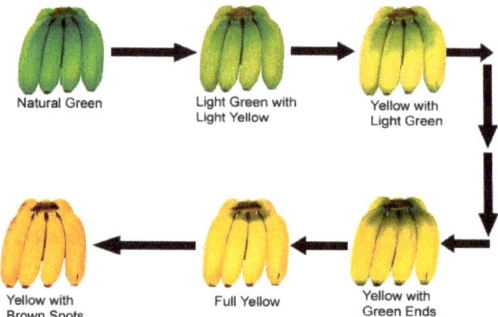

Figure 1. Different stages of the banana ripening process.

Ethylene receptors are embedded in the cells of fruits and the ethylene molecules in the air bind to the receptor sites and help them to ripen [18]. Ethylene performs a series of chemical reactions. These chemical reactions result in fruit ripening by changing the color, aroma, flavor, and composition of fruit (starch, water, and sugar content, etc.) [19]. Table 1, shows the ripening process to measure the condition of banana as follows:

Table 1. Condition measurements for the ripening process of banana [4].

Temperature	16 to 30°C
Comparative Humidity Level	90–95%
Ethylene Concentration	60–100 (pm/kg/h)
Carbon Dioxide (CO_2) Level	Adequate air exchange to prevent CO_2 above 1%

3. Wireless Sensor Network Architecture

In this section, the sensor network architecture is discussed to demonstrate the functionality of the individual sensor nodes and how they work together in the network. The proposed tiered architecture of fruit storage based on a WSN consists of the coordinator sensor node, sink nodes, control unit, and wireless communication system. A node-level intelligent solution is introduced here for significant feature selection and prompt decision at the coordinator level. Many sensors are positioned in the storage container area and a self-organized sensor network architecture is created to monitor behavioral changes in different feature values (including temperature, humidity, ethylene and CO_2, etc.) at different stages of fruit ripening. Figure 2 presents the proposed architecture to of the overall WSN system as follows:

The proposed architecture consists of Xbee sensor nodes that are linked with the router node. To perform a complete and accurate monitoring process, one node in each cluster behaves as a cluster head (router) that is responsible for waking up each neighboring node within the cluster to acquire data and send it to the coordinator for analysis. Rather than the visual inspection of the fruit container condition, every attached node must have aware of their nearest neighboring nodes within the respective cluster and send the values to a router within a specific time frame. Because sometimes sensors are unable to send the right values to the router, the cluster head sets up a mesh network to construct the network backbone and uses relatively more transmit power compared with the other neighboring nodes for better performance.

The role of the coordinator is as a decision-making node that is responsible for deciding the identification of uncertain behavioral areas within the network and passing this decision along with data to the control center. The control center is the brain of the system, which is liable for data logging, data visualization, ANN decision making, and then generating an alarm condition about the fruit ripening process and location to the control administrator.

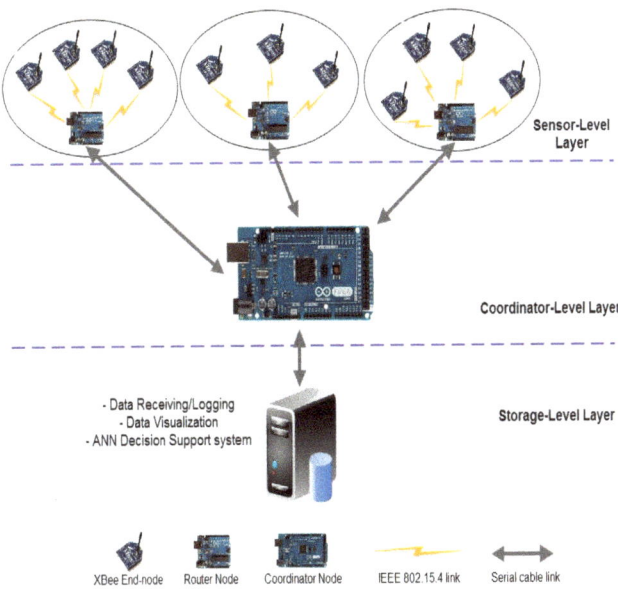

Figure 2. Overall proposed WSN system architecture.

The micro-controller unit controls the operation of the end nodes and stores and deals with the collected feature data along with computational analysis. Figure 3 presents every process of the attached microcontroller that presents a vital task for data fusion in the Arduino board with the sensor and sending the sensor data to the coordinator.

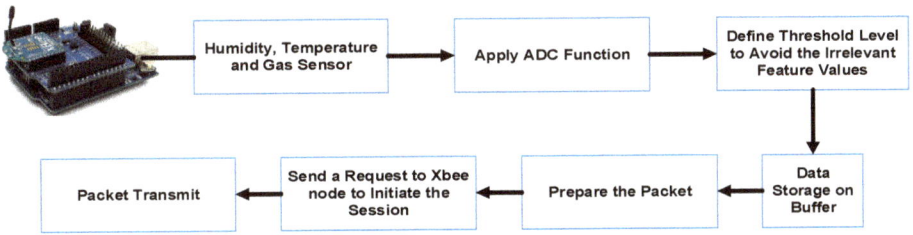

Figure 3. Micro-controller steps for data fusion, Analog-to-Digital Converter (ADC) transformation, packet configuration, and role of Xbee.

At the microcontroller level, the software architecture of sending and receiving the Xbee node is divided into two layers, embedded operating system kernel level and Application Programming Interface (API) level layer, respectively. The first layer provides a low-level transmitting node driver to all attached Xbee devices, and the second layer presents a sensor acquisition component and RF transmitter. The RF transmitter is used to cover the wide area of signal transmission that is attached to the Xbee nodes. Embedded Operating System (OS) provides an efficient software platform of the attached nodes consisting of different libraries and API.

The software architecture flowchart of sensor nodes is presented in Figure 4 including the different steps. In Figure 4, a flowchart of the sensor node is shows the transmission of data and initialization of the Xbee node to register. The software program initializes a request to the Xbee node and a transfer request to the microcontroller, then powers on the sensor node and starts initialization of the protocol stack phase and sends the signal to the network coordinator to assign the network address.

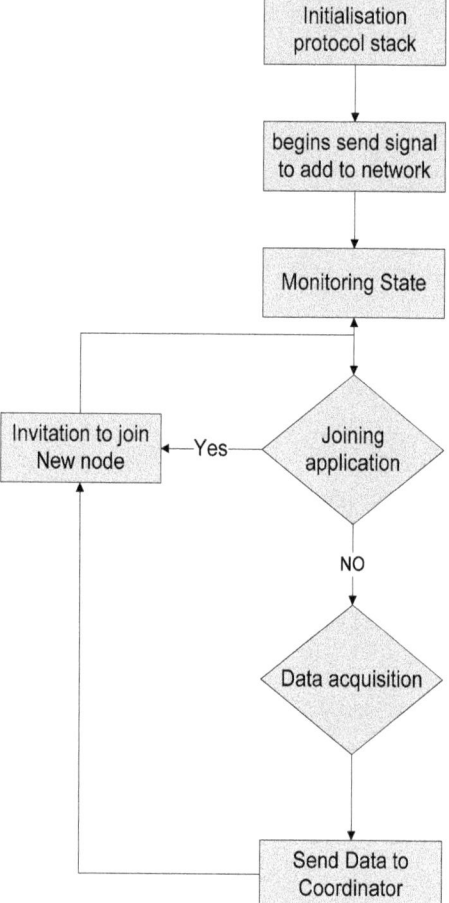

Figure 4. Flowchart of the sensor node.

On the other side, the sink node initializes the protocol stack and the interrupt is released. After that, the software program in the microcontroller instigates configuring the network, and if it is successfully configured, the sink node connects the Xbee node with the coordinator and assigns the physical address, channel number, and network ID and places the nodes into monitoring state. If the receiving node gets some data, it will judge and analyze the sensing node for validation and send feedback to the sending node and a request to the coordinator node for decision making, as shown in Figure 5.

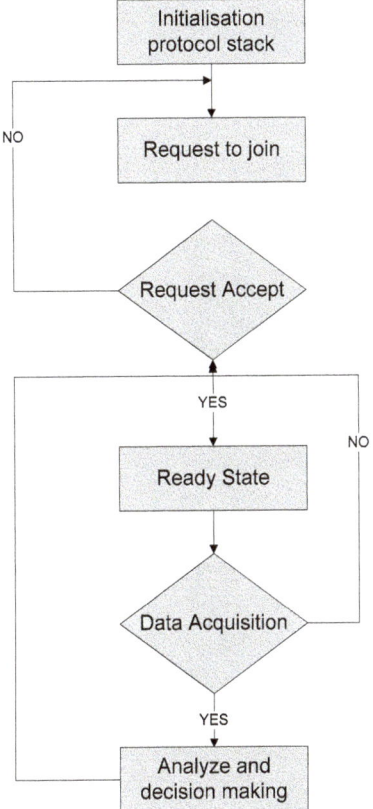

Figure 5. Flowchart of the sink node.

4. WSN-Based Banana Ripening Process Monitoring Experimental Setup

Ethylene is a colorless, odorless, and invisible gas within fruit, in especially high concentrations in banana, with no known harmful consequence on human life [20]. The relatively simple and small ethylene gas molecule contains two carbon atoms along with four hydrogen atoms of the value of 28.05 g/mol^{-1}. As discussed earlier, in the whole progression of banana ripening, ethylene gas is gradually produced and depends on the banana storage time and its weight. Deciding the ethylene concentration level released from banana can be a suitable procedure for evaluating its ripening process. Figure 6 shows the experimental measurement system containing a gas and temperature sensor to detect the current maturity condition of the banana in the container. Measurement of the ethylene gas released from the banana can react with the senor electrolyte that exists inside the sensor voltage. Ethylene concentration is estimated from the electrolyte sensor voltage. It also allows monitoring of the constant flow of ethylene gas emission in the detection system down to 0.01 ppm.

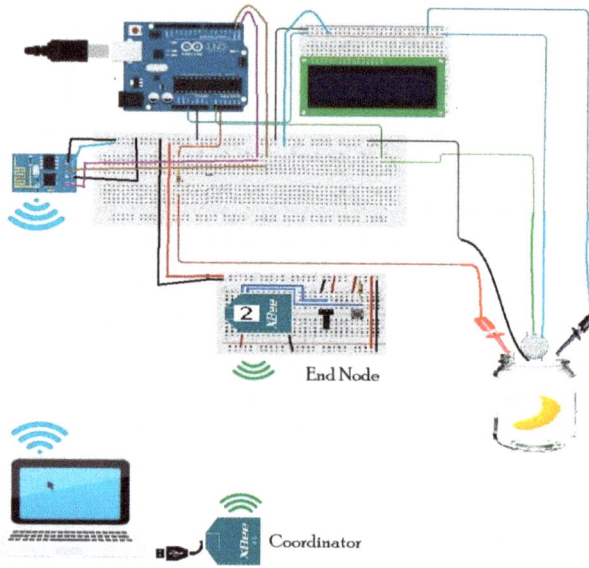

Figure 6. Testbed experimental setup.

5. Neural Network Architecture and Feature Extraction

Significant features are chosen as input values that calculate the fruit temperature, ethylene, and carbon dioxide. The main reason for choosing these features is the relationship with the present condition of the banana ripening process [1]. In this research, Matlab/Simulink script is used to detect the feature values. Feature values are stored in a log file and associated with the microcontroller module for computational analysis. Sometimes, the transform signal method may be difficult to apply with traditional mathematical techniques in the ripening monitoring process [1], while the Feed Forward Neural Network (FFNN) method allows the I/O mapping process with non-linear relationships between all nodes [21]. The NN can recognize the uncharacteristic illustration of transform signals because of the default ability of classification and generalization process, specifically, when the sensitivity of the actual process and response time occur in the repetition of fault sets and create uncertainty in the ripening monitoring process [1]. In the next stage, a multi-layer FFNN is used to identify the uncertainty in sensor values at diverse time slots from the initial point to the ripening process. The proposed architecture of the ANN for banana ripening process monitoring is presented below in Figure 7.

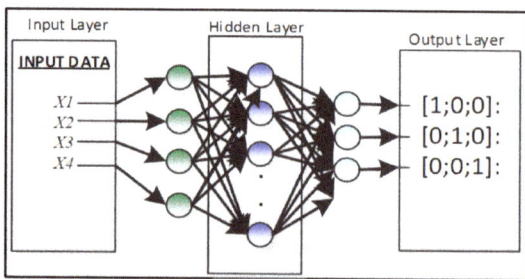

Figure 7. ANN architecture for classification.

The output layer in Figure 8 presents the current state of the banana. It contains a total of four NN nodes, and the hidden layer activation function *logsig* is employed for every proposed output [1]. Three dissimilar forms of architectures ([4 × 8 × 3], [4 × 12 × 3], [4 × 15 × 3]) are practiced to attain the necessary output in an appropriate time frame.

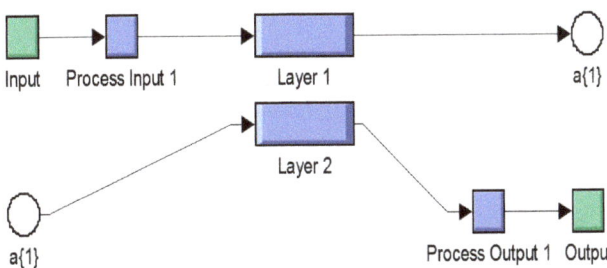

Figure 8. Internal design of NN.

In support of the required target output, classified vector classes are prepared and given by:

- [1;0;0]: Banana Normal Condition,
- [0;1;0]: Banana Rotten Condition,
- [0;0;1]: Banana Unknown Condition.

All feature values (temperature, ethylene, CO_2, and humidity) were stored in text files and allocated values with banana health. Matlab scripts were simulated to combine all the feature sets and produce the range of training data for the testing process and its validation in both healthy and ripening cases. Figure 8 shows the classified internal arrangement of an individual NN for the Xbee node.

Once the NN model is initialized for the non-linear modeling of the overall system, certain NN data have to be measured and targeted node precedents have to be decided for further processes. Hidden layer neurons and the transfer function are initialized to calculate the error criteria and training goal achievement. Then, the initial values of the layers' weight for output is set [1]. The short description and configuration details of the NN layers are defined in Table 2.

Table 2. Details of the implemented ANN.

NN Phases	ANN Configuration for Implementation
Network Type	Feed Forward Neural Network (FFNN)
Learning Scheme	Back Propagation (BP)
Training Target	0.001
Input data of each Xbee node for each experiment.	Four inputs of 1D ANN matrix where all data in each sensing point near node are in a ripening process index.
No. of neurons in the hidden layer.	Diverse N architectures are used with different values of neurons inside the hidden layer. For example, [4 × 8 × 4], [4 × 12 × 4] and [4 × 15 × 4] (see Figure 9).
Vector of classes for the target outputs.	Mathematical matrices refer to the classified vector classes with value 0 or 1.

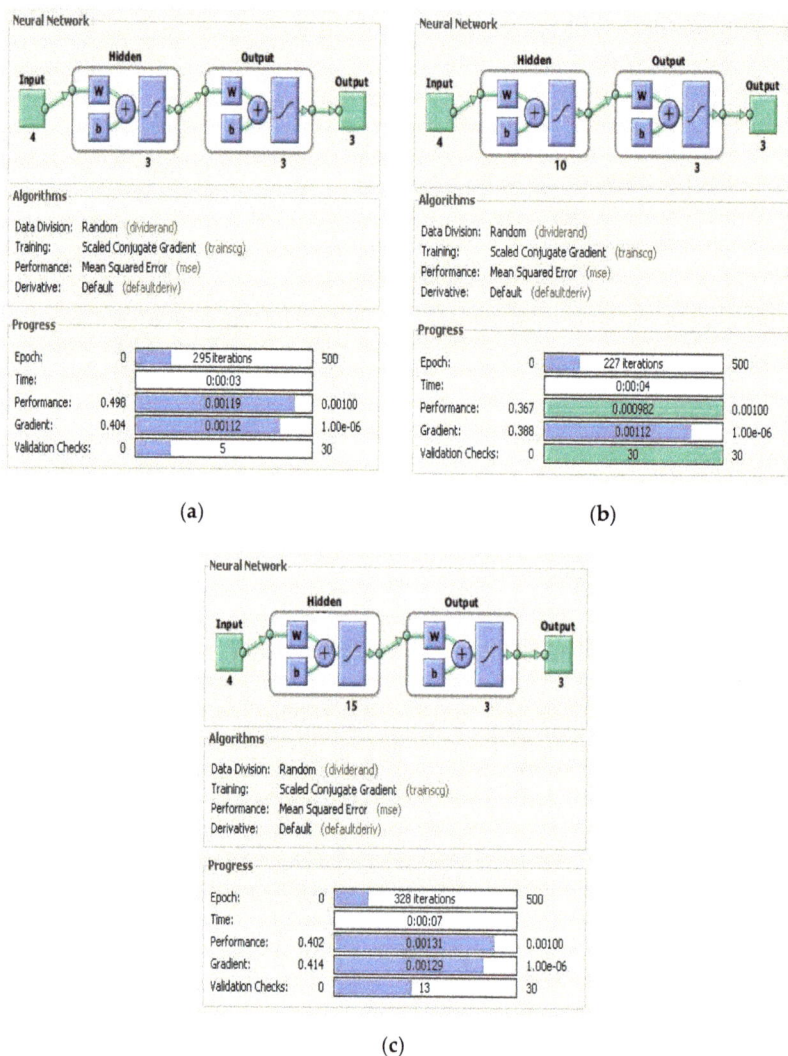

Figure 9. Different ANN architectures chosen for fruit health monitoring: (**a**) [4 × 8 × 3]; (**b**) [4 × 12 × 3]; (**c**) [4 × 15 × 3].

6. Measurements and Results

For the measurement of the sensor values, the ethylene dissolves the electrolyte that counts the electrodes by oxidization at a sampling rate of 50 Hz. A small amount of current is produced by the oxidization reaction. Ethylene gas is measured in ppm under the parched condition of the experimental room and container. Four samples are taken at different time frames according to the banana ripening process. The ethylene sensor measures the gas concentration from 0 to 10 ppm. The practical flow ratio of ethylene gas was measured at 0.4 L/min^{-1} with concentration values of 2.49 ppm (sample 1), 4.89 ppm (sample 2), 8.05 ppm (sample 3), and 10 ppm (sample 4) at high accuracy rate 0.01 ppm. All the data were captured through Xbee nodes and analyzed at the coordinator level. The experimental results demonstrated a dramatic increment of temperature values of fruit when ethylene volume values were high, as shown in Figure 10, as follows:

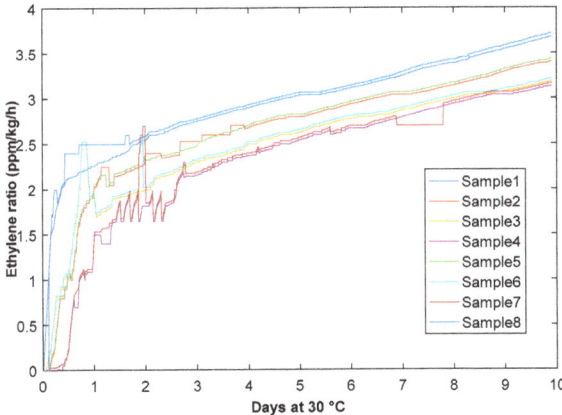

Figure 10. Ethylene production rate during ripening at 30 °C using gas sensor and Xbee mote.

Table 3 shows the different sample values for classification and training purposes that were acquired from sensors and transmitted through Xbee motes.

Table 3. Measured different feature samples.

Sample No.	Temperature	Ethylene	CO_2	Humidity
S1	16	2.49	0.3	60
S2	19	4.89	0.5	69
S3	22	8.05	0.9	75
S4	25	10.0	1.1	84
S5	22	5.5	1.2	81
S6	23	6.2	0.4	79
S7	24	9.4	0.3	89
S8	19	9.9	0.3	79

Figure 11 presents photographs that were taken to show the influence of 1-MCP exposure on the color of ripening bananas. Figure 12 shows the effects of 1-MCP on delaying the ripening color stage of banana using a graphical illustration. All the data shown in Figure 13 were captured from the sensor nodes. Preliminary experiments showed the clear effect of 1-MCP on partially ripened bananas as indicated by a change in color. The 1-MCP treated sample has a better appearance with much less browning and sugar spots. The treated bananas had developed less yellow color even after 7 days of the treatment, whereas the control bananas without any treatment had developed brown spots with a fully developed yellow color.

Figure 11. Influence of 1-MCP behavior on the visual quality of bananas.

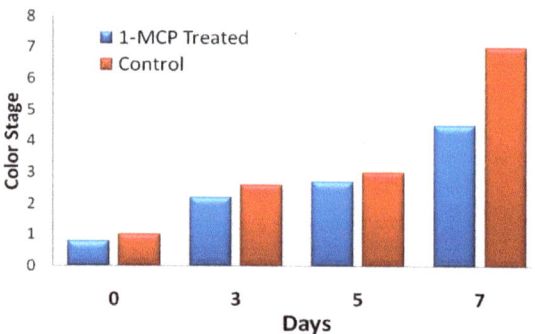

Figure 12. Influence of 1-MCP treatment on color changes of ripening bananas in days.

The following stage is to classify the uncertainty management in Xbee sensor values in diverse frames using multi-layer FFNN from the ripening process. It can be observed in Table 4 that selected NN architecture [4 × 12 × 3] has shown better Mean Squared Error (MSE) performance among other architectures in the classification process. Processing time and reasonable epochs were applied during the training period, which show better efficiency among all the tested NN architectures with less error percentage.

Table 4. Different NN architectures for classification performance.

Arch	Sample	MSE	No. of Epoch	Accuracy	Classification Error
[4 × 8 × 3]	S1	7.79×10^{-2}	72	92.2	7.8
	S2	7.42×10^{-2}	65	93.7	6.3
	S3	7.45×10^{-2}	75	92.4	7.6
	S4	7.99×10^{-2}	101	91.9	8.1
	S5	7.01×10^{-2}	66	90.2	9.8
	S6	6.89×10^{-2}	62	89.1	10.9
	S7	6.91×10^{-2}	84	92.5	7.5
	S8	7.02×10^{-2}	92	94.5	5.5
[4 × 12 × 3]	S1	8.27×10^{-2}	117	96.2	3.8
	S2	9.01×10^{-2}	125	96.3	3.7
	S3	8.98×10^{-2}	132	97.4	2.6
	S4	9.29×10^{-2}	131	97.1	2.9
	S5	7.49×10^{-2}	110	95.9	4.1
	S6	7.33×10^{-2}	98	97.8	2.2
	S7	7.38×10^{-2}	101	96.7	3.3
	S8	7.54×10^{-2}	104	96.6	3.4
[4 × 15 × 3]	S1	7.98×10^{-2}	401	91.9	8.1
	S2	6.45×10^{-2}	310	90.2	9.8
	S3	6.05×10^{-2}	400	83.4	17.6
	S4	7.13×10^{-2}	372	87.2	13.8
	S5	7.13×10^{-2}	400	85.3	14.7
	S6	7.13×10^{-2}	386	88.3	11.7
	S7	7.13×10^{-2}	398	89.1	10.9
	S8	7.13×10^{-2}	402	86.6	13.4

The next step is to measure the data validation coming from the Xbee motes. Figure 13 shows the NN architecture training performance chart of the NN architecture [4 × 12 × 3], which achieved a reasonable and excellent performance result during the neural network testing. After computing the NN testing, the next phase is to measure the combination of the confusion matrix to achieve the training target error. To build the confusion matrix network, test highlight information is provided into the NN system, which is shown in Figure 14. In the graphs, the confusion grid holds the training data regarding the analysis between the target and output classes. Three procedural stages, preparing, testing and approval of the banana maturing process, were tested individually to measure the performance of the system. Four vertical and horizontal classes were used to illustrate the accurate testing of the data validation process to reflect all the sample targeted values of input sets. The green cells show those data groups of trail classes that are classified as accurate and successful testing during the training process. In Figure 14, each corner demonstrates the number of cases that are tested through the NN architecture and again the number of cases to decide the targeted condition of banana ripening measurement data. The red cells represent those data sets that are wrongly classified or might be not validated during testing. The blue cell shows the overall percentage depends on test cases that are classified correctly in green cells and another way around on red cells.

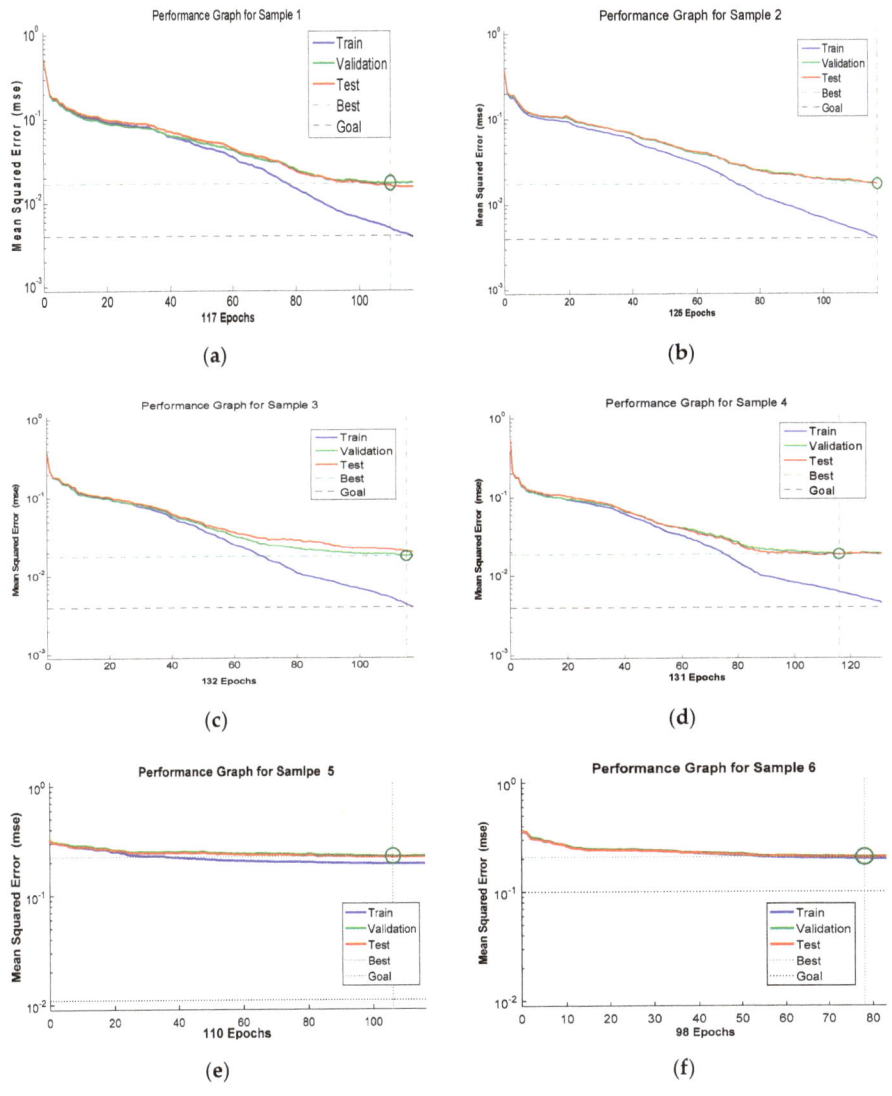

Figure 13. *Cont.*

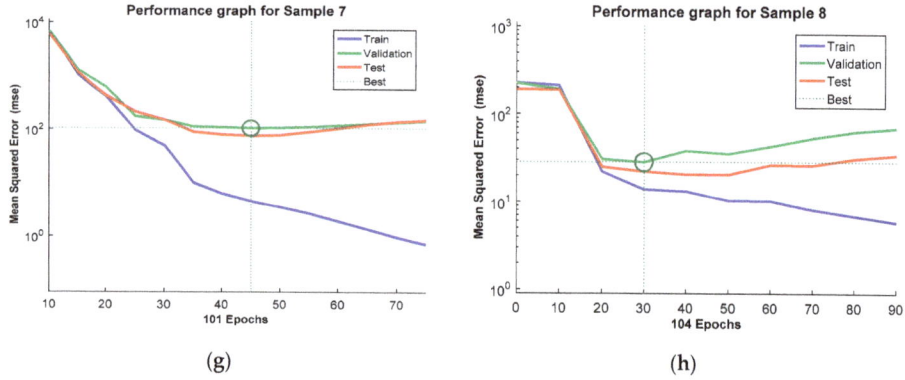

(g) (h)

Figure 13. Performance graphs samples using [4 × 12 × 3] neural network architecture.

It can be easily being observed from Figure 14, each class has a maximum under 1200 testing trails in the green cells to show the accurate validation of datasets to observe the targeted output percentage. If we look at sample 1, a very low number of datasets are incorrectly classified as compared with the green cell. Target class 1 obtained 13 types of incorrectly classified sample trials in output class 2, 35 in class 3, and only 5 in target output class 4. Overall, 94.7 percent accuracy was achieved in the gray cell and a 5.3 percent error rate was identified, which shows the efficiency of the proposed architecture. All the targeted class aggregated output was calculated in the blue cell, which is 96.2% with only a 3.8% error rate, which is the satisfactory ratio. We can observe in sample 3, the cumulative accuracy percentage of all the test classes is 97.4% with only 2.6% error rate that are incorrectly classified as dataset trails within a reasonable processing time frame. This shows the proficiency in the proposed ANN method to decrease the amount of imprecision in analysis and validate the sample trained data.

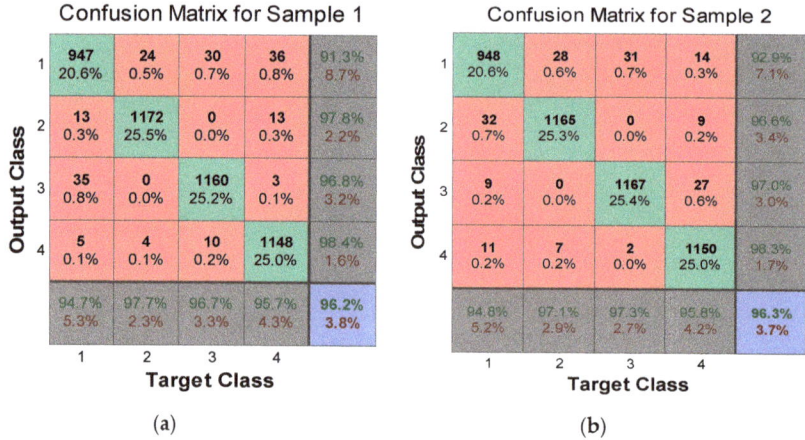

(a) (b)

Figure 14. *Cont.*

Figure 14. Confusion matrices of eight data samples using targeted and output classes.

7. Conclusions and Future Directions

The feasibility of Xbee-based motes was experimentally demonstrated for monitoring the banana ripening process while in storage. The role of the network architecture of the Xbee sensor node and sink end-node was discussed in detail regarding their ability to monitor the condition of all the required diagnosis parameters and stages of banana ripening. Different significant features (temperature, humidity, ethylene, and CO_2) are selected from sample data sets and extracted at different time frames for analysis and training. For classification and training purposes, a supervised ANN architecture is presented to show the efficiency of the network in the diagnosis of the current condition of banana (healthy/rotten). The simulated results showed the precise and general behavior of the ripening process of different parameters, especially ethylene gas, on fruit condition. To improve the mean squared error rate, three types of ANN architecture were tested and [$4 \times 12 \times 3$] demonstrated a reasonable quantity of hidden layers with a high accuracy rate in the classification of features vector.

Future development of this research would be extended toward the utilization of multiple fruit and vegetables for the diagnosis of their type and existing condition in a normal atmosphere environment and cold storage refrigeration. A multilayered structure in which the outer layer is an effective barrier to 1-MCP can be used to prevent loss of 1-MCP gas molecules to the general environment. A comparison of different Xbee motes would be an interesting area with another artificial intelligence technique for better communication between nodes and to precisely predict when different parameter measurements can be taken in parallel with fruits and vegetables to create the complexity.

Author Contributions: Conceptualization, S.A. (Saud Altaf) and S.A. (Shafiq Ahmed); methodology, S.A. (Saud Altaf) and S.A. (Shafiq Ahmed); software, S.A. (Saud Altaf); validation, S.A. (Saud Altaf); formal analysis, S.A. (Saud Altaf), S.A. (Shafiq Ahmed). and M.W.S.; resources, S.A. (Shafiq Ahmed) and M.Z.; data curation, S.A. (Saud Altaf); writing—original draft preparation, S.A. (Saud Altaf) and S.A. (Shafiq Ahmed); writing—review and editing, S.A. (Saud Altaf) and S.A. (Shafiq Ahmed) and M.Z. All authors have read and agreed to the published version of the manuscript.

Funding: Deanship of Scientific Research at King Saud University for funding this work through research group no. (RG- 1438-089).

Acknowledgments: The authors extend their appreciation to the Deanship of Scientific Research at King Saud University for funding this work through research group no. (RG- 1438-089). The authors would like to give thanks to Pir Mehr Ali Shah Arid Agriculture University, Pakistan, for collecting datasets and the testbed environment.

Conflicts of Interest: The authors declare no conflict of interest.

References

1. Chen, L.-Y.; Wu, C.-C.; Chou, T.-I.; Chiu, S.-W.; Tang, K.-T. Development of a Dual MOS electronic nose/camera system for improving fruit ripeness classification. *Sensors* **2018**, *18*, 3256. [CrossRef] [PubMed]
2. Altaf, S.; Soomro, W.; Rawi, M.I.M. Student Performance Prediction using Multi-Layers Artificial Neural Networks: A Case Study on Educational Data Mining. In Proceedings of the 2019 3rd International Conference on Information System and Data Mining—ICISDM, Houston, TX, USA, 6–8 April 2019.
3. Ruan, J.; Shi, Y. Monitoring and assessing fruit freshness in IOT-based e-commerce delivery using scenario analysis and interval number approaches. *Inf. Sci.* **2016**, *373*, 557–570. [CrossRef]
4. Badia-Melis, R.; Ruiz-Garcia, L.; Garcia-Hierro, J.; Villalba, J.I.R. Refrigerated fruit storage monitoring combining two different wireless sensing technologies: RFID and WSN. *Sensors* **2015**, *15*, 4781–4795. [CrossRef] [PubMed]
5. Ouni, S.; Ayoub, Z.T. Cooperative Association/Re-association Approaches to optimize energy consumption for Real-Time IEEE 802.15.4/ZigBee wireless sensor networks. *Wirel. Pers. Commun.* **2013**, *71*, 3157–3183. [CrossRef]
6. Gol, N.B.; Rao, T.V.R. Banana Fruit Ripening as Influenced by Edible Coatings. *Int. J. Fruit Sci.* **2011**, *11*, 119–135. [CrossRef]
7. Matindoust, S.; Baghaei, M.N.; Shahrokh, M.A.; Zou, Z.; Zheng, L. Food quality and safety monitoring using gas sensor array in intelligent packaging. *Sens. Rev.* **2016**, *36*, 169–183. [CrossRef]

8. Chandravathi, C.; Mahadevan, K.; Sheela, S.K.; Hashir, A.R. Detection of Oxytocin in fruits and vegetables using wireless sensor. *Indian J. Sci. Technol.* **2016**, *9*, 16. [CrossRef]
9. Wang, X.; He, Q.; Matetic, M.; Jemric, T.; Zhang, X. Development and evaluation on a wireless multi-gas-sensors system for improving traceability and transparency of table grape cold chain. *Comput. Electron. Agric.* **2017**, *135*, 195–207. [CrossRef]
10. Altaf, S.; Al-Anbuky, A.; Gholam, H.H. Fault Diagnosis in a Distributed Motor Network Using Artificial Neural Network. In Proceedings of the International Symposium on Power Electronics, Electrical Drives, Automation and Motion, Ischia, Italy, 18–20 June 2014; pp. 190–197.
11. Riad, M.; Elgammal, A.; Elzanfaly, D. Efficient Management of Perishable Inventory by Utilizing IoT. In Proceedings of the IEEE International Conference on Engineering, Technology and Innovation (ICE/ITMC), Stuttgart, Germnay, 17–20 June 2018; pp. 1–9.
12. Krairiksh, M.; Varith, J.; Kanjanavapastit, A. Wireless sensor network for monitoring maturity stage of fruit. *Wirel. Sens. Netw.* **2011**, *3*, 318–321. [CrossRef]
13. Altaf, S.; Soomro, M.; Mehmood, M. Fault diagnosis and detection in industrial motor network environment using knowledge-level modelling technique. *Model. Simul. Eng.* **2017**, *2017*, 1–10. [CrossRef]
14. Zhuang, J.; Hou, C.; Tang, Y.; He, Y.; Guo, Q.; Miao, A.; Zhong, Z.; Luo, S. Assessment of external properties for identifying banana fruit maturity stages using optical imaging techniques. *Sensors* **2019**, *19*, 2910. [CrossRef] [PubMed]
15. Gomez, S.M.; Vergara, A.; Ruiz, H.; Safari, N.; Elayabalan, S.; Ocimati, W.; Blomme, G. AI-powered banana diseases and pest detection. *Plant Methods* **2019**, *15*. [CrossRef]
16. Ibba, P.; Falco, A.; Rivadeneyra, A.; Lugli, P. Low-Cost Bio-Impedance Analysis System for the Evaluation of Fruit Ripeness. In Proceedings of the 2018 IEEE Sensors, New Delhi, India, 28–31 October 2018; pp. 1–4.
17. Sabilla, I.A.; Wahyuni, C.S.; Fatichah, C.; Herumurti, D. Determining Banana Types and Ripeness from Image using Machine Learning Methods. In Proceedings of the International Conference of Artificial Intelligence and Information Technology (ICAIIT), Yogyakarta, Indonesia, 13–15 March 2019; pp. 407–412.
18. AboBakr, A.; Mohsen, M.; Said, L.A.; Madian, A.H.; Elwakil, A.S.; Radwan, A.G. Banana Ripening and Corresponding Variations in Bio Impedance and Glucose Levels. In Proceedings of the Novel Intelligent and Leading Emerging Sciences Conference (NILES), Giza, Egypt, 28–30 October 2019; pp. 130–133.
19. Manzoli, A.; Steffens, C.; Paschoalin, R.T.; Correa, A.A.; Alves, W.F.; Leite, F.L.; Herrmann, P.S.P. Low-Cost Gas Sensors Produced by the Graphite Line-Patterning Technique Applied to Monitoring Banana Ripeness. *Sensors* **2011**, *11*, 6425–6434. [CrossRef] [PubMed]
20. Chowdhury, A.; Bera, T.K.; Ghoshal, D.; Chakraborty, B. Studying the Electrical Impedance Variations in Banana Ripening Using Electrical Impedance Spectroscopy (Eis). In Proceedings of the Third International Conference on Computer, Communication, Control and Information Technology (C3IT), Hooghly, India, 7–8 February 2015; pp. 1–4.
21. Sanaeifar, A.; Mohtasebi, S.S.; Ghasemi-Varnamkhasti, M.; Siadat, M. Application of an Electronic Nose System Coupled with Artificial Neural Network for Classification of Banana Samples during shelf-Life Process. In Proceedings of the International Conference on Control, Decision and Information Technologies (CoDIT), Metz, France, 3–5 November 2014; pp. 753–757.

 © 2020 by the authors. Licensee MDPI, Basel, Switzerland. This article is an open access article distributed under the terms and conditions of the Creative Commons Attribution (CC BY) license (http://creativecommons.org/licenses/by/4.0/).

Article

Unidimensional ACGAN Applied to Link Establishment Behaviors Recognition of a Short-Wave Radio Station

Zilong Wu, Hong Chen and Yingke Lei *

College of Electronic Countermeasures, National University of Defense Technology, Hefei 230037, China; wuzilong@nudt.edu.cn (Z.W.); ch2sun@mail.ustc.edu.cn (H.C.)
* Correspondence: 22920142204021@stu.xmu.edu.cn

Received: 25 June 2020; Accepted: 28 July 2020; Published: 31 July 2020

Abstract: It is difficult to obtain many labeled Link Establishment (LE) behavior signals sent by non-cooperative short-wave radio stations. We propose a novel unidimensional Auxiliary Classifier Generative Adversarial Network (ACGAN) to get more signals and then use unidimensional DenseNet to recognize LE behaviors. Firstly, a few real samples were randomly selected from many real signals as the training set of unidimensional ACGAN. Then, the new training set was formed by combining real samples with fake samples generated by the trained ACGAN. In addition, the unidimensional convolutional auto-coder was proposed to describe the reliability of these generated samples. Finally, different LE behaviors could be recognized without the communication protocol standard by using the new training set to train unidimensional DenseNet. Experimental results revealed that unidimensional ACGAN effectively augmented the training set, thus improving the performance of recognition algorithm. When the number of original training samples was 400, 700, 1000, or 1300, the recognition accuracy of unidimensional ACGAN+DenseNet was 1.92, 6.16, 4.63, and 3.06% higher, respectively, than that of unidimensional DenseNet.

Keywords: unidimensional ACGAN; signal recognition; data augmentation; link establishment behaviors; DenseNet; short-wave radio station

1. Introduction

In the field of electronic reconnaissance, only a few Link Establishment (LE) behaviors signals of short-wave radio stations can be detected by non-collaborative sensors. Therefore, unidimensional ACGAN is utilized to get more LE behavior signals, avoiding the problem of lacking a large number of samples to train neural network. Actually, researchers in the military field are very concerned about how to recognize LE behaviors of non-collaborative radio stations, which help the researchers infer network topology of these radio stations. For example, if we find that the behavior of a radio station is Call behavior, some other radio stations that communicate with the radio station appear after a period of time. Hence, all the above radio stations belong to the same communication network. We can also infer how many radio stations are in the current communication network by analyzing the newly emerged electromagnetic signals.

The short-wave radio station refers to wireless communication equipment, of which the working frequency is 3–30 MHz. The most classic way of communication in the military field is to use a short-wave radio station due to its simple equipment, low power, and mature technologies. At present, most of the commands among the brigade, the battalion, and the company are transmitted via a short-wave radio station. Therefore, research on the LE behaviors of short-wave radio stations is of great significance to intelligence reconnaissance. The LE behaviors of a short-wave radio station are a kind of communication behavior of a radio station, which means a radio station starts a specific communication for different purposes. In fact, the seven kinds of LE signals correspond to seven kinds of LE behaviors consisting of Call behavior, Handshake behavior, Notification behavior, Time Offset behavior, Group Time Broadcast behavior, Broadcast behavior, and Scanning Call behavior [1]. Meaningfully, all research in military area desires to acquire the topology of the network where the radio station is located, and the status of the radio station as a node in the communication network, which could be facilitated by LE behavior recognition. For example, if a radio station frequently conducts Call behavior, the tactical status of the radio station is very important, and the owner is likely to be the commander in their organization. On the other hand, if a radio station seldom conducts Call behavior, the radio station may have a low tactical status. However, only radio LE behavior signals can be collected by non-collaborative sensors, which means there is no help for protocol standard of the radio station. Moreover, only a small number of LE behavior signals can be acquired from enemy radio stations, which increases the difficulty of LE behavior recognition. Limited to the unknown communication protocol standard and only a few collected labeled signals, LE behaviors can still be recognized by using the proposed algorithm in this work to directly process physical layer signals.

At present, the research on LE behavior recognition of radio stations at home and abroad is in the initial stage. Research has been done [2–5] on the communication behaviors of radio signals, but they all differ from a short-wave radio station's LE behavior recognition. The research done in [6] uses a novel improved unidimensional DenseNet to recognize the LE behaviors of a short-wave radio station, and the whole recognition process does not need the help of communication protocol standard, avoiding complex signal feature transformation. However, when there are only a few samples with labels, the improved unidimensional DenseNet recognition accuracy needs to be further improved. In the actual electronic countermeasure environment, only a small amount of LE behavior signals can be obtained. In view of this special case, further research is needed. In terms of the implementation issue, it is possible to generalize current research results to the short wave radio station with constraints by combining the technologies in intelligent control and ideas in this work [7,8].

The Generative Adversarial Network (GAN) [9] can generate fake samples that are very similar to the real samples and then achieve the purpose of data augmentation, which indicates that GAN has great potential to solve the problem of a few LE behavior signals. The role of GAN can be roughly divided into style transfer [10–13] and data augmentation [14–16] according to application scenarios. In the field of style transfer, the pix2pix [17] is of epoch-making significance. Isola et al. realized the style transfer of paired images through pix2pix rather than simple pixel-to-pixel mapping. BicycleGAN further improved the performance of pix2pix [18]. CycleGAN [19] and DiscoGAN [20] are able to realize style transfer without using pairs of images. In the field of data augmentation, the research of GAN mainly focuses on the improvement of network structure. On the basis of the GAN model, Conditional Generative Adversarial Network (CGAN) [21,22] adds additional conditional information to the generator and discriminator to guide the training of the network model, and finally CGAN can generate samples corresponding to the specified labels. Energy-based Generative Adversarial Network (EBGAN) [23] introduced the concept and method of energy into GAN and regarded the discriminator as an energy function. ACGAN [24,25] added an auxiliary classifier to the output of the discriminator to improve the performance of GAN, and ACGAN also proposed using the class of each sample to update and improve the loss function, which significantly improved the performance of the network model. In the field of LE behavior recognition of short-wave radio stations, ACGAN can generate some labeled signals according to a small number of labeled signals, which achieves the purpose of data augmentation. However, the original ACGAN model is only applicable to the field of computer vision. Therefore, in this paper, unidimensional ACGAN is proposed to achieve data augmentation of unidimensional LE behavior signals.

Aiming at solving the problem that there are only a few LE behavior signals with labels of short-wave radio stations, a new unidimensional ACGAN is proposed to acquire more LE behavior signals. The following is the overall idea of this work: According to the short-wave communication protocol standard (MIL-STD-188-141B), seven kinds of LE behavior signals are simulated, and then these signals are used to verify the feasibility and effectiveness of the proposed algorithm model. Firstly, a small number of real LE behavior signals were randomly selected to train unidimensional ACGAN. In order to explore the effectiveness of unidimensional ACGAN, a new unidimensional Convolutional Auto-Encoder (CAE) which was used to demonstrate the deep features distribution of these generated samples was proposed. Then, the generated samples were mixed with the initial training samples to form a new training set, and the new training set was used to train the recognition network model. Finally, the LE behaviors of a short-wave radio station were recognized based on a small number of labeled samples. The whole training and recognition process of algorithm model did not need the help of communication protocol, which met the demand of real electronic countermeasures. Meanwhile, it also showed that the proposed algorithm model had the value of practical application.

Our main contributions are as follows:

- A method based on ACGAN+DenseNet was proposed to recognize radio stations' LE behaviors without the communication protocol standard, which means a lot in the filed in the military field;
- A new ACGAN called unidimensional ACGAN was presented to generate more LE behavior signals. The presented ACGAN was able to directly process and generate unidimensional electromagnetic signals, while the original ACGAN is mostly used in the field of computer vision rather than unidimensional signals;
- We used a unidimensional Convolutional Auto-Encoder to represent deep features of the generated samples, which provided a novel way to verify the reliability of ACGAN when applied in the generation of electromagnetic signals.

The idea adopted in this work provides a reference for research on communication behaviors of non-collaborative radio stations. Once we have mastered the communication behaviors of radio stations belonging to a communication network, we can effectively infer the topological relationships between the radio stations. We hope that more people will be interested in research on LE behavior recognition of non-collaborative radio stations.

The remainder of this paper is organized as follows: Section 2 introduces the recognition algorithm model in detail and Section 3 introduces the experimental results and analysis. Finally, Section 4 shows our conclusion.

2. Methods

2.1. ACGAN

ACGAN, as a variant of Conditional GAN (CGAN), is widely used to generate fake "real" data. There are two modules in all different GAN models, which include a generator module and discriminator module. In the game against each other between the generator and discriminator during the training, the generator and discriminator can reach the ideal state. Then the generator can generate fake samples which are very similar to the real samples. The differences between ACGAN and CGAN are that ACGAN not only uses information of data's labels for training, but also provides the category judgment of different samples. The structures of GAN and ACGAN are shown in Figure 1.

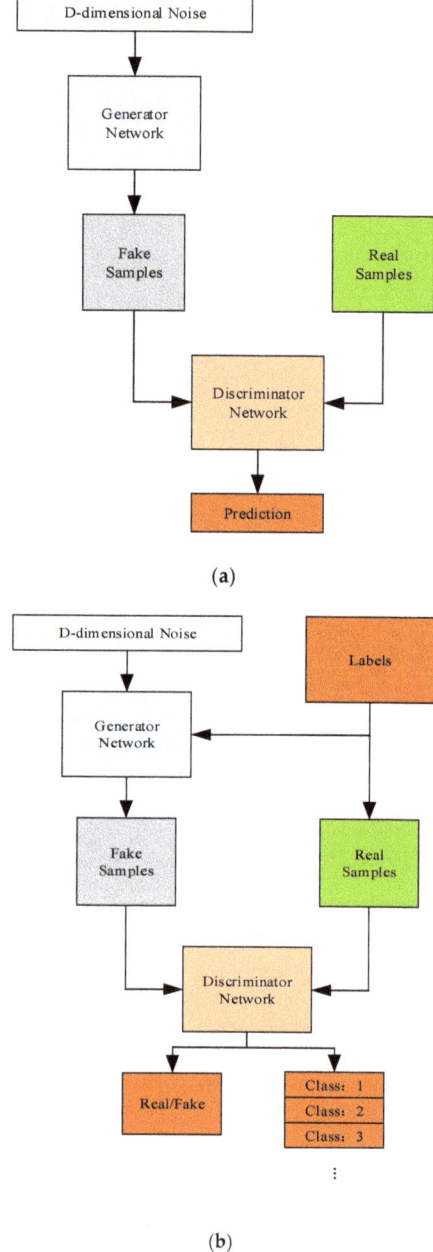

Figure 1. Structures of GAN and ACGAN. (**a**) Structure of original GAN; (**b**) structure of original ACGAN.

Compared with GAN, ACGAN generates more similar samples and it can also generate many samples with labels at a time. Therefore, ACGAN is very suitable for data augmentation and thus we are able to easily acquire more LE behavior signals with different labels.

During the training of ACGAN, the loss function of discriminator is expressed as:

$$L_D = L_S + L_C \tag{1}$$

$$L_S = E[\log P(S = real|x_{real})] + E[\log P(S = fake|x_{fake})] \tag{2}$$

$$L_C = E[\log P(C = c|x_{real})] + E[\log P(C = c|x_{fake})] \tag{3}$$

The loss function of the generator is expressed as:

$$L_D' = L_C' - L_S' \tag{4}$$

$$L_C' = E[\log P(C = c|x_{fake})] L_C = E[\log P(C = c|x_{fake})] \tag{5}$$

$$L_S' = E[\log P(S = fake|x_{fake})] \tag{6}$$

During the training, the discriminator and generator alternately update parameters in the network model. The goals of network optimization are to maximize the L_D of discriminator and L_D' of generator. In other words, the discriminator tries to distinguish the real samples from the fake samples, and the generator tries to make the generated samples be judged by the discriminator as the real samples. Finally, the loss function of the network model tends to be stable and the process of training is over.

2.2. Unidimensional ACGAN

Because it is often necessary to transform the electromagnetic signals into their deep features, the features are then treated as images. The LE behavior signals of a short-wave radio station cannot be directly put into a traditional ACGAN. The traditional method of signal recognition is to transform signals into their characteristic domain, and then these signals are processed as two-dimensional images.

However, the LE behavior signals of a short-wave communication station have few differences and their modulation is almost the same, except for the difference of the valid 26 bits. Thus, the characteristic transformations of LE behavior signals are incapable of getting better performance in recognizing LE behavior signals. Therefore, a new unidimensional ACGAN is proposed in this work. The unidimensional ACGAN was trained directly by the unidimensional LE behavior signals to achieve the purpose of getting more LE behavior signals with labels.

The structure of generator in unidimensional ACGAN is shown in Table 1.

Table 1. The structure of generator in unidimensional ACGAN.

Layer	Input Size	Output Size
Input	500(Noise)+1(Label)	500
Fully Connected	500	1472 * 30
Reshape	1472 * 30	(1472, 30)
BN(0.8)	(1472, 30)	(1472, 30)
UpSampling1D	(1472, 30)	(2944, 30)
Conv1D(KS = 3,1(s))	(2944, 30)	(2944, 30)
Activation("ReLU")+BN(0.8)	(2944, 30)	(2944, 30)
UpSampling1D	(2944, 30)	(5888, 30)
Conv1D(KS = 3,1(s))	(5888, 30)	(5888, 20)
Activation("ReLU")+BN(0.8)	(5888, 20)	(5888, 20)
Conv1D(KS = 3,1(s))	(5888, 20)	(5888, 1)
Output(Activation("tanh"))	(5888, 1)	(5888, 1)

As shown in Table 1, BN (0.8) represents the Batch Normalization (BN) layer, and the momentum is equal to 0.8. UpSampling1D means the data is upsampled by 2 times. Conv1D (Kernel Size (KS) = 3,1 (s)) denotes unidimensional convolution operation, the Kernel Size (KS) of which is 3 and convolutional stride is 1. Activation ('*') denotes that the activation function is *.

The structure of the discriminator in unidimensional ACGAN is shown in Table 2.

Table 2. The structure of discriminator in unidimensional ACGAN.

Layer	Input Size	Output Size
Input	(5888, 1)	(5888, 1)
Conv1D(KS = 3,2(s))	(5888, 1)	(2944, 20)
Actication("LeakyReLU(0.2)")	(2944, 20)	(2944, 20)
Dropout(0.25)	(2944, 20)	(2944, 20)
Conv1D(KS = 3,2(s))	(2944, 20)	(1472, 20)
Actication("LeakyReLU(0.2)")	(1472, 20)	(1472, 20)
Dropout(0.25)	(1472, 20)	(1472, 20)
Conv1D(KS = 3,2(s))	(1472, 20)	(736, 30)
Actication("LeakyReLU(0.2)")	(736, 30)	(736, 30)
Dropout(0.25)	(736, 30)	(736, 30)
Conv1D(KS = 3,2(s))	(736, 30)	(368, 30)
Actication("LeakyReLU(0.2)")	(368, 30)	(368, 30)
Dropout(0.25)	(368, 30)	(368, 30)
Flatten	(368, 30)	368 * 30
Output	368 * 30	1(real/fake) 7(class)

As shown in Table 2, Conv1D (KS = 3,2 (s)) denotes unidimensional convolution operation, the Kernel Size (KS) of which is 3 and convolutional stride is 2. Activation ("LeakyReLU(0.2)") means that the activation function is LeakyReLU(γ = 0.2). 7(class) represents the class of sample output by the discriminator.

Up to this point, the proposed unidimensional ACGAN has been shown in detail in Tables 1 and 2. For LE behavior signals of a short-wave radio station, unidimensional ACGAN could be used to generate some data samples with different labels, and the original samples and generated samples could be combined to obtain a new training set. Then unidimensional DenseNet could be trained to effectively recognize different LE behaviors of a radio station. The method in this work is able to improve the accuracy of signals recognition.

2.3. LE Behavior Recognition Algorithm Based on Unidimensional ACGAN+DenseNet

When the number of LE behavior signals with labels of a short-wave radio station is relatively small, ta combination of unidimensional ACGAN and unidimensional DenseNet can improve the accuracy of LE behavior recognition. Firstly, the powerful generative adversarial ability of unidimensional ACGAN was used to generate fake "real" samples according to original training samples. Then real samples and generated fake "real" samples were mixed to form a new training set. The unidimensional DenseNet we proposed in [6] was able to effectively automatically extract the deep features of LE behavior signals, and finally Softmax classifier was used to realize the recognition of a radio station's different LE behaviors. The whole process of the algorithm's recognition did not need the help of the communication protocol, which provided a new idea for electronic reconnaissance. The framework of the recognition algorithm is shown in Figure 2.

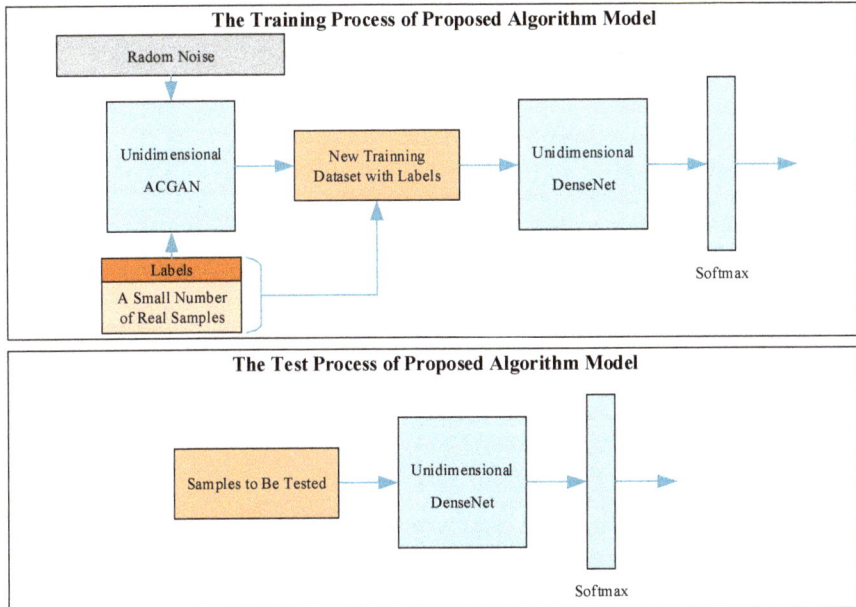

Figure 2. The framework of a short-wave radio station's LE behavior recognition algorithm.

The steps of the recognition algorithm are as follows:

Step 1: Firstly, a small number of samples with labels expressed as **X** are randomly selected from the real samples. Another sample expressed as **Y**, the dimensions of which are (500, 1), are generated by random noise, and these samples are randomly labeled as 0, 1, 2, 3, 4, 5, and 6, corresponding to seven kinds of LE behaviors.

Step 2: The **X** and the **Y** are put into unidimensional ACGAN as training data set. Unidimensional AGCAN begins to be trained. These parameters in generators and discriminators are updated alternately.

Step 3: According to the epoch and batch size, which we have set, repeat **Steps 1 and 2**.

Step 4: Randomly generate some noise sequences with specific labels, and then input these sequences into unidimensional ACGAN that have been already trained. Some fake samples with specific labels, expressed as **Z**, are also generated. Then a new training set is formed by mixing the real samples **X** with the generated samples **Z**.

Step 5: The batch size and epoch are set properly, and the new training set is used to train unidimensional DenseNet.

Step 6: A short-wave radio station's LE behaviors are recognized by the trained unidimensional DenseNet.

3. Experimental Results and Analysis

Experimental environment: Intel (R) Core (TM) i9-9900K CPU, NVIDIA RTX TITAN×1, TensorFlow 1.12.0, and Keras 2.2.5.

3.1. LE Behavior Signals Dataset

According to the third-generation short-wave communication protocol standard (MIL-STD-188-141B), seven kinds of LE behavior signals only differ in the valid bits (26 bits) in their data frame. The LE behavior signals used in experiments are generated as shown in Figure 3.

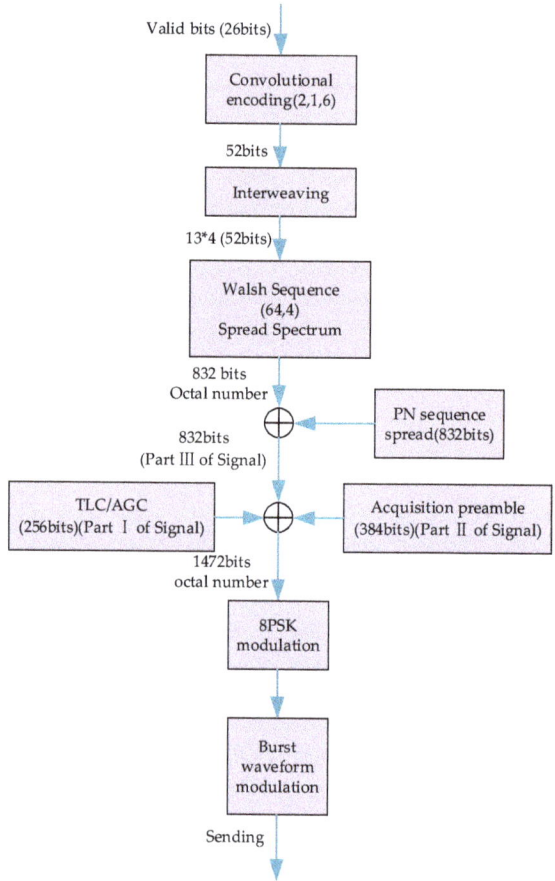

Figure 3. How the LE behavior signals used in experiments are generated.

As shown in Figure 3, TLC/AGC represent the Transmit Level Control process and Automatic Gain Control process. The short-wave communication protocol standard (MIL-STD-188-141B) stipulates that the carrier frequency is 1800 Hz, and the raised cosine filter is used to form waveform. The seven kinds of LE behavior signals, the dimensions of which were (5888, 1), were obtained, and the size of each behavior signal's dimension was (5888, 1). In fact, 14,000 LE behavior signals were simulated and the number of every kind of LE behavior signal was 2000. Finally, the LE behavior signal dataset was ready for experiments.

3.2. Unidimensional ACGAN Generates LE Behavior Signals

There were a total of 14,000 signals that we simulated, which belonged to seven kinds of different LE behavior signals. Seven hundred signals were randomly selected from 14,000 signals, and they were treated as training sets of unidimensional ACGAN. We also selected 700 real samples as a validation set, and another 6300 samples formed the test set. An Adam optimizer was used in the experiments, the initial learning rate was 0.0002, momentum was 0.5, and batch size was 32. These LE behavior signals with SNR = 0 dB were put into unidimensional ACGAN.

The epoch was set as 20,000 when the unidimensional ACGAN was already trained adequately. The value of the discriminator's loss and the value of the generator's loss changed with training time going, as shown in Figure 4.

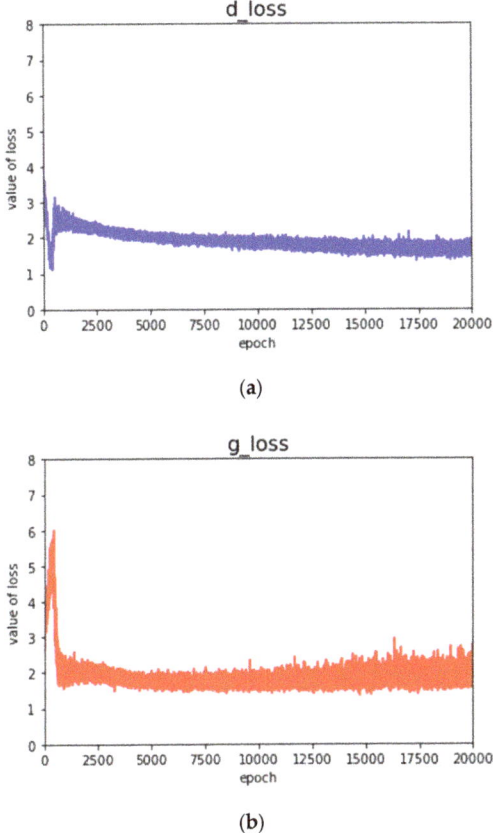

Figure 4. The curve of the loss function' change. (a) Discriminator; (b) generator.

As shown in Figure 4, in the initial stage of network training, the loss of the discriminator is in the descending state, while the loss function of the generator is in the rising state. This indicates that the fake samples generated by the generator could not deceive the discriminator, while the discriminator could learn the deep features of the real samples and correctly distinguish the real samples from the fake samples. As a network model going to be trained adequately, the loss of the discriminator started to rise and the loss of the generator started to decline, which indicates that the fake samples generated by the generator were approaching the distribution of the real samples, and it was difficult for the discriminator to distinguish the real samples from the generated samples correctly. Then, the loss of the discriminator was in the descending state again and the loss of the generator was in the rising state again, indicating that the generator and the discriminator of the network model were fighting with each other in the training process. The generator gradually generated more real fake samples. Finally, the loss of the discriminator and the generator gradually became stable. The loss of the discriminator decreased slightly and the loss of the generator increased slightly, indicating that the samples generated by the generator could better approach the distribution of the real samples.

The Call behavior signal was one of the seven kinds of LE behavior signals of a short-wave radio station. As the number of epochs increases, the changes of the Call behavior signal generated by the unidimensional ACGAN's generator are roughly shown in Figure 5.

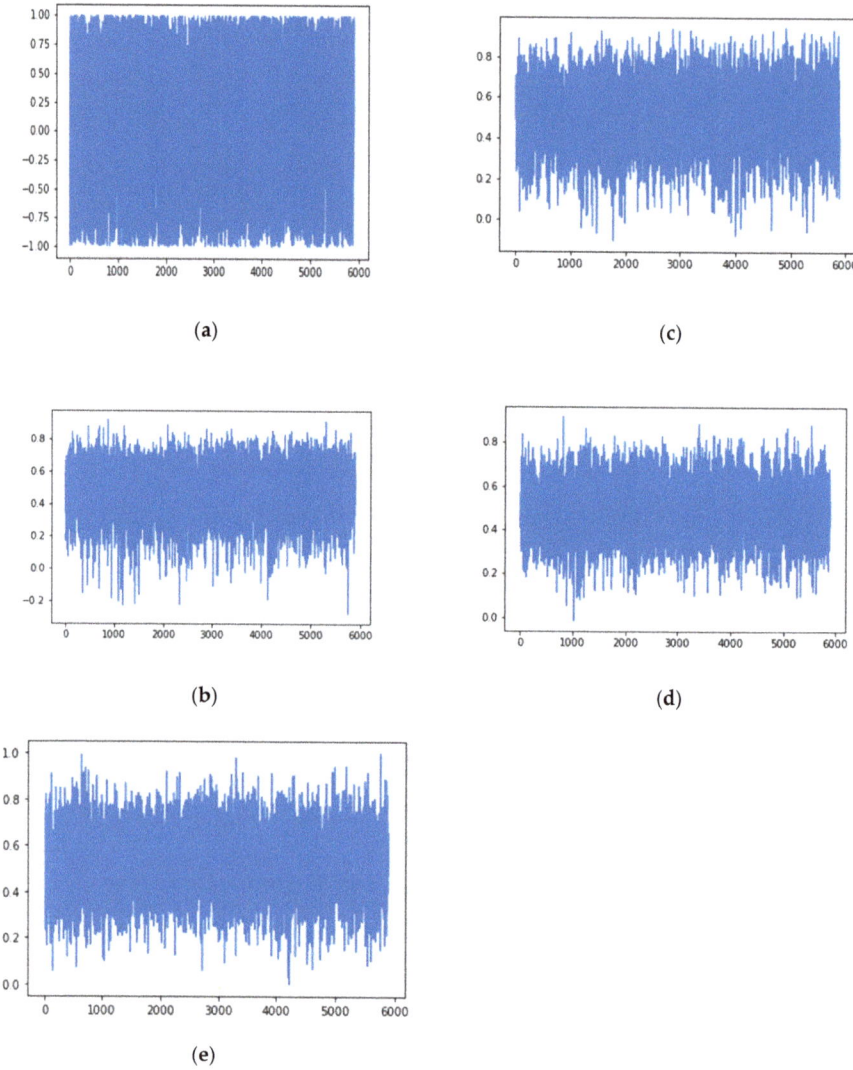

Figure 5. With different training epoch, the changes of the Call behavior signal generated by generator. (a) When epoch is 200; (b) when epoch is 1000; (c) when epoch is 2000; (d) when epoch is 20,000; (e) a real Call behavior signal.

According to Figure 5, with the process of training going on, after the value of epoch is greater than 2000, the Call behavior signal generated by unidimensional ACGAN is close to the real Call behavior signal. When the value of epoch was equal to 20,000, unidimensional ACGAN could generate fake "real" LE behavior signals. When the value of epoch is equal to 20,000, seven kinds of LE behavior signals generated by the network are shown in Figure 6.

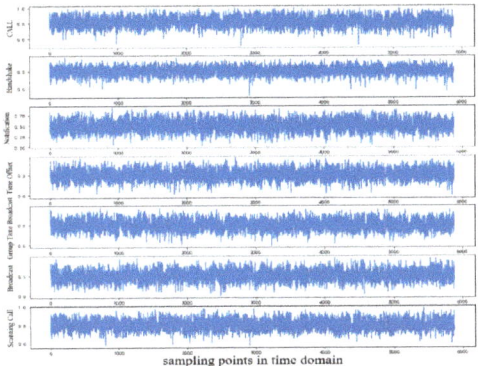

Figure 6. Seven kinds of LE behavior signals generated by unidimensional ACGAN.

As shown in Figures 5 and 6, these samples generated by unidimensional ACGAN are similar to those real signals. However, whether there were differences between each kind of behavior signal requires more study on characteristic distribution of each kind of sample signal generated by our network. The unidimensional Convolutional Auto-Encoder (CAE) can effectively show the deep characteristic differences between different samples [26–28]. Therefore, the output of the auto-encoder was set as a two-dimensional vector. Differences between each kind of behavior signals are shown through the visualization of the two-dimensional vectors. The structures of unidimensional encoder and unidimensional decoder we proposed are shown in Table 3.

Table 3. The structures of encoder and decoder.

Encoder		
Layers	Input Size	Output Size
Input Layer	(5888, 1)	(5888, 1)
Conv1D, S = 1, KS = 3	(5888, 1)	(5888, 20)
MaxPooling1D(2)	(5888, 20)	(2944, 20)
Conv1D, S = 1, KS = 3	(2944, 20)	(2944, 20)
MaxPooling1D(2)	(2944, 20)	(1472, 20)
Flatten	(1472, 20)	29, 440
Fully Connected	29440	256
(Out Layer)Fully Connected	256	2
Decoder		
Layers	Input Size	Output Size
Input Layer	2	2
Fully Connected	2	256
Fully Connected	256	29440
Reshape	29440	(1472, 20)
UpSampling1D	(1472, 20)	(2944, 20)
Conv1D, S = 1, KS = 3	(2944,20)	(2944,20)
UpSampling1D	(2944,20)	(5888,20)
Conv1D, S = 1, KS = 3	(5888,20)	(5888,20)
(Out Layer)Conv1D, S = 1, KS = 3	(5888, 20)	(5888, 1)

In Table 3, Conv1D represents one-dimensional convolution operation, S = 1 represents the stride of convolutional operation is 1, and KS represents that the size of the convolution kernel is 3. MaxPooling1D(2) represents that the stride of max-pooling is 2. And UpSampling1D represents that data is upsampled by two times.

The LE behavior signals whose SNR = 0 dB make up our dataset in our experiment. Firstly, 700 real signals selected at random were used to train the encoder, and then the fake signals generated by unidimensional ACGAN were put into the trained encoder. When the epoch of training unidimensional ACGAN was different, we received fake signals corresponding to different epoch with different labels (0, 1, 2, 3, 4, 5, 6) by unidimensional ACGAN. Then we put those fake samples into the CAE. The output of CAE is shown in Figure 7.

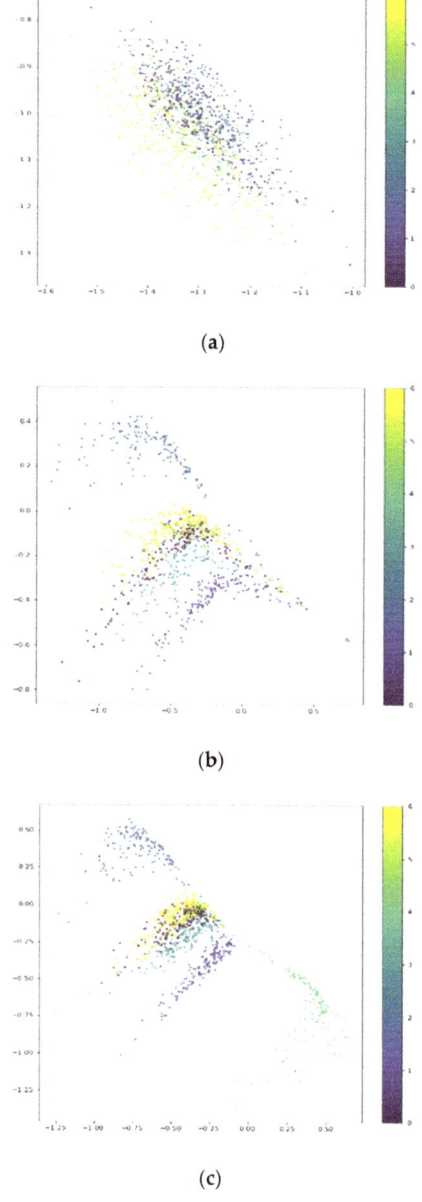

(a)

(b)

(c)

Figure 7. *Cont.*

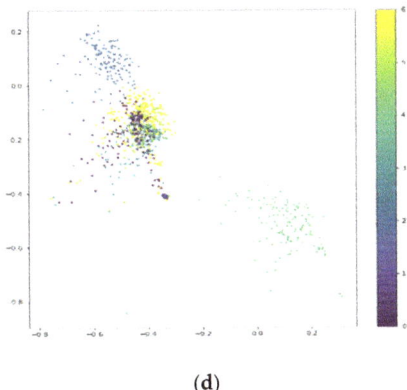

(d)

Figure 7. The outputs of unidimensional Convolutional Auto-Encoder (CAE) corresponding to generated signals with different training epochs. (**a**) When epoch = 500; (**b**) when epoch = 2000; (**c**) when epoch = 5000; (**d**) when epoch = 20,000.

As shown in Figure 7, with the increase of the value of training epoch, the differences in characteristic distribution of the seven kinds of LE behavior signals become more and more obvious. According to Figure 7d, the distribution of yellow slightly overlaps the distribution of other colors because in the process of simulating signals labeled by 6 (6 means yellow), the 11 valid bits representing the called radio station's address are generated randomly. This results in the similarities between Scan Call behavior signals and other LE behavior signals, and different LE behavior signals with SNR = 0 dB were only a little different. When the epoch was greater than 5000, the loss g_loss of the generator was still going up and down within a certain range, which made the characteristic distribution of the generated signals have a certain contingency. Comparing Figure 7c,d, with the training going on, the characteristic distribution of samples belonging to same class becomes more clustered. The experimental results show that unidimensional ACGAN could effectively generate different kinds of LE behavior signals, and unidimensional ACGAN could also learn the deep features of different kinds of behavior signals.

3.3. The LE Behavior Recognition Performance Based on Unidimensional ACGAN + DenseNet

In order to explore the performance of the unidimensional ACGAN+DenseNet algorithm for LE behavior recognition, we still randomly selected 700 samples from the simulated 14,000 LE behavior signals whose SNR was 0 dB. These 700 samples as the training set were put into unidimensional ACGAN. Then we selected 700 samples as the validation set and the 6300 samples were regarded as the test set. The original 700 samples were combined with fake samples generated by unidimensional ACGAN and then a new training set was acquired. The unidimensional DenseNet was be trained by the new training set. The value of the epoch for training the unidimensional DenseNet was set as 10, and the batch was set as 8. According to all the 50 Monte Carlo experiments, as the number of generated samples, which are used to train unidimensional ACGAN, changed, the average accuracy of unidimensional DenseNet changed, as shown in Figure 8.

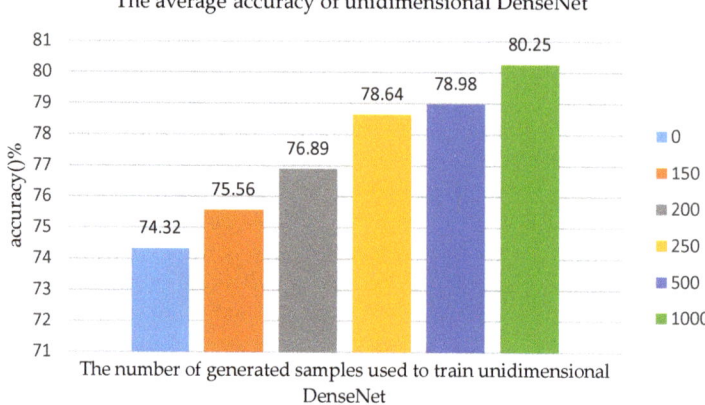

Figure 8. As the number of generated samples used to train unidimensional ACGAN changes, the average accuracy of unidimensional DenseNet changes.

As shown in Figure 8, when the number of the generated sample is 0, which means we used all real samples to train unidimensional DenseNet, the accuracy of recognition algorithm is 74.32%. As the unidimensional ACGAN generated more samples, the performance of the recognition algorithm was better. When the number of generated samples was 150, 200, 250, 500 or 1000, the accuracy of the algorithm model was improved by 1.24, 2.57, 4.32, 4.66, and 5.93%, respectively, compared to when the number of generated samples was 0. The experimental results show that the recognition performance of the network model could be improved if the samples generated by unidimensional ACGAN were combined with the real training data set.

In order to adequately verify the performance of our algorithm based on unidimensional ACGAN+DenseNet, especially under the condition that there were only few samples with labels, we set SNR = 0 dB and we used a different number of real samples to train unidimensional ACGAN. Seven hundred fake samples were generated by unidimensional ACGAN and then a new training set was acquired by combining 700 fake samples and the original training samples. There were 6300 samples still used as test set. According to all the 50 Monte Carlo experiments, as the number of original real samples which were used to train unidimensional ACGAN changed, the average recognition accuracy of unidimensional ACGAN+DenseNet algorithm changed, as shown in Figure 9.

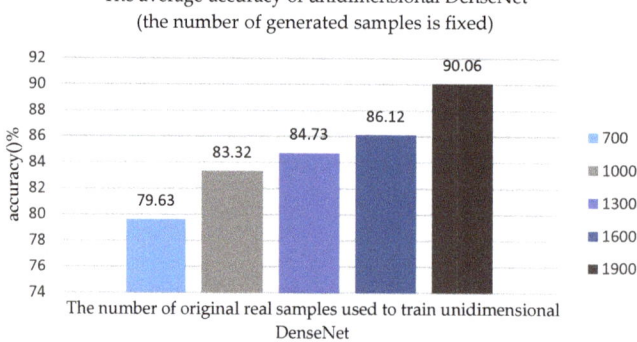

Figure 9. As the number of original real samples used to train unidimensional ACGAN changes, the average accuracy of unidimensional DenseNet changes, and the number of generated samples is fixed.

As shown in Figure 9, if the number of generated samples is fixed at 700, and the recognition accuracy is 79.63% when the number of original training samples is 700. When the number of original training samples increased to 1000, 1300, 1600, or 1900, the recognition accuracy of the network increased to 83.32, 84.73, 86.12, and 90.06%. The experimental results show that the more original training samples we used, the better performance of unidimensional DenseNet would be, which is consistent with the real scene.

In order to further verify that more original training samples, which are used to train unidimensional ACGAN, were able to improve the performance of the recognition algorithm, the number of generated samples was no longer fixed at 700, and the number of generated samples was the same as the number of original training samples. Finally, according to all the 50 Monte Carlo experiments, as the number of original real samples, which are used to train unidimensional ACGAN, changed, the average recognition accuracy of the algorithm changed, as shown in Figure 10.

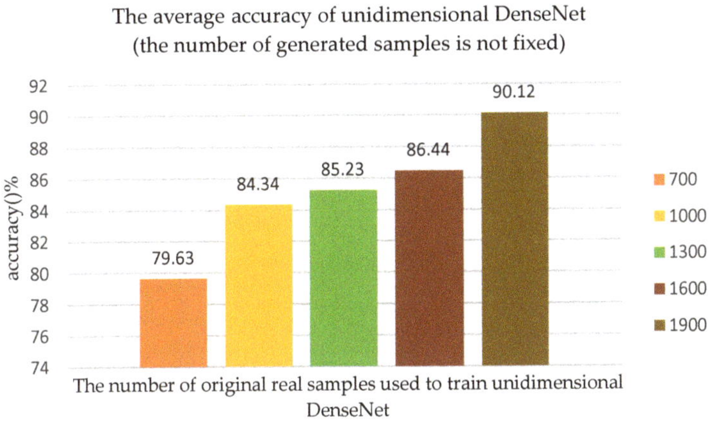

The average accuracy of unidimensional DenseNet
(the number of generated samples is not fixed)

[Bar chart showing accuracy(%) vs the number of original real samples used to train unidimensional DenseNet: 700 → 79.63, 1000 → 84.34, 1300 → 85.23, 1600 → 86.44, 1900 → 90.12]

Figure 10. As the number of original real samples used to train unidimensional ACGAN changes, the average accuracy of unidimensional DenseNet changes, and the number of generated samples is the same as the number of original training samples.

As shown in Figure 10, when the number of original training samples is 1000, 1300, 1600, or 1900, unidimensional ACGAN generates 1000, 1300, 1600, and 1900 LE behavior signals, respectively. Then, the new training set was used to train the network model and the final recognition accuracy increased as the number of training samples increased. When the number of original training samples was 1900, compared to when the number of original training samples was 700, the recognition accuracy of network model was improved by 10.49%.

Comparing Figures 9 and 10, we can see that when the number of original training samples are the same, the number of generated samples will affect the final recognition accuracy of the network model. When the number of original samples was small, the difference in the number of generated samples will have a great influence on the recognition accuracy of the network model. For example, when the number of original training samples is 1000, the recognition accuracy in Figure 9 is 83.32%, while the recognition accuracy in Figure 10 is 84.34%. However, when the number of original training samples was large, the number of generated samples had a small impact on the final recognition accuracy of the network model, such as the number of original training samples being 1900. In summary, when training samples were sufficient, these samples could fully represent the essential features of these samples so that there was no need to utilize too many generated samples to train the network. However, it was necessary to generate more samples to train the network when the number of training samples was small.

3.4. Comparison Experiment

In order to illustrate the advantages of the unidimensional ACGAN+DenseNet algorithm, the proposed method was compared with Multilayer Perceptron (MLP), Lenet-5 [29,30], and unidimensional DenseNet [6]. They were all used to recognize different LE behaviors.

The internal structure of MLP is as follows: Input layer–512–256–7–Output layer; "512", "256", and "7" represent the number of neurons in each hidden layer. The internal structure of Lenet-5 is as follows: Input layer–Conv1D(6,5,1(s))–MaxPooling(2)–Conv1D(16,5,1(s))–Maxpooling(2)–flatten–256–7–Output layer; "6" and "16" represent the number of convolution kernel; "5" represents the size of convolution kernel; "1(s)" represents that the stride of convolutional operation is 1; "MaxPooling(2)" represents that the stride of max-pooling is 2; "flatten" represents the flatten layer. The structure of unidimensional DenseNet is the same as the DenseNet used in this work.

From 14,000 LE behavior signals with SNR = 0 dB, 400, 700, 1000, and 1300 samples were randomly selected as the training sets, respectively, and 6300 LE behavior signals were selected as the test set. The number of original real training samples used to train unidimensional ACGAN and the recognition accuracy of each algorithm are shown in Figure 11.

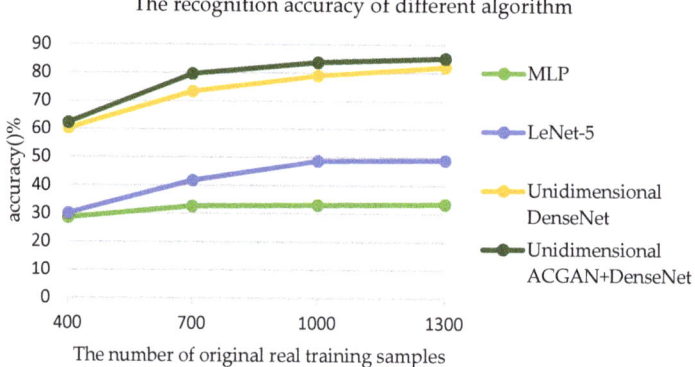

Figure 11. The number of original real training samples used to train unidimensional ACGAN changes and the recognition accuracy of the four algorithms.

As shown in Figure 11, when the number of original training samples is 400, 700, 1000, or 1300, the recognition accuracy of the algorithm in this work is higher than that of unidimensional DenseNet, LeNet-5, and MLP. In particular, when there were insufficient samples with labels, the recognition accuracy of LeNet-5 and MLP was less than 50%. When the number of training samples was 400, 700, 1000, or 1300, the recognition accuracy of unidimensional DenseNet was 1.92, 6.16, 4.63, and 3.06% lower than that of unidimensional ACGAN+DenseNet, respectively. It can also be known from Figure 11 that when the number of original training samples is large, such as 1300, the recognition accuracy of unidimensional ACGAN+DenseNet begins to be close to that of unidimensional DenseNet.

4. Conclusions

In the field of electronic countermeasures, not only is there no help of short-wave communication protocol standard, but it also is difficult to obtain a large number of LE behavior signals with labels. Therefore, it is a difficult problem to recognize the LE behaviors of a short-wave radio station. Firstly, according to the third-generation short-wave communication protocol standard, we simulated a radio station's LE behavior signals with labels on behalf of labeled 3G ALE signals. Then a novel unidimensional ACGAN was proposed. A small number of samples with labels were used to train unidimensional ACGAN and some fake samples were generated. Then a new training set was formed by combining original real samples with generated samples, and the new training set was used to train the unidimensional DenseNet. Finally, the unidimensional DenseNet that had been trained was used to recognize different LE behaviors of a short-wave radio station. The experimental results show that the more training samples and the more fake samples generated were used to train unidimensional ACGAN, the better performance the network model had. Meanwhile, the experimental results showed that the performance of unidimensional ACGAN+DenseNet was much better than that of LeNet-5 and MLP. Moreover, when the number of original training samples was 400, 700, 1000, or 1300, the recognition accuracy of the algorithm in this work was 1.92, 6.16, 4.63, and 3.06% higher than that of unidimensional DenseNet.

In summary, the proposed algorithm can recognize a non-collaborative radio station's LE behaviors without the help of the communication protocol standard. In particular, when the number of labeled training samples was very small, such as 700, 1000, 1300, 1600, and 1900, the recognition accuracy of the algorithm could reach 79.63, 84.34, 85.23, 86.44, and 90.12%, respectively. In addition, a new unidimensional CAE was presented to explain the reliability of samples generated by ACGAN. Further, in terms of the application of the proposed algorithm in real time, we used a few LE behavior signals collected by sensors to train networks and the trained networks were used to recognize the new detected LE signals. When applying the algorithm, we did not need to consider the training time of networks, but the testing time of networks which was very short in reality.

In future work, the algorithm model of unidimensional ACGAN+DenseNet can continue to be optimized. In particular, the unidimensional ACGAN needs to be improved so that it can generate more real unidimensional electromagnetic signals, and other better data-augmentation techniques should be utilized to generate LE behavior signals; hence, there would be more contribution in terms of signals' data augmentation. In addition, the unidimensional ACGAN was adopted in this work to directly generate time-domain signals. ACGAN was used to process features after signals' characteristic transformation and then directly generate "real" features of signals to train the recognition network, which may have made the algorithm more productive under the condition that there were only a few labeled samples. Moreover, some study on real samples should be presented, as experiments are currently all based on simulated samples. As soon as conditions permit, the LE behavior signals of a short-wave radio station should be collected in a real environment by non-collaborative sensors and then the real samples should bde utilized to verify the performance of the proposed algorithm. Although the structure of the proposed algorithm should be further improved, it already has the capability to recognize radio signals, which demonstrates that we can recognize LE behaviors of radio station even without the communication protocol. Finally, research should be done on whether our idea could be applied to analyze communication behaviors and LE behaviors of other types of radio stations.

Author Contributions: Conceptualization, Z.W. and Y.L.; methodology, Z.W.; software, Z.W.; validation, H.C. and Y.L.; formal analysis, H.C.; investigation, Y.L.; resources, Y.L.; data curation, Z.W.; writing—original draft preparation, Z.W.; writing—review and editing, H.C.; visualization, Z.W.; supervision, H.C.; project administration, Y.L.; funding acquisition, Y.L. All authors have read and agreed to the published version of the manuscript.

Funding: This research was funded by National Defense Science and Technology Key Laboratory Fund Project of China, grant number 6142106180402.

Conflicts of Interest: The authors declare no conflict of interest.

References

1. U.S. Standard MIL-STD-188-141B. *Interoperability and Performance Standards for Medium and High Frequency Radio Systems*; US Department of Defense: Arlington County, VA, USA, 1999.
2. Yuan, Y. Network Communication Behaviors Modeling Method on Data Classification Research. Master's Thesis, Dept. Computer Applied Technology, UESTC, Chengdu, China, 2015.

3. Liu, C.; Wu, X.; Zhu, L.; Yao, C.; Yu, L.; Wang, L.; Tong, W.; Pan, T. The Communication Relationship Discovery Based on the Spectrum Monitoring Data by Improved DBSCAN. *IEEE Access* **2019**, *7*, 121793–121804. [CrossRef]
4. Liu, C.; Wu, X.; Yao, C.; Zhu, L.; Zhou, Y.; Zhang, H. Discovery and research of communication relation based on communication rules of ultrashort wave radio station. In Proceedings of the 2019 IEEE 4th International Conference on Big Data Analytics (ICBDA), Suzhou, China, 15–18 March 2019; pp. 112–117.
5. Xiang, Y.; Xu, Z.; You, L. Instruction flow mining algorithm based on the temporal sequence of node communication actions. *J. Commun. China* **2019**, *40*, 51–60.
6. Wu, Z.; Chen, H.; Lei, Y.; Xiong, H. Recognizing Automatic Link Establishment Behaviors of a Short-Wave Radio Station by an Improved Unidimensional DenseNet. *IEEE Access* **2020**, *8*, 96055–96064. [CrossRef]
7. Qiu, J.; Sun, K.; Rudas, I.J.; Gao, H. Command filter-based adaptive NN control for MIMO nonlinear systems with full-state constraints and actuator hysteresis. *IEEE Trans. Cybern.* **2019**, *50*, 1–11. [CrossRef] [PubMed]
8. Qiu, J.; Sun, K.; Wang, T.; Gao, H.-J. Observer-based fuzzy adaptive event-triggered control for pure-feedback nonlinear systems with prescribed performance. *IEEE Trans. Fuzzy Syst.* **2019**, *27*, 2152–2162. [CrossRef]
9. Goodfellow, I.; Pouget-Abadie, J.; Mirza, M.; Xu, B.; Warde-Farley, D.; Ozair, S.; Courville, A.; Bengio, Y. Generative adversarial networks. *Adv. Neural Inf. Process. Syst.* **2014**, *3*, 2672–2680.
10. Azadi, S.; Fisher, M.; Kim, V.; Wang, Z.; Shechtman, E.; Darrell, T. Multi-content GAN for few-shot font style transfer. In Proceeding of the Conference on Computer Vision and Pattern Recognition 2018 IEEE/CVF, Salt Lake, UT, USA, 18–23 June 2018; pp. 7564–7573. [CrossRef]
11. Chang, H.; Lu, J.; Yu, F.; Finkelstein, A. PairedCycleGAN: Asymmetric style transfer for applying and removing makeup. In Proceedings of the 2018 IEEE/CVF Conference on Computer Vision and Pattern Recognition, Salt Lake, UT, USA, 18–23 June 2018; pp. 40–48.
12. Niu, X.; Yang, D.; Yang, K.; Pan, H.; Dou, Y. Image Translation Between High-Resolution Remote Sensing Optical and SAR Data Using Conditional GAN. In *Intelligent Tutoring Systems*; Springer Science and Business Media LLC: Hudson Square, NY, USA, 2018; pp. 245–255.
13. Yang, S.; Wang, Z.; Wang, Z.; Xu, N.; Liu, J.; Guo, Z. Controllable Artistic Text Style Transfer via Shape-Matching GAN. In Proceedings of the 2019 IEEE/CVF International Conference on Computer Vision (ICCV), Seoul, Korea, 27 October–2 November 2019; pp. 4441–4450.
14. Sandfort, V.; Yan, K.; Pickhardt, P.J.; Summers, R.M. Data augmentation using generative adversarial networks (CycleGAN) to improve generalizability in CT segmentation tasks. *Sci. Rep.* **2019**, *9*, 16884. [CrossRef] [PubMed]
15. Luo, Y.; Lu, B.-L. EEG data augmentation for emotion recognition using a conditional Wasserstein GAN. In Proceedings of the 2018 40th Annual International Conference of the IEEE Engineering in Medicine and Biology Society (EMBC), Honolulu, HI, USA, 17–21 July 2018; Volume 2018, pp. 2535–2538.
16. Tang, B.; Tu, Y.; Zhang, Z.; Lin, Y.; Zhang, S. Digital signal modulation classification with data augmentation using generative adversarial nets in cognitive radio networks. *IEEE Access* **2018**, *6*, 15713–15722. [CrossRef]
17. Isola, P.; Zhu, J.-Y.; Zhou, T.; Efros, A.A. Image-to-image translation with conditional adversarial networks. In Proceedings of the 2017 IEEE Conference on Computer Vision and Pattern Recognition (CVPR), Honolulu, HI, USA, 21–26 July 2017; pp. 5967–5976. [CrossRef]
18. Zhu, J.-Y.; Zhang, R.; Pathak, D.; Darrell, T.; Efros, A.A.; Wang, O.; Shechtman, E. Toward multimodal image-to-image translation. *Neural Inf. Process. Syst.* **2017**, 465–476.
19. Cho, S.W.; Baek, N.R.; Koo, J.H.; Arsalan, M.; Park, K.R. Semantic segmentation with low light images by modified CycleGAN-based image enhancement. *IEEE Access* **2020**, *8*, 93561–93585. [CrossRef]
20. Kim, T.; Cha, M.; Kim, H.; Lee, J.K.; Kim, J. Learning to discover cross-domain relations with generative adversarial networks. Computer vision and pattern recognition. *arXiv* **2017**, arXiv:1703.05192.
21. Deng, J.; Pang, G.; Zhang, Z.; Pang, Z.; Yang, H.; Yang, G. cGAN based facial expression recognition for human-robot interaction. *IEEE Access* **2019**, *7*, 9848–9859. [CrossRef]
22. Liu, J.; Gu, C.; Wang, J.; Youn, G.; Kim, J.-U. Multi-scale multi-class conditional generative adversarial network for handwritten character generation. *J. Supercomput.* **2017**, *75*, 1922–1940. [CrossRef]
23. Zhao, J.; Mathieu, M.; Lecun, Y. Energy-based generative adversarial network. Machine Learning. *arXiv* **2016**, arXiv:1609.03126.

24. Zhikai, Y.; Leping, B.; Teng, W.; Tianrui, Z.; Fen, W. Fire Image Generation Based on ACGAN. In Proceedings of the 2019 Chinese Control and Decision Conference (CCDC), Nanchang, China, 3–5 June 2019; pp. 5743–5746. [CrossRef]
25. Yao, Z.; Dong, H.; Liu, F.; Guo, Y. Conditional image synthesis using stacked auxiliary classifier generative adversarial networks. In *Future of Information and Communication Conference*; Springer: Cham, Switzerland, 2018.
26. Zhang, C.; Cheng, X.; Liu, J.; He, J.; Liu, G. Deep Sparse Autoencoder for Feature Extraction and Diagnosis of Locomotive Adhesion Status. *J. Control. Sci. Eng.* **2018**, *2018*, 1–9. [CrossRef]
27. Azarang, A.; Manoochehri, H.E.; Kehtarnavaz, N. Convolutional autoencoder-based multispectral image fusion. *IEEE Access* **2019**, *7*, 35673–35683. [CrossRef]
28. Karimpouli, S.; Tahmasebi, P.; Tahmasebi, P. Segmentation of digital rock images using deep convolutional autoencoder networks. *Comput. Geosci.* **2019**, *126*, 142–150. [CrossRef]
29. Liu, N.; Xu, Y.; Tian, Y.; Ma, H.; Wen, S. Background classification method based on deep learning for intelligent automotive radar target detection. *Futur. Gener. Comput. Syst.* **2019**, *94*, 524–535. [CrossRef]
30. Wen, L.; Li, X.; Gao, L.; Zhang, Y. A New Convolutional Neural Network-Based Data-Driven Fault Diagnosis Method. *IEEE Trans. Ind. Electron.* **2018**, *65*, 5990–5998. [CrossRef]

 © 2020 by the authors. Licensee MDPI, Basel, Switzerland. This article is an open access article distributed under the terms and conditions of the Creative Commons Attribution (CC BY) license (http://creativecommons.org/licenses/by/4.0/).

Article

Control System for Vertical Take-Off and Landing Vehicle's Adaptive Landing Based on Multi-Sensor Data Fusion

Hongyan Tang [1], Dan Zhang [1,2,*] and Zhongxue Gan [1]

1. Institute of AI and Robotics, Academy for Engineering & Technology, Fudan University, Shanghai 200433, China; hytang@fudan.edu.cn (H.T.); ganzhongxue@fudan.edu.cn (Z.G.)
2. Lassonde School of Engineering, York University, Toronto, ON M3J 1P3, Canada
* Correspondence: dzhang99@yorku.ca; Tel.: +1-6472090959

Received: 29 June 2020; Accepted: 4 August 2020; Published: 7 August 2020

Abstract: Vertical take-off and landing unmanned aerial vehicles (VTOL UAV) are widely used in various fields because of their stable flight, easy operation, and low requirements for take-off and landing environments. To further expand the UAV's take-off and landing environment to include a non-structural complex environment, this study developed a landing gear robot for VTOL vehicles. This article mainly introduces the adaptive landing control of the landing gear robot in an unstructured environment. Based on the depth camera (TOF camera), IMU, and optical flow sensor, the control system achieves multi-sensor data fusion and uses a robotic kinematical model to achieve adaptive landing. Finally, this study verifies the feasibility and effectiveness of adaptive landing through experiments.

Keywords: landing gear; adaptive landing; data fusion

1. Introduction

Compared with fixed-wing aircraft, vertical take-off and landing (VTOL) vehicles benefit from its multirotor power mode and have much fewer requirements for take-off and landing sites. VTOL vehicles are widely used in reconnaissance, search and rescue, logistics and other fields. The reduced requirements for the landing site lower the design requirements of the landing gear, but also limit the aircraft's ability to take off and land on non-structural terrain.

To further expand the aircraft's landing and landing environment, that is, to take off and land on complex unstructured terrain, more and more scholars have begun paying attention to the design of the adaptive landing gear of VTOL vehicles. The Mission Adaptive Rotor (MAR) project of the Defense Advanced Research Projects Agency (DARPA) organization was the first one to propose adaptive landing [1]. They adopted a legged mechanism to enable aircraft to adapt to different terrains. Subsequently, based on the difference in power output, two different types (active and passive) of adaptive landing gear were developed.

The active adaptive landing gear has developed into rigid-body landing gear and flexible-body landing gear. Rigid landing gear [2] mainly uses rigid connectors as the joints of the landing gear. The main representatives are the plane hinged robot landing gear of Edinburgh Napier University, Edinburgh, UK [3], the leg landing gear from Russia [4], the articulated leg landing gear of Kanazawa Institute of Technology in Japan [5], etc. All these landing gears apply legged mechanism to adjust touching points on the ground. The legged structure makes the robot's modelling and controlling easier, but also make a challenge to driving motors on their hip joints. The driving motor keeps working all the time to keep the joints' position, which may waste power energy. Flexible landing gear replaces joint motion by the deformation of flexible rods. The main representatives are the cable-driven landing gear of Georgia Institute of Technology [6], the avian landing gear [7] of UTHA University. The cable-driven landing gear uses spring dampers and cables to adjust its posture, and absorbs landing shock by the cable and spring damper. It is complicated but can be applied in heavy unmanned aerial vehicles (UAVs) and crewed aircrafts. The avian landing gear uses a soft gripper instead of a gild-body link, which can grip a rod and help vehicle standing on the rod. The passive landing gear is mainly powered by the weight of the robot and uses an under-actuated mechanism to achieve passive balance adjustment of the robot during the landing process, such as soft shock absorbers [8] of Imperial College London, a Four-bar linkage-based landing mechanism [9], and flexible landing gear [10] of China University of Petroleum.

Both positive and active landing gear can adjust posture by their mechanism. These mechanisms are the base hardware for adaptive landing. To complete the automatic landing, it also needs a control system to drive the mechanism. To realize the adaptive landing function, the aircraft needs to be based on mechanism configuration and the design of the control algorithm [11]. According to the different sensors used by robots, adaptive landing controllers can be divided into three categories. The first is the contact sensor, such as the tact switch [12,13], the pressure sensor [3,14], etc. These sensors are usually placed at the contact point of the landing gear on the ground and use a passive control method, which requires the relatively high real-time performance of the system. The second is visual sensors, which use three-dimensional visual scanning [15] to determine the terrain of the landing point, calculate the driving joint of the landing gear, and achieve adaptive landing. The third is the inertial measurement unit (IMU) [2,16]. IMU achieves adaptive landing by the different attitude control laws of the landing gear during landing. This control method requires higher design requirements for aircraft control algorithms.

On the other hand, computer vision is widely used in robotics. In UAV field, computer vision is applied in vision position [17,18] and visual recognition [19,20]. Vision position can calculate linear velocity with video data streams, and scan the 3D target with dual-camera. Vision recognition is applied in target and obstacle recognition. A depth camera is a novel vision sensor which can output both RGB image and depth image. The depth camera is a kind of low-cost 3D scanning approach comparing to 3D laser scanners, which is widely used in robot motion feedback [21], motion measurement [22], UAV obstacle avoidance [23], and other fields.

This article tries to apply the depth camera in VTOL UAV's adaptive landing. Based on the hardware development foundation of the early landing gear robot of the laboratory team, this article combined with multi-sensor information, such as depth vision sensor, optical flow sensor, IMU, etc., data fusion, motion control of the landing gear robot to realize the aircraft adaptive control on complex unstructured terrain.

This article first introduces the main structure of the landing gear robot developed by authors. The next section introduces the mathematical basis of the control algorithm. Then the article proposes an adaptive landing control algorithm based on multi-sensor data fusion, and finally verifies the feasibility and effectiveness of the algorithm through experiments.

2. Adaptive Landing Gear Robot

The landing gear robot analyzed in this paper is based on the tripod robot designed by authors. The amphibious robot (as shown in Figure 1a) can fly in the sky, dive underwater, and run on the ground. It is mainly composed of flying robots and landing gear robot. This article focuses on the control system in which landing gear robots land adaptively on unstructured terrain during the landing. This section introduces the components of the landing gear robot, including the mechanical structure, the power module, and the sensor module. This section lays the foundations for the following mathematical analysis and introduction of the control algorithm design.

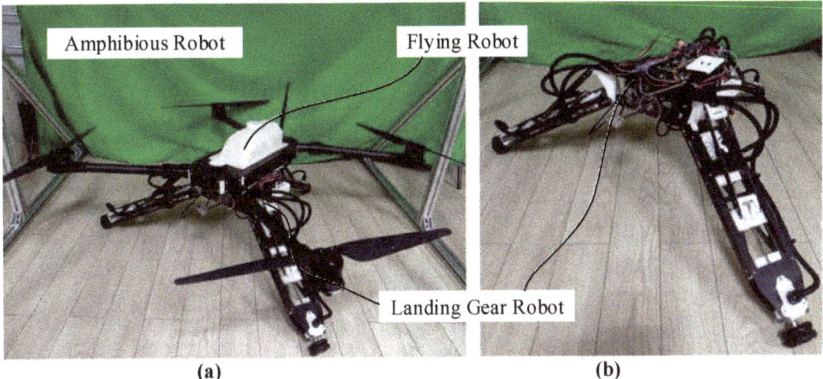

Figure 1. (a) The amphibious robot and (b) landing gear robot.

2.1. Mechanical Structure

Figure 1b shows the landing gear robot studied in this article, which is mainly composed of a base and three limbs. The base is the main bearing part of the landing gear robot. The top of the base is connected and fixed with the flying robot through bolts. The base is mainly composed of three parts: Diving component, sensing component, and structural components. The structural components mainly use carbon fiberboard, and photosensitive resin, which are produced by cutting and 3D printing, and the parts are fastened and connected by screw nuts. The main function of the sensor component is to install the sensors required for the robot system, including the depth camera and optical flow sensor. The diving component is to achieve the robot's diving function in the water, which is not the focus of this article.

The top view of the base is as shown in Figure 2a. The geometric center of the base is defined as point O. The forward direction is the X axis direction of the base. The left direction is the Y axis. According to the right-hand rule, the direction of the vertical top surface is the Z axis direction. The center points of the six rotation axes connected with limbs are points A_i and points E_i (i = 1, 2, 3). Points A_i and points E_i are centrosymmetric around the Z axis. In the horizontal direction, the distance from Point A_i and Point E_i to the center point is r_A and r_E, respectively.

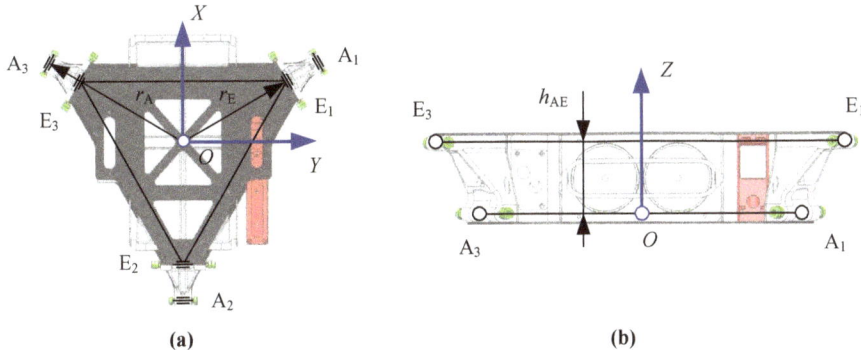

Figure 2. (a) Geometry structural and (b) dimensions of the base.

The three limbs use the design of the slider link mechanism, shown in Figure 3. By driving the translation of slider B_i, the swing rod A_iC_i is rotated, thereby controlling the height of the landing point C_i of the landing gear robot. The schematic diagram of the mechanism is shown in Figure 3b.

Point A_i is the connecting point of the connecting rod D_iE_i, and the base and Point E_i is the connecting point of the swing rod A_iC_i and the base. Point D_i is the connecting point of the connecting rod and the slider. The length of the connecting rod is l_{DE}. The length and eccentricity of the swing rod are l_{AC} and d_C, respectively. The displacement and eccentricity of the slider are l_{Si} and d_S, respectively.

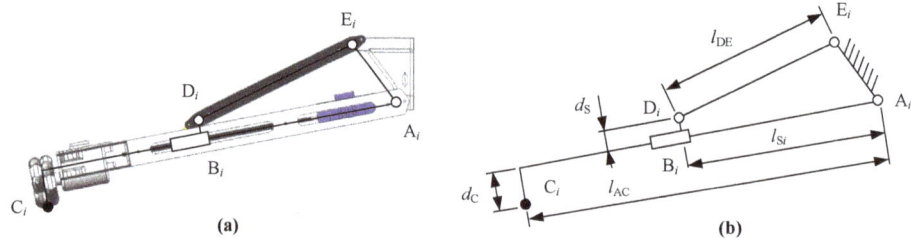

Figure 3. (a) Geometry structural and (b) dimensions of the limb.

The structure size parameters of the mechanism are shown in Table 1. The optimization of the structure size parameters of the structure is not the focus of this article and will not be discussed.

Table 1. Dimension parameters of the landing gear robot.

Parameter	Value	Parameter	Value
r_A	170 mm	l_{AC}	415 mm
r_E	220 mm	d_C	40 mm
h_{AE}	65 mm	l_{DE}	195 mm
d_S	22 mm		

2.2. Power System

The design and layout of the power system are related to the functional requirements of the landing gear robot. In this study, the landing gear robot mainly has two functions: Adaptive landing and omnidirectional motion. Therefore, this study designed two sets of power systems, as shown in Figure 4.

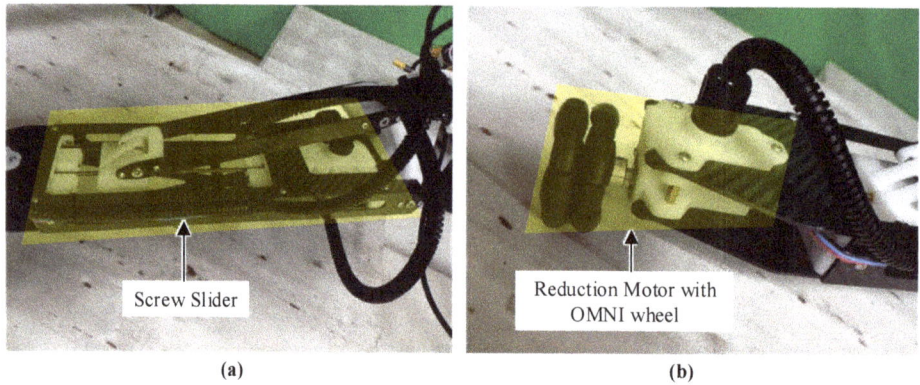

Figure 4. (a) Screw slider and (b) reduction motor with omnidirectional (OMNI) wheel.

Figure 4a shows the screw slider assembly. The motor uses a DC reduction motor with an encoder. The structure of the leading screw allows the motor to keep the position of the slider (self-locking) when it stops rotating. Limit switches are installed at both ends of the leading screw to prevent the slider from locking beyond the stroke (locked-rotor). Figure 4b demonstrates an omnidirectional wheel assembly, which is mainly composed of a reduction motor and an omnidirectional wheel. The omnidirectional wheel uses an omnidirectional (OMNI) wheel with a diameter of 56 mm, and the reduction motor uses a DJI M2006 motor (made by SZ DJI Technology CO., Ltd., Shenzhen, China), which can realize the feedback of position, speed, and torque of the motor.

2.3. Sensor and Control System

The sensors of the robot mainly include depth cameras and optical flow sensors, show in Figure 5c. Depth cameras use Intel Realsense435i (made by Intel Corporation, Santa Clara, CA, USA), which can output image signals, depth signals, gyro signals, and acceleration signals. Depth cameras are mainly used for the detection of terrain during the descent of the robot. Optical flow sensor uses the lc302 optical flow (made by Youxiang Corporation, Changsha, China), which can detect the moving speed of the aircraft in the horizontal direction.

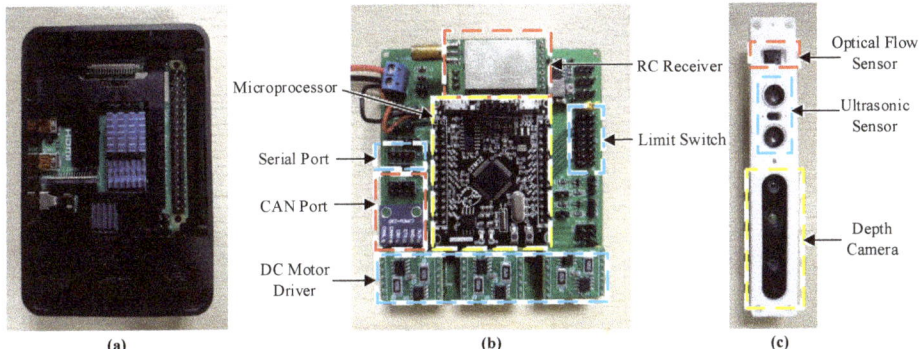

Figure 5. (a) Control computer, (b) driving board, (c) and sensor system.

The control system of the robot is mainly composed of a control computer (show in Figure 5a) and a driving board (show in Figure 5b). The control computer uses the Raspberry Pi 4B (made by RS Components Ltd., Northants, UK) as the main carrier and is operating with the Ubuntu system. The control computer is mainly used for the analysis of image signals, the resolution of attitude signals, and the solution of robot kinematics (kinematical analysis). Once the driving joint variables are calculated, the control computer inputs it to the driving board to control the movement of the landing gear robot.

The driving board is mainly composed of a microprocessor, three DC motor drivers, and communication circuits. The microprocessor uses the STM32F405 chip (made by STMicroelectronics, Agrate, Catania, Italy) as the main processor for signal conversion and motor control. The DC motor drivers use the MOS chip for converting the processor's electrical signal into the current required by the motor. The communication circuits, mainly integrated serial signal and controller area network (CAN) signal converter, are used to convert different signal modules.

3. Mathematical Analysis

This study is based on the mathematical modeling of the robot and the multi-sensor data fusion algorithm to achieve the adaptive landing function of the landing gear robot. The mathematical model of the robot is the mathematical foundation for realizing the robot motion control. The multi-sensor data fusion algorithm is the basis for realizing robot adaptive adjustment.

The basis for landing gear robot to achieve adaptive landing is terrain detection and analysis. Traditional mobile robots that perform three-dimensional reconstruction of environmental information can scan the terrain in a relatively stable state (variable movement oscillations are small). However, it is difficult for landing gear robots to keep stable because they are fixed to flying-submarine robots when they land. In order to maintain the attitude balance and fixed-point flight, it is difficult for the flying robot to ensure that the aircraft's posture is in an ideal static state, which affects the detection of the terrain and the analysis of the landing point by the landing gear robot. Therefore, the landing gear robot's judgment of the landing point needs to comprehensively consider the robot's attitude angle, flight speed and other state information. This state information is collected by multiple sensors, and different sensor data have different characteristic information. In order to obtain complete, accurate, and real-time target state information, this study uses a complementary filter and Kalman filter to perform data fusion on multi-sensor signals.

3.1. Robot Mathematical Model

The mathematical model of the robot is the basis for realizing robot motion control. In the previous section, the article introduces the robot mechanism and the simplified structure of the robot, as shown in Figure 6.

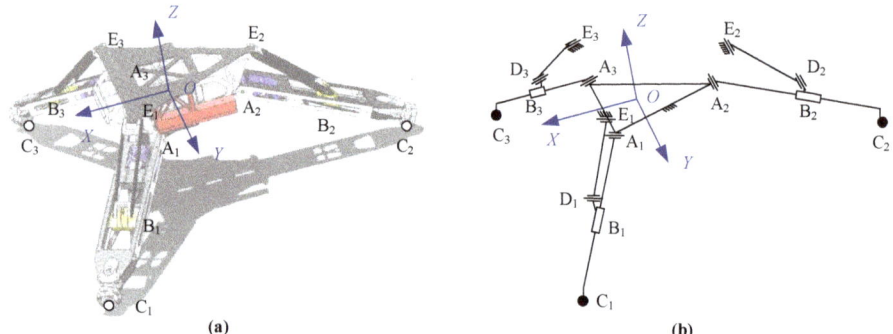

Figure 6. (a) The structural and (b) kinematic diagram of the landing gear robot.

The main motion pair of the mechanism is the rotating pair and the moving pair. Our focus is on the correlation between the joint variable of the slider driving motor and the three landing points. The sketch map of the branch chain can clearly show the constraint relationship between the geometric structures of the mechanism:

$$\begin{array}{c} \overrightarrow{E_iD_i} = \overrightarrow{E_iA_i} + \overrightarrow{A_iB_i} + \overrightarrow{B_iD_i} \\ \overrightarrow{O_iC_i} = \overrightarrow{O_iA_i} + \overrightarrow{A_iC_i} \end{array} \quad (i = 1, 2, 3) \tag{1}$$

The geometric constraints of the robot are converted into mathematical relations as follows:

$$\begin{array}{c} \left| \overrightarrow{E_iD_i} \right| = norm(\overrightarrow{E_iA_i} + l_{Si} \cdot \overrightarrow{es_i} + d_S \cdot \overrightarrow{es_i} \times \overrightarrow{eA_i}) = l_{DE} \\ \overrightarrow{O_iC_i} = \overrightarrow{O_iA_i} + l_{AC} \cdot \overrightarrow{es_i} + d_C \cdot \overrightarrow{eA_i} \times \overrightarrow{es_i} \end{array} \tag{2}$$

where $\overrightarrow{es_i}$ is the unit vector of the axis $\overrightarrow{A_iB_i}$; $\overrightarrow{eA_i}$ is the unit vector of the rotation axis A_i.

The forward kinematics and inverse kinematics of the robot can be solved by Equations (1) and (2). In this study, we are mainly concerned with how to estimate the robot's driving joint variables l_{Si} through the height h_{Ci} of the three contact points of the robot. Here we will introduce the solution process of inverse kinematics.

First, through the height of the contact point, the position and unit vector of the contact point C_i can be obtained:

$$\overrightarrow{OC_i} = R_{AC,i} \begin{pmatrix} d_{OCi} & d_{OCi} & -h_{Ci} \end{pmatrix}^T$$
$$\overrightarrow{es_i} = \frac{1}{\sqrt{2}} R_{AC,i} \cdot \begin{pmatrix} \cos(\alpha_i) & \cos(\alpha_i) & \sin(\alpha_i) \end{pmatrix}^T \quad (3)$$
$$R_{AC,1} = \begin{pmatrix} 1/2 & 0 & 0 \\ 0 & \sqrt{3}/2 & 0 \\ 0 & 0 & 1 \end{pmatrix}, R_{AC,2} = \begin{pmatrix} -1 & 0 & 0 \\ 0 & 0 & 0 \\ 0 & 0 & 1 \end{pmatrix}, R_{AC,3} = \begin{pmatrix} 1/2 & 0 & 0 \\ 0 & -\sqrt{3}/2 & 0 \\ 0 & 0 & 1 \end{pmatrix}$$

where $d_{OCi} = \sqrt{l_{AC}^2 - h_{Ci}^2} + r_A$ is the projected length of the vector $\overrightarrow{OC_i}$ on the horizontal plane; α_i is the angle between the axis $\overrightarrow{A_iC_i}$ and the horizontal plane; $R_{AC,i}$ is the coordinate conversion matrix.

Then, the obtained substitution $\overrightarrow{es_i}$ is substituted into Equations ((1) and (2)), and the driving rod length l_{Si} can be obtained by solving the unary quadratic equation.

3.2. Complementary Filter

In this study, the robot's attitude angle was calculated mainly by acquiring data from accelerometers and gyroscopes. The attitude angle can be obtained by integrating the angular velocity of the gyroscope. However, the gyro sensor has an integral drift, and the angle obtained by direct integration may contain errors. In order to eliminate the error of the sensor as much as possible, this study uses accelerometer data to correct the gyroscope data, and uses a complementary filter to solve the attitude angle.

The solution process of the complementary filter is mainly divided into five processes:

Normalize the acceleration data:

$$a = \begin{pmatrix} a_x & a_y & a_z \end{pmatrix}^T = \frac{a_z}{norm(a_z)} \quad (4)$$

where a is the normalized acceleration value, and a_z is the value directly output by the accelerometer.

Convert the gravity vector g_Z in the global coordinate system to body coordinates:

$$g_Z = \begin{pmatrix} g_{Z,x} \\ g_{Z,y} \\ g_{Z,z} \end{pmatrix} = \begin{pmatrix} 2(q_1 \cdot q_3 - q_0 \cdot q_2) \\ 2(q_0 \cdot q_1 + q_2 \cdot q_3) \\ q_0^2 - q_1^2 - q_2^2 + q_3^2 \end{pmatrix} \quad (5)$$

Compensate the error e by doing a vector cross product in the body coordinate system:

$$e = \begin{pmatrix} e_X \\ e_Y \\ e_Z \end{pmatrix} = a \times g_Z = \begin{pmatrix} a_Y \cdot g_{Z,z} - a_Z \cdot g_{Z,y} \\ a_Z \cdot g_{Z,x} - a_X \cdot g_{Z,z} \\ a_X \cdot g_{Z,y} - a_Y \cdot g_{Z,x} \end{pmatrix} \quad (6)$$
$$e_I = \sum K_I \cdot e$$

Calculate the proportional-integral (PI) of the error and compensate the angular velocity for compensate for the angular velocity:

$$gyo_k = \begin{bmatrix} gyo_{x,k} & gyo_{y,k} & gyo_{z,k} \end{bmatrix}^T = gyo_{k-1} + K_P \cdot e + e_I \quad (7)$$

Update quaternions q_k:

$$q_k = \frac{\Delta t}{2} \begin{pmatrix} \frac{2}{\Delta t} & -gyo_{x,k} & -gyo_{y,k} & -gyo_{z,k} \\ gyo_{x,k} & \frac{2}{\Delta t} & gyo_{z,k} & -gyo_{y,k} \\ gyo_{y,k} & -gyo_{z,k} & \frac{2}{\Delta t} & gyo_{x,k} \\ gyo_{z,k} & gyo_{y,k} & -gyo_{x,k} & \frac{2}{\Delta t} \end{pmatrix} \cdot q_{k-1} = \frac{\Delta t}{2} M_q \cdot q_{k-1} \tag{8}$$

Convert quaternions to angles θ_X and θ_Y:

$$\theta_X = \arctan\left(\frac{2q_2q_2 + 2q_0q_1}{1 - 2q_1^2 - 2q_2^2}\right) \tag{9}$$

$$\theta_Y = \arcsin(-2q_1q_3 + 2q_0q_2)$$

3.3. Kalman Filter

The Kalman filter is a commonly used multi-information fusion method. It is an optimal estimation algorithm, which calculates the state variable of the system in a stepwise recursive manner to solve the minimum amount of estimated variance. In this study, the posture and speed of the robot are used as state variables, and the data of the gyroscope and optical flow sensor are used as the measured values.

To accurately determine the location of the robot's landing point, this study will fuse the gyroscope and optical flow sensor data to calculate the robot's position in the horizontal direction to further determine the robot's landing point. When using a depth camera to scan the landing terrain, the camera is fixed at the center position of the robot. Therefore, the center point of the acquired depth image is used as the depth position corresponding to the center of the robot. Hence this study takes the relative displacement in the horizontal direction as one of the state variables of the filter. The state variables related to the relative displacement of the robot in the horizontal direction also include the angular velocity and linear velocity of the robot. The state variables of the filter can be expressed as:

$$x_k = [\ \Delta P_{X,k} \quad \Delta P_{Y,k} \quad \omega_{X,k} \quad \omega_{Y,k} \quad v_{X,k} \quad v_{X,k}\ ]T \tag{10}$$

where $\Delta P_{i,k}(i = X, Y)$ represents the relative displacement of the robot at the k moment, $\omega_{i,k}$ represents the angular velocity of the robot at the k moment, and $v_{i,k}$ represents the linear velocity of the robot at the k moment.

When the robot is in the landing state, under ideal conditions, the robot's horizontal speed and displacement approach zero, and the robot is in a relatively stable state. Therefore, in this study, the horizontal movement of the robot is approximated to a uniform speed for calculation, in other words, the state variable at the k moment can be predicted from the state variable at the previous moment:

$$\hat{x}_k = A * x_{k-1} = \begin{pmatrix} 0 & 0 & 0 & 0 & \Delta t & 0 \\ 0 & 0 & 0 & 0 & 0 & \Delta t \\ 0 & 0 & 1 & 0 & 0 & 0 \\ 0 & 0 & 0 & 1 & 0 & 0 \\ 0 & 0 & 0 & 0 & 1 & 0 \\ 0 & 0 & 0 & 0 & 0 & 1 \end{pmatrix} * x_{k-1} \tag{11}$$

where $\hat{x}_k$ are the predicted state variables and A is the state transition matrix.

According to the covariance at the $k-1$ moment, the predicted covariance $\hat{P}_k$ at the k moment can be calculated:

$$\hat{P}_k = A * P_k * A^T + Q_k \tag{12}$$

where Q_k is the white noise that the system is interfered with by the outside world, and it is assumed to follow the standard normal distribution $N(0, Q)$.

On the other hand, this study uses the data collected by the gyroscope and optical flow sensor to correct the current estimated state. The collected data and sensor errors are expressed as follows:

$$\mu_z = h_k = \begin{bmatrix} h\omega_X & h\omega_Y & hv_X & hv_Y \end{bmatrix}^T$$
$$\sigma_z^2 = R_k = \begin{bmatrix} R\omega_X & R\omega_Y & Rv_X & Rv_Y \end{bmatrix}^T \quad (13)$$

The relationship between sensor observations and state variables can be expressed by the following formula:

$$\mu_H = H_k * \hat{x}_k = \begin{pmatrix} 0 & 0 & 1 & 0 & 0 & 0 \\ 0 & 0 & 0 & 1 & 0 & 0 \\ 0 & 0 & 0 & h_{C,k} & 1 & 0 \\ 0 & 0 & h_{C,k} & 0 & 0 & 1 \end{pmatrix} * \hat{x}_k \quad (14)$$

$$\sigma_H^2 = H_k * \hat{P}_k * H_k^T$$

where H_k represents the conversion matrix between the state variable and the observed variable; $h_{C,k}$ represents the height of the robot at the current moment.

Both the measured value μ_z obtained and the predicted value μ_H obey the Gaussian distribution. Based on these two values, the optimal estimated value at the current moment can be calculated, which also satisfies the Gaussian distribution. According to the nature of the mean and variance of the Gaussian distribution, the mean and covariance of the best quality values can be respectively obtained as:

$$\mu_k = \mu_H + \frac{\sigma_H^2}{\sigma_H^2 + \sigma_z^2}(\mu_z - \mu_H) = \mu_H + K(\mu_z - \mu_H)$$
$$\sigma_k^2 = \sigma_H^2 - \frac{\sigma_H^2}{\sigma_H^2 + \sigma_z^2}\sigma_H^2 = \sigma_H^2 - K\sigma_H^2 \quad (15)$$

where K is the Kalman gain.

Sorting Equations (10)–(15), the update function of the Kalman filter can be obtained:

$$\begin{aligned}
\hat{x}_k &= A * x_{k-1} \\
\hat{P}_k &= A * P_{k-1} * A^T + Q_k \\
K &= H_k * \hat{P}_k * H_k^T \left(H_k * \hat{P}_k * H_k^T + R \right)^{-1} \\
x_k &= \hat{x}_k + H_k^{-1} * K(h_k - H_k * \hat{x}_k) \\
P_k &= \hat{P}_k - H_k^{-1} * K * H_k * \hat{P}_k
\end{aligned} \quad (16)$$

4. Control Algorithm Design

4.1. Adaptive Landing Process

In the traditional automatic landing process of a drone, the different stages of the drone landing are mainly determined based on the altitude information of the drone. The adaptive landing also needs to collect the environmental information of the landing site for comprehensive decision. In this study, adaptive landing is divided into two stages: The preparation stage and the descending stage.

During the preparation for the landing phase, the system predicts the landing posture of the landing gear robot by collecting terrain information. The control computer collects the sensor information, analyzes and calculates it, and then transmits the robot's driving joint variables to the driving board through the serial port. Then the driving board drives the motors to change the robot's attitude. During the descending phase, the flying robot controls the entire system to descend, and the control computer continues to analyze the terrain and make small adjustments to the robot's landing attitude.

4.2. Algorithm Design

The key to realizing adaptive landing function is to convert the sensor information into the driving joint variables required by the robot. In this study, the sensor information mainly includes the depth image of the depth camera, the triaxle acceleration of the accelerometer, the triaxle angular velocity of the gyroscope, and the biaxial velocity of the optical flow module. The driving joint variables of the robot refer to the displacement of the driving sliders of the robot's three limbs.

In this study, the information conversion is implemented in three steps: Center position analysis, contact position analysis, and drive variable analysis. The whole process is shown in Figure 7.

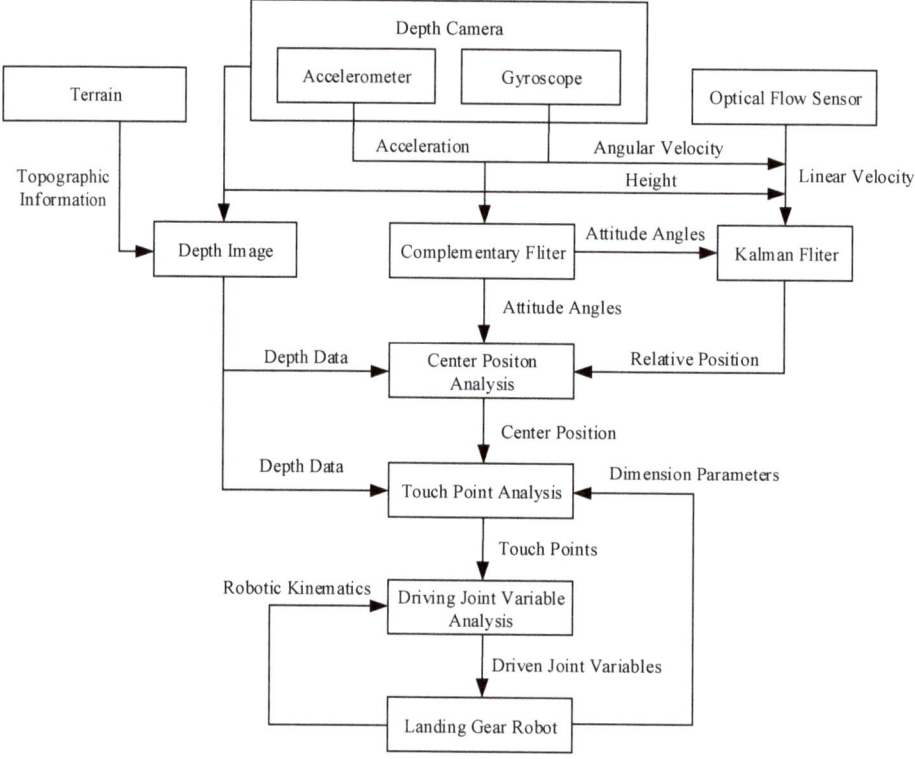

Figure 7. Flow chart of the adaptive landing control algorithm.

4.2.1. Center Position Analysis

The center position refers to the position where the center of the robot landed on the ground, which is, the intended landing point. The determination of the location of the landing point is key to the robot's adaptive terrain. In this study, we mainly discuss how to calculate the position of the center position in depth images.

This study uses two steps to calculate the center position. First, calculate the vertical position of the robot center on the ground, based on the attitude angle of the robot, as shown in Figure 8a:

$$C_V = C_C + K_{img} \cdot \Theta \tag{17}$$

where $C_C = \begin{pmatrix} Pix_X/2 & Pix_Y/2 \end{pmatrix}^T$ is the pixel coordinates of the center position of the depth image; $K_{img} = \begin{pmatrix} k_{img,X} & 0 \\ 0 & k_{img,Y} \end{pmatrix}$ is the ratio parameter between the depth image pixels and the angle; $\Theta = \begin{pmatrix} \theta_X & \theta_Y \end{pmatrix}^T$ is the attitude angle of the robot in the X-axis and Y-axis directions.

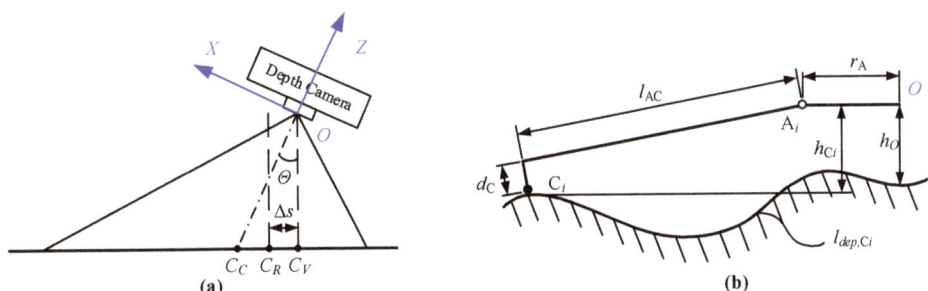

Figure 8. (a) Illustrations of center position analysis and (b) touch point analysis.

Then, according to the speed of the robot in the horizontal direction Δs, correct the center position C_R of the robot at the next moment:

$$C_R = C_V + \Delta s = C_V + K_{img} \cdot \Theta_s$$
$$\Theta_s = \begin{pmatrix} \arctan\left(\dfrac{\Delta P_{X,k}}{f_{dep}(C_V)}\right) & \arctan\left(\dfrac{\Delta P_{Y,k}}{f_{dep}(C_V)}\right) \end{pmatrix}^T \quad (18)$$

where f_{dep} is the depth surface function obtained by surface fitting according to the depth image at the k moment.

4.2.2. Touch Point Analysis

Touch point refers to the contact point between the three limbs of the robot and the ground. Based on the center position C_R of the robot, this study uses the geometric relationship of the robot limbs (as shown in Figure 8b) to calculate the robot's touch point position.

First, calculate the depth curve $l_{dep,Ci}$ through the center of the robot:

$$l_{dep,Ci}(t_i) = \begin{pmatrix} f_{len}(P_{img,i}) \\ f_{dep}(P_{img,i}) \end{pmatrix}, P_{img,i} = R_{AC,i} \cdot t_i + C_R \quad (19)$$

where $P_{img,i}$ is the coordinate position of the i-th limb in the depth image, t_i is the parameter variable; $R_{AC,i}\left(R_{AC,1} = \begin{pmatrix} 1/2 \\ -\sqrt{3}/2 \end{pmatrix}, R_{AC,2} = \begin{pmatrix} -1 \\ 0 \end{pmatrix}, R_{AC,3} = \begin{pmatrix} 1/2 \\ \sqrt{3}/2 \end{pmatrix}\right)$ is rotated matrix; f_{len} is the distance function of the pixel parameter $P_{img,i}$ and the center position in the horizontal direction:

$$f_{len}(P_{img}) = norm\left(f_{dep}(C_V) \cdot \tan\left(\dfrac{P_{img,x}}{k_{img,X}}\right), f_{dep}(C_V) \cdot \tan\left(\dfrac{P_{img,y}}{k_{img,Y}}\right)\right) \quad (20)$$

Then according to the following geometric relationship, the position of the contact point C_i can be obtained:

$$nrom(l_{dep,Ci} - A_{img}) = \sqrt{l_{AC}^2 + d_C^2}$$
$$A_{img} = \begin{pmatrix} r_A \\ f_{dep}(C_R) - h_O \end{pmatrix} \quad (21)$$
$$h_{Ci} = f_{dep}(P_{img,i}) - f_{dep}(C_R) + h_O$$

where A_{img} represents the coordinate position of point A_i in the depth image; h_O is the height of the robot center from the ground after landing.

4.2.3. Driving Variable Analysis

When substituting the centrifugal height h_{Ci} of the touch point (which is calculated by the touch point position) into the robot inverse kinematics (analyzed above), the joint variables of each driving joint can be solved:

$$l_{Si} = F_{inverse}(h_{Ci}) \qquad (22)$$

Finally, the calculated driving joint variables are input to the driving board. The driving board uses the displacement of the slider as the target value to perform PID control on the slider motor to realize the motion control of the robot.

5. Experimental Test

To verify the feasibility of the algorithm proposed in this article, this section will introduce the construction of the experimental platform, the experimental process, and the analysis of the experimental results in detail.

5.1. Experiment Platform and Process

In this study, the verification experiment was mainly carried out on an indoor experimental platform. The experimental platform is shown in Figure 9: It is mainly composed of three parts—the frame, the cable, and the terrain platform. One end of the cable is fixed on the robot, and the other end is a free end that passes through the fixed pulley on the main bracket. By controlling the expansion and contraction of the cable, the robot's landing process is simulated. The terrain platform is composed of modular terrain modules, which are combined with modules of different heights to form different terrains.

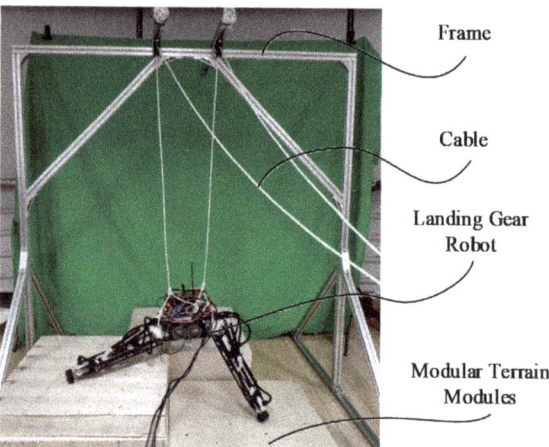

Figure 9. Experiment platform.

In this study, the robot's take-off and landing process was simulated by releasing and pulling back cables. The entire adaptive landing process was monitored in real-time. The experimental process is shown in Figure 10.

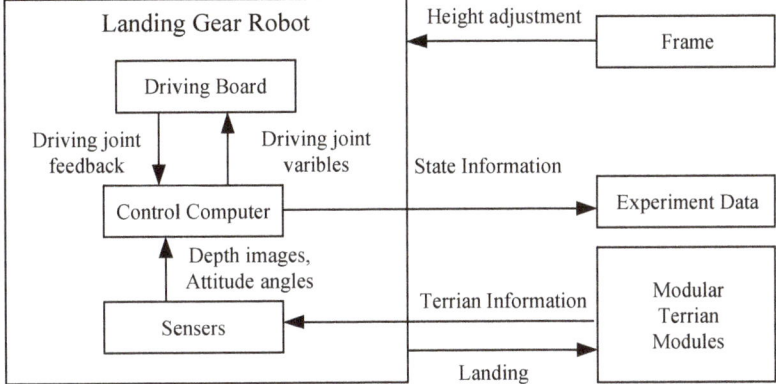

Figure 10. Flow chart of experiment data collection.

When the robot is at the standby altitude, the sensor collects terrain depth information and the status information of the robot and transmits the information to the control computer for processing. The control computer calculates the driving joint variables of the robot according to the algorithm designed above, and outputs them to the driving board of the robot. The driving board executes the motion of the motors according to the driving joint variables, and feeds back the position parameters of the driving motors to the control computer in real-time.

The height of the robot is lowered by slowly releasing the cable, until the robot landed on the ground. The control computer will collect and record the robot attitude information, expected driving joint variable, actual driving joint variable, and center height of the entire experiment. It is used to analyze the adaptive landing function of robots.

In order to test the influence of terrain structure, ambient light and terrain color on the designed algorithm, three groups of contrast experiments were designed. The first group of experiments are built in a normal environment, but with different terrain structures. The second group is designed with different terrain colors. The last group is set with different ambient lights.

5.2. Results and Discussion

This section presents the results and analysis of this experiment. Figure 11 shows the four process moments of the experiment: Initialization, adjustment, descend, and completeness.

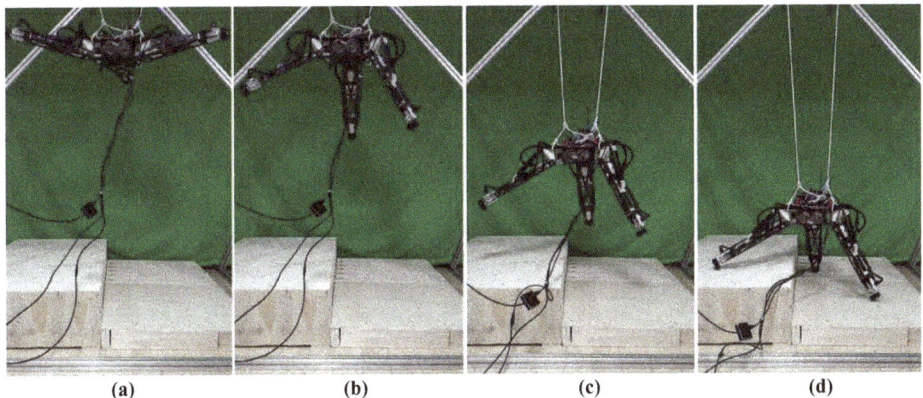

Figure 11. Adaptive landing process: (**a**) Initialization, (**b**) adjustment, (**c**) descend, (**d**) completeness.

Figure 12 shows the robot state variables collected during the entire experiment. The sub-figures in Figure 12 record the relationship between the driving joint variables, the attitude angle, and the height of the robot center position. In Figure 12c, according to the height curve of the robot, the landing process of the robot can be divided into three stages. The first stage (0–20 s, yellow region) is the landing preparation stage, in this stage, the drone will maintain a fixed altitude and fixed-point flight. The second stage (20–33 s, blue region) is the descent stage, in this stage, the drone lands vertically at a low speed. The third stage (33–40 s, green region) is the landing stabilization stage. At this time, the robot system has completed the entire landing process while keeping the height and attitude of the robot stable.

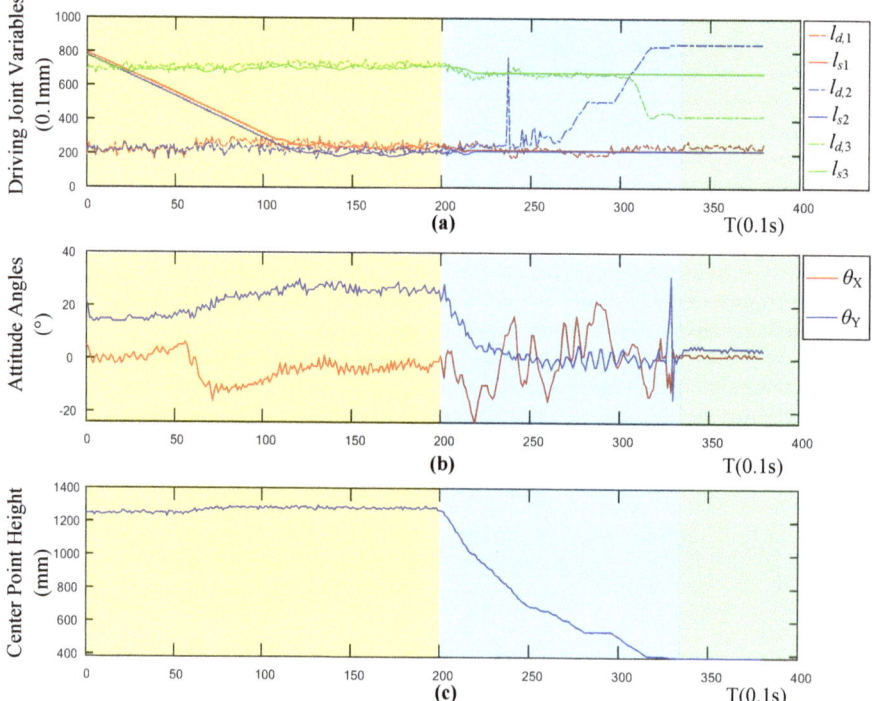

Figure 12. (a) Diagram of driving joint variables, (b) attitude angles, and (c) center point height.

Figure 12b records the changes of the robot's pitch θ_Y and roll angle θ_X during the descending phase. Corresponding to the three stages of the robot's landing process, the angle of the robot changes greatly during the descending phase, due to the disturbance caused by the manual release of the cable. This is similar to the disturbance phenomenon when the drone is performing position control and attitude control during the landing at a fixed point. The robot has a small angle fluctuation during the preparation phase as the robot is influenced by a reaction force generated when the robot rotates its robotic arms.

Figure 12a shows the relationship between the expected driving joint variables $l_{d,i}$ ($i = 1, 2, 3$) and the actual joint variables $l_{s,i}$ throughout the landing process. The dotted line represents the driving joint variable curve calculated by the control computer through the control algorithm, and the solid line represents the driving joint variable fed back by the robot in real-time. In the preparation stage, the robot's driving joint gradually approaches the expected position from the initial position, and finally reaches the curve of the expected position. In the descending phase, when the height is lower than 60 cm, the value of the drive variable calculated by the control computer begins to change significantly.

This is because when the robot is landed to a certain height, the viewing angle of the depth camera is limited, and the landing point of the robot exceeds the calculation range of the viewing angle, thereby causing a disturbance. Therefore, in this study, the height of 100 cm is set as the threshold. When the height of the robot is higher than the threshold, the calculated drive value of the control computer is believed to be reliable, and the robot follows the calculated value. When the height is lower than the threshold, the robot maintains the last trusted value. The resulting curve trend is shown in Figure 12a. When the height is less than 100 cm, the actual joint variables no longer follow the expected joint variable movement.

To further explore the effectiveness of the adaptive landing algorithm, this study calculated the error curve of the driving joint variable and the curve of the robot's tilt angle. It can be seen from Figure 13a that the variation trend of the error curve of the driving joint variables is basically consistent with the variation trend of the drive variable fed back by the robot. In the preparation stage, when the system is stable, there is zero error associated with the three driving joint variables, which shows the effectiveness of the robot motor driving algorithm. Figure 13b shows the deflection angle θ_{err} between the Z axis of the robot's coordinate axis and the Z axis of the geodetic coordinate system. The calculation formula is as follows:

$$\theta_{err} = \arccos(\cos(\theta_X) \cdot \cos(\theta_Y)) \qquad (23)$$

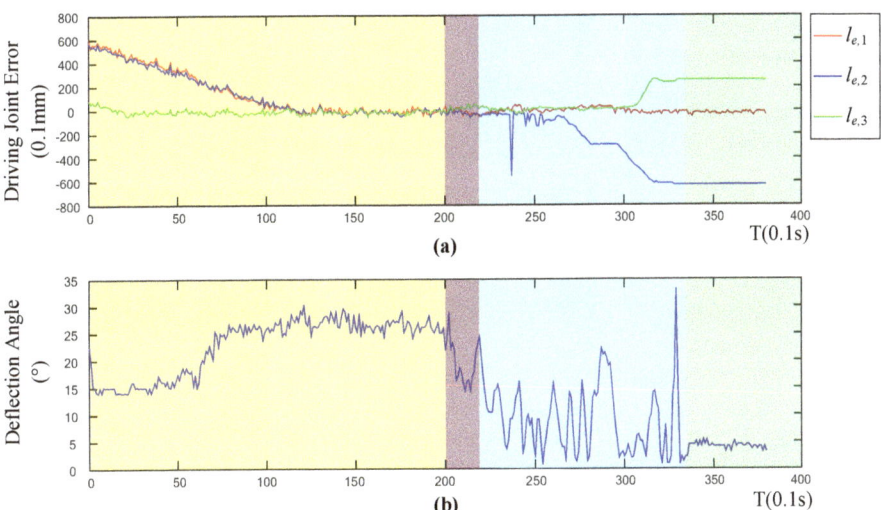

Figure 13. (a) Diagram of driving joint error, and (b) deflection angle.

It can be seen from Figure 13 that during the landing stabilization phase, the tilt angle of the robot is less than five degrees, and the landing can be considered as stable and horizontal. In the preparation phase and the landing phase, the tilt angle of the robot is larger, while in the final stabilized phase, the tilt angle is smaller. This proves that the adaptive control algorithm is less affected by the flight attitude change and speed change, which verifies the effectiveness of the control algorithm.

The changing curve of the robot's posture angle in Figure 12 shows that the posture angle will vibrate with the movement of the robot. However, when the robot nearly reaches the expected position, it will still move in small amplitude with the change of the expected driving joint variable, and this high-frequency, small-amplitude vibration motion will have a greater impact on the flight stability of the flying robot.

Figure 14 demonstrates the results of the experiments in different terrains. The sub-figures in the first row shows the 3D reconstructions in 1-m height. The second and third row present the RGB images and the depth images separately. The curves of the driving joint variables, attitude angles, and center point height are shown in fourth, fifth, and sixth rows. Each column represents the corresponding experiments. In the first experiment, the terrain is flat, and all three driving joint variables are almost the same, which satisfies the flat terrain. In the second experiment, a bulge is set on the ground. One driving joint variable is bigger than the others to satisfy the terrain with one bulge. In the last experiment, one high bugle and one low bugle are set on the ground. All three driving joint variables are different owing to the terrain. These results show that the designed control systems can work well. On the other hand, when the robot landed, the attitude angle errors were small and between plus and minus 10 degrees, also shows the effectiveness of the designed system. The testing results show that the designed system work well in different rigid terrains.

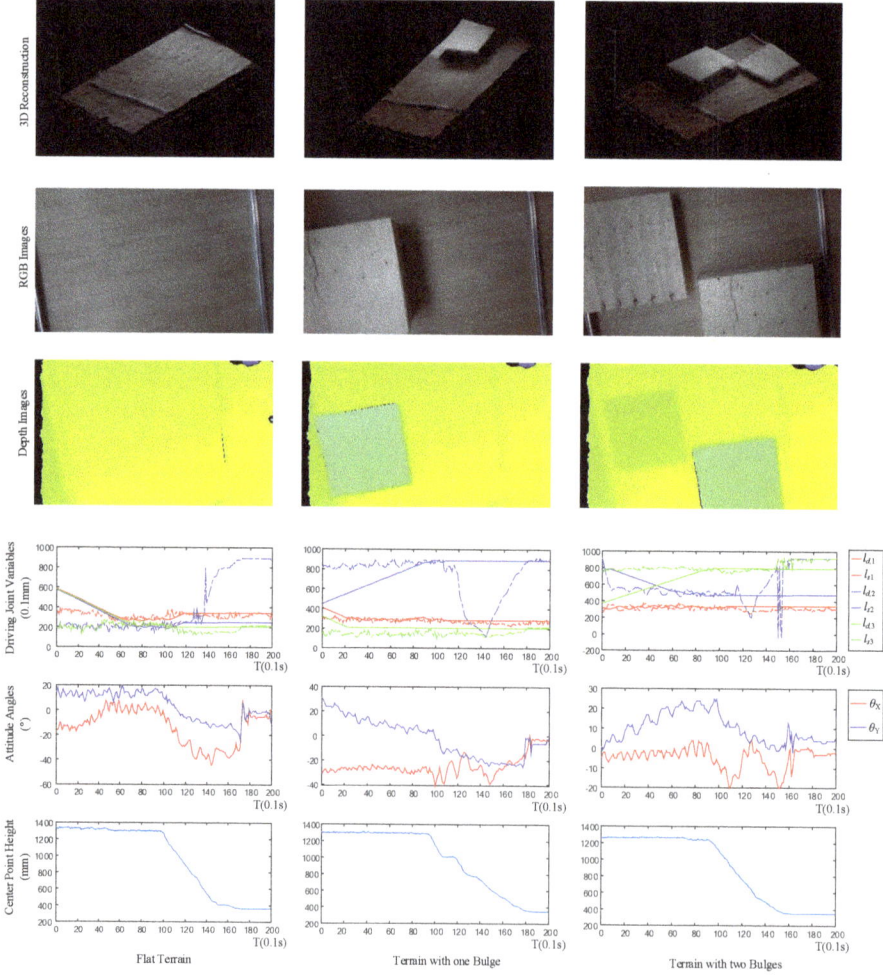

Figure 14. Experiments in different terrains.

Figure 15 shows the results of the experiments in different environmental conditions. Just like Figure 14, the sub-figures in different rows also show the results of 3D reconstructions, RGB images, depth images, driving joint variables, attitude angles, and center point height separately, while each column represents the corresponding experiments in different environmental conditions. In the first column, the figures present the results of the experiment with bright sunlight. The bright sunlight and shadow are clear in the RGB image, and the bright sunlight does influence the depth camera, which causes noise in the 3D reconstruction and the depth image. However, the attitude angle curves show that the angle error are still small, which means the effect of bright sunlight is restricted. In the second experiment, the terrain is covered by black foams. However, the change of terrain color has no impact on the depth camera nor the control system. In the last column, the experiment is tested in the dark environment. The terrain cannot be distinguished from the RGB image, but are still clear in the 3D reconstruction and the depth image. Moreover, the attitude angle curves also show darkness has no impact on the control system. All the results show that the ambient light and the terrain color have a limited impact on the designed control system.

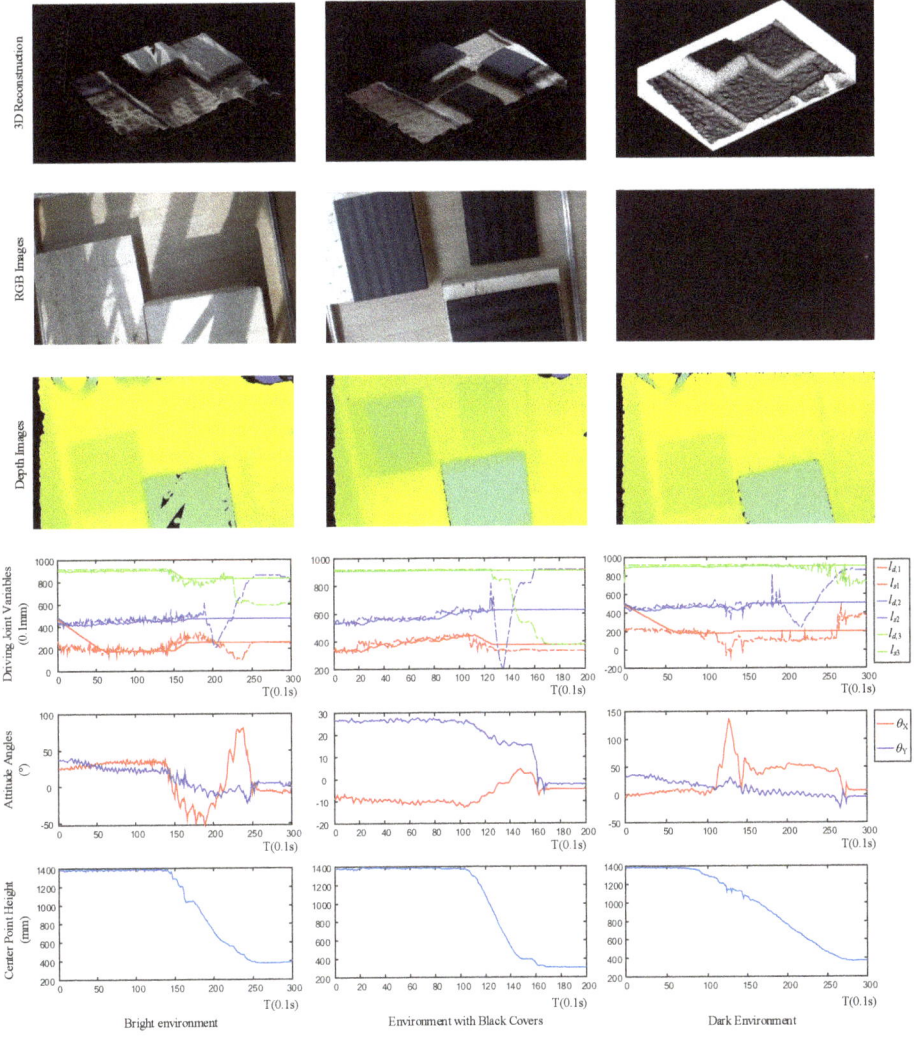

Figure 15. Experiments in different environmental conditions.

In order to determine the specific cause of the expected joint jitter, this study continues to analyze the relationship between the expected driving joint variable and the depth information and attitude angle. In this study, the expected driving joint variables, depth of the center point, and attitude angle curves are regarded as three kinds of signals, and the jitter of the signals is caused by noise. Therefore, in this article, the noise signals of the three signals are separated by wavelet filtering, and a comparative analysis is performed to determine the source of the expected driving joint variable noise. Figure 16 shows the filtered signal and noise signal of the driving joint signal, center position height signal, and inclination signal during the preparation stage. It can be seen from the figure that the filtered signal retains the movement trend of the original signal, but the change is relatively smooth. To analyze the correlation between various noise signals, this paper analyzes the Pearson correlation coefficient [24] among the three signals.

$$\rho M, N = \frac{\mathrm{cov}(M,N)}{\sigma M \sigma N} = \frac{E(MN) - E(M)E(N)}{\sqrt{E(M^2) - E^2(M)}\sqrt{E(N^2) - E^2(N)}} \quad (24)$$

where $\mathrm{cov}(M,N)$ represents the covariance of the signal $M, N = h_{nO}, l_{n,1}, l_{n,2}, l_{n,3}, \theta_{n,X}, \theta_{n,Y}$, and σ represents the standard deviation of the signal.

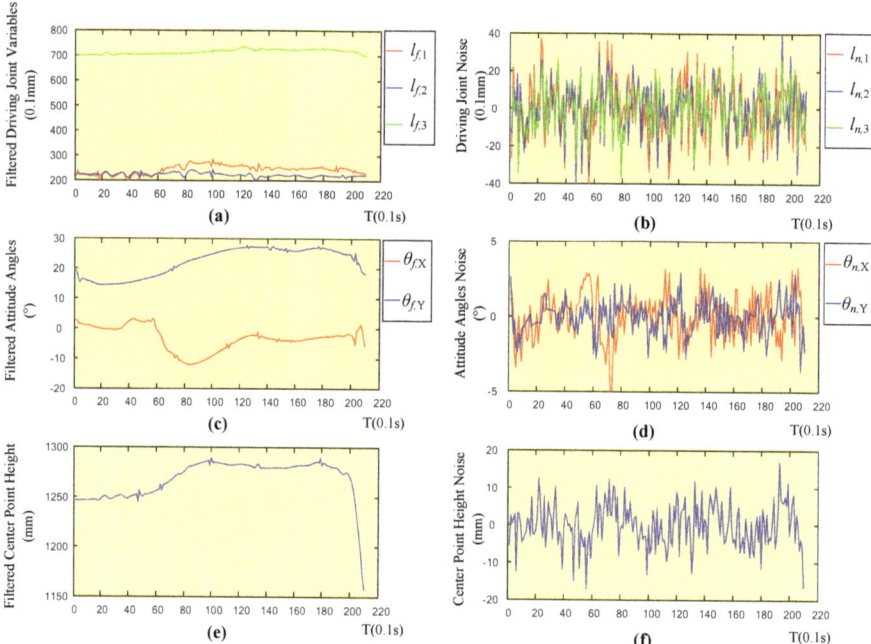

Figure 16. (**a**) The filtered curve of driving joint variables, (**c**) attitude angles, (**e**) center point height, and (**b**) noise curve of driving joint variables, (**d**) attitude angles, (**f**) center point height.

As shown in Table 2, it can be seen from the results that the driving joint variable is highly correlated with the noise signal at the center position height, which are, 0.7871, 0.7602, and 0.8064, respectively, but the noise signal correlation with the corner signal is relatively small. The height data of the center position comes from the depth sensor, so the main cause for the vibration of the driving variable is the signal noise of the depth sensor. By optimizing the depth solution algorithm of the depth sensor or filtering the depth image, the vibration of the driving joint can be alleviated.

Table 2. Pearson correlation coefficient.

Coefficient	$l_{n,1}$	$l_{n,2}$	$l_{n,3}$
h_{nO}	0.7871	0.7602	0.8064
$\theta_{n,X}$	−0.1398	−0.0615	0.0247
$\theta_{n,Y}$	0.1228	0.0421	0.1124

From the experiments, the results verify that the designed control algorithm can achieve adaptive landing in different rigid terrains, and the ambient light and the terrain color have a limited impact on the designed control system. However, some problems are also explored in the experiments. First, noise signals from the depth camera cause an error in robot's kinematical analysis. Second, visual angles of the depth camera limit the control algorithm when vehicles go down below the threshold height. Third, ambient light causes the noise in the depth camera. We will keep working on solving these three problems and testing the system on more terrains in the future.

6. Conclusions

In order to expand the take-off and landing functions of uncrewed aerial vehicles in a non-structural responsible environment, this study designs an adaptive landing control system for our self-developed robots. The control system uses multi-sensor fusion technology to realize three-dimensional terrain scanning of the landing area. To accurately calculate the robot's landing position, the control system uses a Kalman filter and complementary filter algorithms to fuse the robot's inclination and speed information. Then, based on the robot kinematics model, the robot's landing posture and driving joint variables are determined. Finally, this article verifies the feasibility and effectiveness of the adaptive landing control system through experiments. Based on the analysis of experimental data, the causes of the robot's lower abdominal concussion are discussed, which provides a basis for further optimization of the robot control algorithm design.

Author Contributions: Conceptualization, D.Z., H.T., Z.G.; Methodology, D.Z., H.T.; Software, H.T., Z.G.; Formal Analysis, D.Z., H.T., Z.G.; Investigation, D.Z., H.T.; Writing-Original Draft Preparation, H.T.; Writing—Review & Editing, D.Z., Z.G. All authors have read and agreed to the published version of the manuscript.

Funding: This research received no external funding.

Conflicts of Interest: The authors declare no conflict of interest.

References

1. Robotic Landing Gear Could Enable Future Helicopters to Take Off and Land Almost Anywhere. Available online: https://www.darpa.mil/news-events/2015-09-10 (accessed on 9 October 2015).
2. Stolz, B.; Brödermann, T.; Castiello, E.; Englberger, G.; Erne, D.; Gasser, J.; Hayoz, E.; Müller, S.; Muhlebach, L.; Löw, T. An adaptive landing gear for extending the operational range of helicopters. In Proceedings of the 2018 International Conference on Intelligent Robots and Systems (IROS 2018), Lanzhou, China, 24–27 August 2018; pp. 1757–1763.
3. Boix, D.M.; Goh, K.; McWhinnie, J. Modelling and control of helicopter robotic landing gear for uneven ground conditions. In Proceedings of the 2017 Workshop on Research, Education and Development of Unmanned Aerial Systems (RED-UAS 2017), Linköping, Sweden, 3–5 October 2017; pp. 60–65.
4. Sarkisov, Y.S.; Yashin, G.A.; Tsykunov, E.; Tsetserukou, D. DroneGear: A Novel Robotic Landing Gear with Embedded Optical Torque Sensors for Safe Multicopter Landing on an Uneven Surface. *IEEE Robot. Autom. Lett.* **2018**, *3*, 1912–1917. [CrossRef]
5. Miyata, K.; Sasagawa, T.; Doi, T.; Tadakuma, K. A Study of Leg-Type Landing Gear for Aerial Vehicles-Development of One Leg Model. *J. Robot. Mechatron.* **2011**, *23*, 266–270. [CrossRef]
6. Di Leo, C.V.; Leon, B.; Wachlin, J.; Kurien, M.; Rimoli, J.J.; Costello, M. Cable-driven four-bar link robotic landing gear mechanism: Rapid design and survivability testing. In Proceedings of the 2018

AIAA/ASCE/AHS/ASC Structures, Structural Dynamics, and Materials Conference, Kissimmee, FL, USA, 8–12 January 2018; pp. 0491:1–0491:25.

7. Doyle, C.E.; Bird, J.J.; Isom, T.A.; Kallman, J.C.; Bareiss, D.; Dunlop, D.J.; King, R.J.; Abbott, J.J.; Minor, M.A. An Avian-Inspired Passive Mechanism for Quadrotor Perching. *IEEE-Asme Trans. Mechatron.* **2013**, *18*, 506–517. [CrossRef]
8. Zhang, K.; Chermprayong, P.; Tzoumanikas, D.; Li, W.; Grimm, M.; Smentoch, M.; Leutenegger, S.; Kovac, M. Bioinspired design of a landing system with soft shock absorbers for autonomous aerial robots. *J. Field Robot.* **2019**, *36*, 230–251. [CrossRef]
9. Tieu, M.; Michael, D.M.; Pflueger, J.B.; Sethi, M.S.; Shimazu, K.N.; Anthony, T.M.; Lee, C.L. Demonstrations of bio-inspired perching landing gear for UAVs. In Proceedings of the Bioinspiration, biomimetics, and Bioreplication 2016, Las Vegas, NV, USA, 20–24 March 2016; pp. 9797:1–9797:8.
10. Luo, C.; Zhao, W.; Du, Z.; Yu, L. A neural network based landing method for an unmanned aerial vehicle with soft landing gears. *Appl. Sci.* **2019**, *9*, 2976. [CrossRef]
11. Hu, D.; Li, Y.; Xu, M.; Tang, Z. Research on UAV Adaptive Landing Gear Control System. In Proceedings of the 2018 2nd International Conference on Artificial Intelligence, Automation and Control Technologies (AIACT 2018), Osaka, Japan, 26–29 April 2018; Volume 1061, p. 012019.
12. Arns, M. *Novel Reconfigurable Delta Robot Dual-Functioning as Adaptive Landing Gear and Manipulator*; York University Press: Toronto, ON, Canada, 2019.
13. Jia, R.; Jizhen, W.; Xiaochuan, L.; Yazhou, G. Terrain-adaptive Bionic Landing Gear System Design for Multi-Rotor UAVs. In Proceedings of the 2019 Chinese Control and Decision Conference (CCDC 2019), Nanchang, China, 3–5 June 2019; pp. 5757–5762.
14. Huang, M. Control strategy of launch vehicle and lander with adaptive landing gear for sloped landing. *Acta Astronaut.* **2019**, *161*, 509–523. [CrossRef]
15. Cabecinhas, D.; Naldi, R.; Silvestre, C.; Cunha, R.; Marconi, L. Robust Landing and Sliding Maneuver Hybrid Controller for a Quadrotor Vehicle. *IEEE Trans. Control. Syst. Technol.* **2016**, *24*, 400–412. [CrossRef]
16. Goh, K.; Boix, D.M.; Mcwhinnie, J.; Smith, G.D. Control of rotorcraft landing gear on different ground conditions. In Proceedings of the 2016 IEEE International Conference on Mechatronics and Automation, Harbin, China, 7–10 August 2016; pp. 181–186.
17. Ho, H.W.; De Croon, G.C.H.E.; Van Kampen, E.; Chu, Q.P.; Mulder, M. Adaptive Gain Control Strategy for Constant Optical Flow Divergence Landing. *IEEE Trans. Robot.* **2018**, *34*, 508–516. [CrossRef]
18. Miller, A.B.; Miller, B.M.; Popov, A.K.; Stepanyan, K. UAV Landing Based on the Optical Flow Videonavigation. *Sensors* **2019**, *19*, 1351. [CrossRef] [PubMed]
19. Ho, H.W.; De Wagter, C.; Remes, B.D.W.; De Croon, G.C.H.E. Optical-flow based self-supervised learning of obstacle appearance applied to MAV landing. *Robot. Auton. Syst.* **2018**, *100*, 78–94. [CrossRef]
20. Cheng, H.; Chen, T.; Tien, C. Motion Estimation by Hybrid Optical Flow Technology for UAV Landing in an Unvisited Area. *Sensors* **2019**, *19*, 1380. [CrossRef] [PubMed]
21. Lin, H.; Chiang, M. The Integration of the Image Sensor with a 3-DOF Pneumatic Parallel Manipulator. *Sensors* **2016**, *16*, 1026. [CrossRef] [PubMed]
22. Bilal, D.K.; Unel, M.; Yildiz, M.; Koc, B. Realtime Localization and Estimation of Loads on Aircraft Wings from Depth Images. *Sensors* **2020**, *20*, 3405. [CrossRef] [PubMed]
23. Gao, M.; Yu, M.; Guo, H.; Xu, Y. Mobile Robot Indoor Positioning Based on a Combination of Visual and Inertial Sensors. *Sensors* **2019**, *19*, 1773. [CrossRef] [PubMed]
24. Benesty, J.; Chen, J.; Huang, Y.; Cohen, I. *Noise Reduction in Speech Processing*; Pearson Correlation Coefficient; Springer: Berlin/Heidelberg, Germany, 2009; pp. 1–4. ISBN 978-3-642-00295-3.

© 2020 by the authors. Licensee MDPI, Basel, Switzerland. This article is an open access article distributed under the terms and conditions of the Creative Commons Attribution (CC BY) license (http://creativecommons.org/licenses/by/4.0/).

Article

Vehicle Classification Based on FBG Sensor Arrays Using Neural Networks

Michal Frniak, Miroslav Markovic, Patrik Kamencay *, Jozef Dubovan, Miroslav Benco and Milan Dado

Faculty of Electrical Engineering and Information Technology, University of Zilina, 01026 Zilina, Slovakia; michal.frniak@uniza.sk (M.F.); miroslav.markovic@uniza.sk (M.M.); jozef.dubovan@uniza.sk (J.D.); miroslav.benco@uniza.sk (M.B.); milan.dado@uniza.sk (M.D.)
* Correspondence: patrik.kamencay@uniza.sk; Tel.: +421-41-513-2225

Received: 17 June 2020; Accepted: 7 August 2020; Published: 10 August 2020

Abstract: This article is focused on the automatic classification of passing vehicles through an experimental platform using optical sensor arrays. The amount of data generated from various sensor systems is growing proportionally every year. Therefore, it is necessary to look for more progressive solutions to these problems. Methods of implementing artificial intelligence are becoming a new trend in this area. At first, an experimental platform with two separate groups of fiber Bragg grating sensor arrays (horizontally and vertically oriented) installed into the top pavement layers was created. Interrogators were connected to sensor arrays to measure pavement deformation caused by vehicles passing over the pavement. Next, neural networks for visual classification with a closed-circuit television camera to separate vehicles into different classes were used. This classification was used for the verification of measured and analyzed data from sensor arrays. The newly proposed neural network for vehicle classification from the sensor array dataset was created. From the obtained experimental results, it is evident that our proposed neural network was capable of separating trucks from other vehicles, with an accuracy of 94.9%, and classifying vehicles into three different classes, with an accuracy of 70.8%. Based on the experimental results, extending sensor arrays as described in the last part of the paper is recommended.

Keywords: vehicle classification; FBG; artificial intelligence; smart sensors

1. Introduction

The issue of traffic monitoring and management has arisen due to a growing number of personal vehicles, trucks, and other types of vehicles. Due to existing road capacities being based on historic designs, the condition of these road communications deteriorates with a lack of growing financial investment to maintain and expand the road network. With these requirements, vehicle visual identification is not sufficient for traffic management and the prediction of the future state of traffic and road conditions. For this purpose, existing monitoring areas are being innovated with new sensor platforms, not only for the statistical purpose of monitoring areas. Additional information such as traffic density, vehicle weight distribution, overweight vehicles, and trucks could be included in automatic warning systems for the prediction of possible critical traffic situations. There are several technological approaches based on different principles. Each of them has various advantages and disadvantages, such as operating duration, traffic density, meteorological condition limits, resistance to chemical and mechanical damage from maintenance vehicles, etc.

All motor vehicles are classified into 11 base classes by current legislation in the states of the central European Union. Meanwhile, according to the Federal Highway Administration under the United States Department of Transportation, there are even 13 classes. These classes consist of personal vehicles, trucks, technical vehicles, public transport vehicles, and their subclasses. For decades, the only sufficient method to classify vehicles was by visual recognition. This method was strongly limited by meteorological conditions. In the last two decades, several different technical designs for classifying vehicles without a visual part of classifications have been proposed. At first, based on metallic vehicle chassis and axle parts, there were designs to measure magnetic field parameters of crossing vehicles. This included inductive loops or anisotropic magneto-resistive sensors built into the road pavement [1–4]. These technological designs achieved accurate results for specific vehicle classes with magnetic signatures. A different approach was by vehicle weight signature. Technological solutions based on piezoelectric sensors [5] and bending plate sensors are widely used in road traffic monitoring and vehicle measurements [6]. There were also experimental solutions such as the usage of hydro-electric sensors [7] with a bending metal plate at the top of the vessel filled with a specific liquid. Weigh-in-motion technologies measuring specific parameters such as the weight signature could be used, as well as other technologies including fiber optic sensors [8], wireless vibration sensors [9], or using embedded strain gauge sensors [10]. As an additional capability, this could be measured by smart pavements based on conductive cementitious materials [11]. Optic sensors based on Fiber Bragg Grating (FBG) were also successfully tested on different types of transport, such as railways. It was in Naples in Italy where this type of sensor was used for speed and weight measurements with the detection of train wheel abrasions as additional information for transport safety [12].

Vehicle classification and the measurement of vehicle parameters, such as weigh-in-motion, were the aim of several international research projects. The weighing-in-motion of road vehicles was a research aim in the European research project COST323 over two decades ago [13]. In the last decade, research ideas relating to infrastructure monitoring including road traffic have been studied, e.g., by COST projects TU1402 for structural health monitoring and TU1406 for roadway bridges [14–16].

Optical fiber sensors are becoming a very important part of smart Internet of Things (IoT) infrastructures, also on roads and highways. They can additionally perform different functions in critical infrastructure protection and monitoring. There is a broad spectrum of technological solutions of fiber optic sensors and optical sensors systems. For our investigation, we used the FBG sensor network built into the entry road into the campus of our university. Fiber Bragg Grating (FBG) sensors are classified as passive optical fiber components that are compatible with existing types of telecommunication fiber systems and can operate directly with incident light (most commonly in the 1550 nm range). Thus, they can be directly incorporated into the optical transmission chain. The fundamental principle on which the FBG work is based is Fresnel diffraction and interference. The propagating optical field may be refracted or reflected at the interface in the transmission medium with different refractive indices. The FBG operates as a light reflector for a specific (desired) spectrum of wavelengths, to ensure that the phase-matching condition is met. Other (undesirable) wavelengths are only slightly influenced by the Bragg grating [17–19].

In recent years, the different Convolutional Neural Network (CNN) architectures [20–22] applied to image processing constitute the current dominant computer vision theory, especially in tasks such as image classification (vehicle classification). The main goal of these networks is to transform the input image layer-by-layer from the input image to the final class scores. The input image is processed by a series of convolution layers with filters (kernels), max pooling layers, and Fully Connected (FC) layers. Finally, the activation function, such as softmax or sigmoid to classify the outputs (small cars, sedans, crossovers, family vans, or trucks), is used. In our case, the AlexNet [20] and GoogLeNet [23] convolutional neural networks were chosen. The basic architecture of the AlexNet consists of some convolutional layers (five layers), followed by max pooling layers, FC layers (three layers), and a softmax layer [20,24,25]. On the other hand, the architecture of GoogLeNet consists of 22 layers (nine inception modules). The main motivation for the inception modules' (layers') creation is to make a deeper CNN network so that highly accurate results could be achieved [23,26,27]. For vehicle classification, several works using deep learning and convolutional neural networks were described in [20].

The aim of this article is vehicle classification with FBG sensor arrays using artificial intelligence from partial records. The proposed neural network was trained using a dataset with a lack of information on the vehicle's speed, which we created by visual recognition of the vehicle passing through our testing platform. The majority of recorded vehicles were detected only through their left wheels, which reduces records from a 3D vehicle surface to one line of deformation. These records simulated situations where the vehicle's driver tried to avoid detection with a changed trajectory through the roadside or an emergency line without visual recognition.

2. Materials and Methods

The main goal of the research is the use of optical sensor networks for the classification of passing vehicles through a test platform based on neural networks for car recognition using an industrial camera. For this purpose, a test platform was built, which is described in Section 2.1.

2.1. Experimental Platform

The test platform for the measurement of additional vehicle characteristics is located at the University of Zilina campus on the entry road to the main parking lot. This monitoring area consists of several sensor arrays based on two technological applications of FBG sensors. All these sensors are built in the 2nd asphalt pavement layer covered with a top asphalt layer with a height of 6 cm above the sensors. Two electric loops were installed for the initialization of measurements, but the main goal was to use only optic-based sensors as FBGs. Those were realized in two different placements and numbers, as shown in Figure 1.

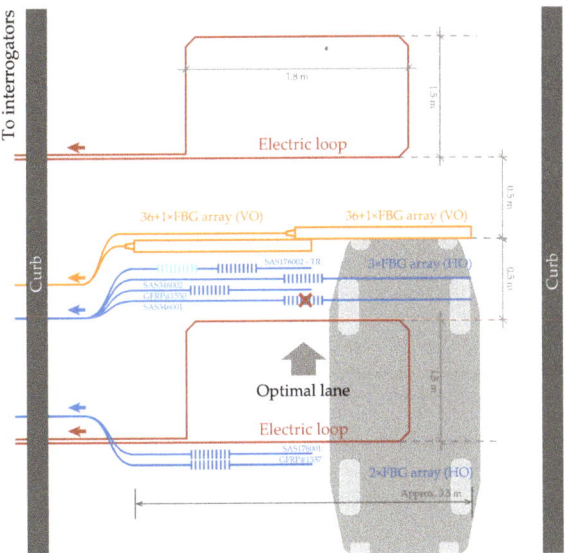

Figure 1. Real test platform scheme with multiple Fiber Bragg Grating (FBG) sensors. Some are connected as an FBG sensor array. The red cross indicates a dysfunctional FBG sensor (destroyed when the test platform was created). This test platform was built on the road into the university campus.

2.1.1. Vertically Oriented FBG Sensors

The 1st type of FBG was attached vertically on a perforated aluminum chassis with approximately a 10 cm distance between these Vertically Oriented (VO) sensors (orange sensors in Figure 1) positioned orthogonally to the direction of the vehicle, as shown in Figure 2. Based on the configuration of these sensors and their placement, there are several limitations. One of them is the distance between vertical

FBG sensors. Each vehicle's wheel is captured in a range from 3 to 4 vertical FBG sensors. Due to the construction of the aluminum chassis with these sensors in a partially liquid material such as asphalt, it is problematic to determine wheel width. This is a necessary parameter for calculating the weight distribution area through measuring the wheel and accurately determining the vehicle class.

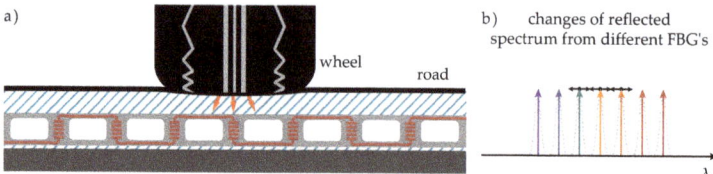

Figure 2. (**a**) Wheel pressure is applied to vertically oriented FGB sensors; (**b**) reflected optical spectrum shift is given by pressure change (every FBG reflects light on the other'scentral wavelength in idle status—the FBG's position is also known).

2.1.2. Horizontally Oriented FBG Sensors

The second type of FBG sensor was horizontally placed orthogonally (blue sensors in Figure 1) at different distances from vertical FBG sensors in the direction of the vehicle. Horizontally Oriented (HO) FBG sensors were installed with two different active lengths of sensors (measured on the whole fiber length using one FBG sensor). The first sensor had a length of 3460 mm, and the second had a length of 1760 mm. One of the optical fibers with shorter sensors contained another FBG for temperature compensation. Both horizontal sensor lengths had a passive length of 300 mm and an operative temperature range from −40 to +80 °C. All horizontal sensors were attached to the bottom asphalt layer by asphalt glue. This allowed for the measurement of exact flexibility and strength changes of the top asphalt layers during the measurements of overpassing vehicles. Due to the vehicle wheel trajectory over those sensors and their type, we observed both compression and tension, as shown in Figure 3.

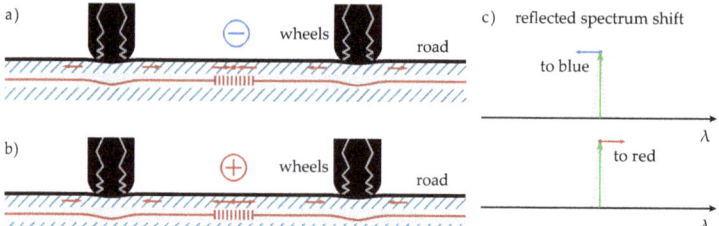

Figure 3. Illustration of some scenarios of wheel pressure to horizontally oriented FGB sensors (**a**) when the wheel's pressure in the horizontal line is negative (compressive stress) or (**b**) positive (tensile stress), (**c**) and the appropriate reflected light spectrum change in wavelength.

2.1.3. Measurement Units

Each set of measurement data was from the FBG sensor arrays consisting of 2 lines of 36 vertical sensors orthogonal to the vehicle direction and 2 sensors for the temperature compensation of the vertical sensors. From the horizontal FBG sensors, there were 3 horizontal sensors at a different level. Two of those sensors had an active length of 1760 mm, and one had a length of 3460 mm. One fiber with a shorter length contained an FBG sensor for temperature compensation created for different wavelengths. The sampling rate of the two interrogators connected to the FBG sensor arrays was 500 samples/s.

Output matrix data of each measurement had 2000 time samples (4 s) of the 34 vertically oriented FBG sensors used. This output matrix was extended by measurements from 4 horizontally oriented

FBG sensors with a dimension of 2000 time samples (4 s). We used only 34 of 36 vertical FBG sensors because the last 2 peaks of reflected intensity on specific wavelengths were too low for processing in the interrogator, and this caused problems with measured data consistency, as shown in Figure 4.

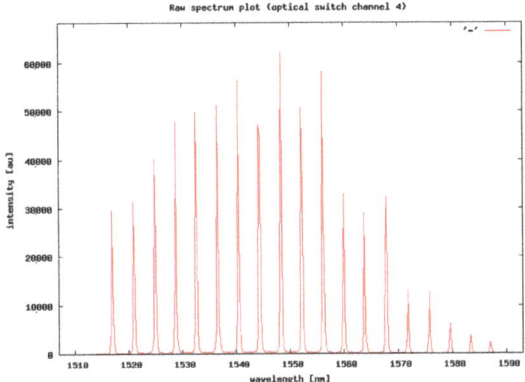

Figure 4. Time sample of the reflected optical spectrum from the FBG array received by the interrogator.

The 1st peak value of the FBG sensor, set at 1517 nm, was dedicated to temperature compensation. The last 2 unused vertical FBG sensors were preset at wavelengths of 1583.74 and 1587.68 nm. Both matrices for 2000 measurements were synchronized into the same time range. This format and size of data were applicable only in one direction of the vehicles due to the position of each sensor array.

2.2. Proposed Methodology

The block diagram of the proposed methodology is shown in Figure 5. Firstly, datasets based on FBG sensor data and Closed-Circuit Television (CCTV) were created. Next, the modified neural networks for visual classification using a CCTV camera system for FBG dataset annotation were used. This classification was used for the verification of measured and analyzed data from the sensor arrays. Finally, the newly proposed neural network for vehicle classification from the sensor array dataset was created.

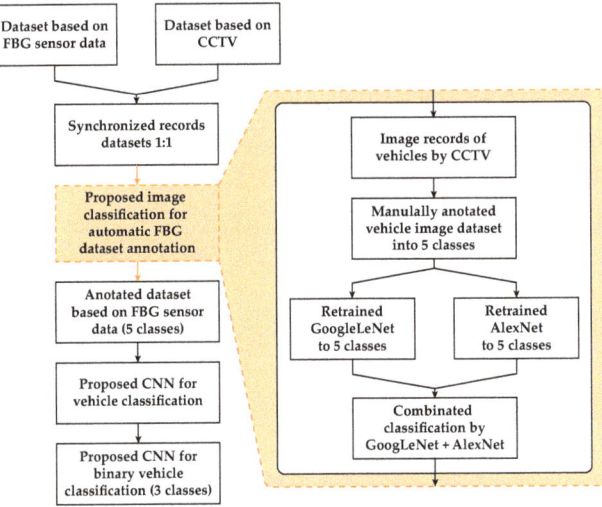

Figure 5. Block diagram of the proposed methodology.

Two separate datasets were created. Firstly, an image dataset based on CCTV was created for the acceleration of the automatized learning process for vehicle classification based on FBG sensor data. Secondly, a dataset based on FBG sensor data was created for final vehicle classification by the proposed CNN.

2.2.1. Dataset Based on FBG Sensor Data

Each vehicle's record from the test platform was created with a matrix from vertical FBG sensors with a size of 2000 measurements by 34 sensors. With a sampling rate of 500 samples/s, this represents a period of 4 s per each vehicle. The record detail of the full pressure map of the vehicle with a wheelbase of 2570 mm is presented in Figure 6.

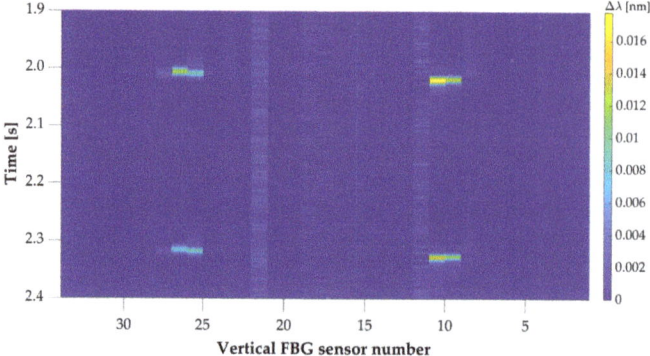

Figure 6. Record detail of the pressure map for a vehicle with an optimal line. The colormap represents the values of the wavelength change of the reflected optical spectrum by FBG in nm.

The shift in samples for each axle between the wheels in Figure 6 is caused by the installation shift of aluminum strips for vertical FBG sensors, shown in Figure 2 with orange color. The partial pressure map (only left wheels) of the vehicle with a wheelbase of 2511 mm is in Figure 7. Both vehicle's details show the detection of the 1st axle at time position 2 s. This was based on two way detection.

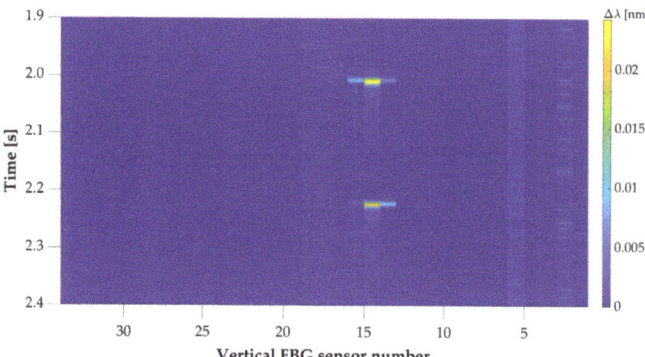

Figure 7. Record detail of the pressure map for the overpassing vehicle with left wheels. The colormap represents the values of the wavelength change of the reflected optical spectrum by FBG in nm.

For speed determination without information on the specific wheelbase of the vehicle from visual recognition, there were built-in horizontal FBG sensors of 2 lengths. Those sensors were placed asymmetrically towards the left side of the road. Record details from the overpassing vehicle recognized by both lines are shown below in Figure 8 and the overpassing vehicle recognized by one line in Figure 9.

Figure 8 is a record detail of the same vehicle's record as shown in Figure 6. A vehicle with the optimal line was captured with vertical and horizontal sensors; thus, we were able to determine vehicle speed and wheelbase distances. In Figures 7 and 9 is shown the same overpassing vehicle recognized by only one line of wheels by vertically oriented FBG sensors.

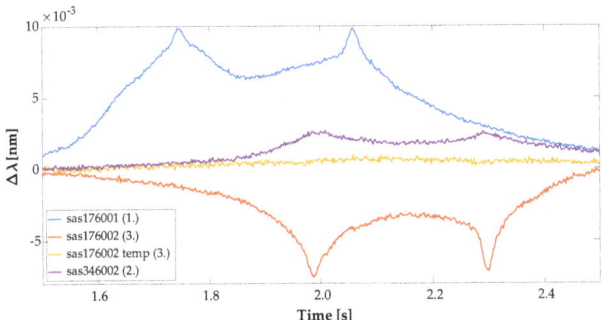

Figure 8. Record detail of axle detection from horizontal FBG sensors from the overpassing vehicle with an optimal traffic line as reflected in the optical spectrum wavelength change detected by the FBGs.

Records with only one line (footprint) of wheels of the vehicle recognized by vertically oriented FBG sensors, and those vehicles that were not recognized by horizontally oriented FBG sensors and measured data seem to be akin to a Nothing-on-Road state (NoR).

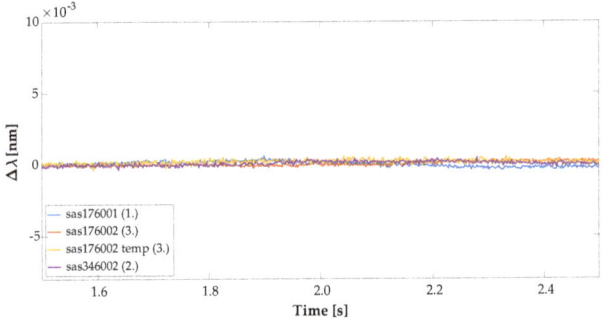

Figure 9. Record detail of axle detection from horizontal FBG sensors from the overpassing vehicle with a non-optimal traffic line as reflected in the optical spectrum wavelength change detected by the FBGs.

For the simplification of vehicle detection, we summed all wavelength shifts of all vertical FBG sensors per each timestamp. The summed wavelength shift for all k sensors in specific time t_n was compared with the summed wavelength shift for all k sensors in previous time t_{n-i}. Reference value $\Delta \lambda_R$ was added to this value, which corresponds to the minimum recorded pressure on the sensors from one vehicle's wheel detection. The reference value of $\Delta \lambda_R$ for the 1st axle detection was 0.015 nm with an air temperature over the test platform in the range from 15 to 30 °C. The equation for the 1st axle's detection is:

$$\sum_k \left| \Delta \lambda_{k, t_{n-i}} \right| + \Delta \lambda_R \leq \sum_k \left| \Delta \lambda_{k, t_n} \right|. \quad (1)$$

The record details of the summed values per 2 strips with vertical FBG sensors shifted by NoR values are shown below in Figure 10. The right wheels of vehicles shown by the blue curve for summed sensors with Positions 1 to 18 were detected. The left vehicle wheels are shown by the orange curve for summed sensors with Positions 19 to 34 by the left strip with vertical FBG sensors. The record detail shown is for the same vehicle as in Figures 6 and 8.

On the graph of the overpassing vehicle recognized only by one line of vehicle wheels in Figure 11, there was a partial record with no detection of the vehicle's right wheels. Only left wheels were detected by the sensors in Positions 1 to 18 with a blue curve.

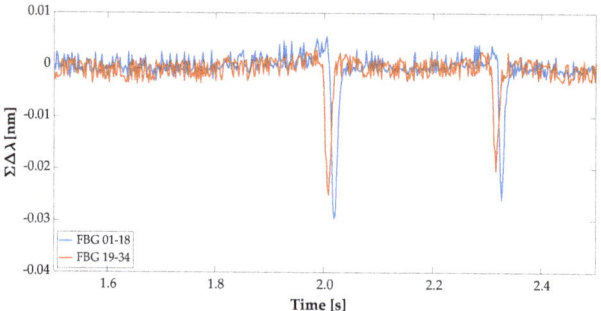

Figure 10. Record detail of the summary values of the wavelength changes (of the reflected optical spectrum by FBG) shifted by Nothing-on-Road (NoR) values.

Figures 7, 9 and 11 depict the same partially recognized vehicle, where it was not possible to determine the vehicle's speed and wheelbase distances from the minimal two lines of the FBG sensors. This information could only be used in combination with visual identification of the vehicle's model with technical parameters.

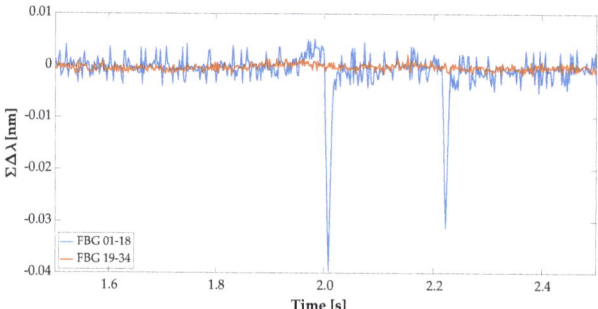

Figure 11. Record details of the summed wavelength shifts (of the reflected optical spectrum) from the vertical FBG sensors of the overpassing vehicle recognized only by the left wheels.

2.2.2. Dataset Based on CCTV

Our test platform is incapable of accurately determining wheel width and other additional parameters based on it. For this reason, we decided to define each vehicle class by wheelbase and weight ranges in combination with visual recognition. For this, we used security CCTV monitoring the entry ramp used to access the road with the testing platform. This entry ramp serves as a measurement separator in the direction of monitored vehicles, as shown in Figure 12.

The input images from CCTV were at a resolution of 1920 × 1080 px. The area of interest, with an image size of 800 × 800 px (red rectangle), is shown in Figure 12.

Figure 12. Entry ramp view from CCTV with the area of interest (red rectangle with a resolution of 800 × 800 px) with a timestamp.

2.2.3. Synchronized Records' Datasets

All vehicles were monitored with CCTV and measured using FBG sensor arrays for 1 month. Per each overpassing vehicle's record, there was 1 synchronized vehicle image. These images were classified by 2 CNNs for image classification, validated as shown in Figure 5, and integrated with records from FBG sensor arrays. Those records were impossible to classify only from vertically oriented FBG sensor arrays without image classification. For the next vehicle's classification using FBG sensors, there were only relevant data from the chassis of vertical FBG sensors from Positions 1 to 18.

2.2.4. Proposed Image Classification for Automatic FBG Dataset Annotation

For the visual verification of the 5 determined classes, we tested the dataset on 3 different CNNs in the MATLAB® workspace in Version 2019b. We decided to use AlexNet [12], GoogLeNet [13], and ResNet-50 [28,29]. Each pre-trained network was modified in the final layers for specific class number outputs.

The architecture modification of the pretrained CNN AlexNet from 1000 classes to 5 classes is shown in Figure 13. The modification of pretrained Directed Acyclic Graph (DAG) CNN GoogLeNet with the same number of pretrained classes as AlexNet to 5 classes is shown in Figure 14.

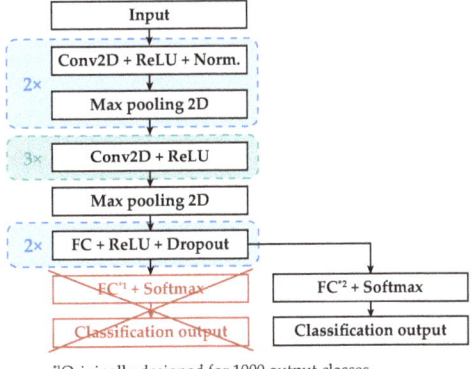

*1 Originally designed for 1000 output classes
*2 Redesigned to 5 output classes

Figure 13. Architecture modification of the pretrained AlexNet.

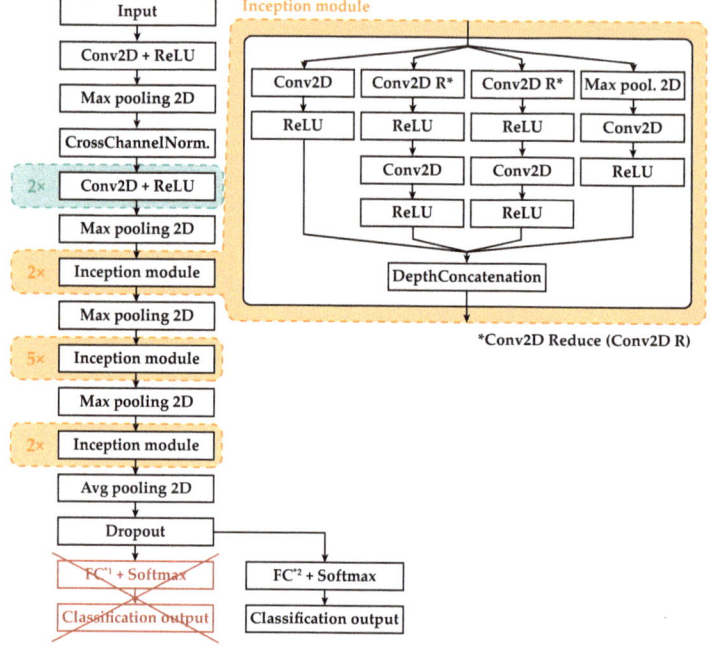

Figure 14. Architecture modification of the pretrained GoogLeNet.

The training phase consisted of 650 vehicle images for each class. The test phase consisted of a minimum of 100 vehicle images for each class. Those images were next resized to the necessary input size to each CNN [20,23].

For this reason, we decided to create 5 vehicle classes. The 1st class was small cars with hatchback bodyworks with a weight up to 1.5 t and up to a 2650 mm axle spacing. The 2nd class was vehicles such as sedans and their long versions or combo bodyworks. The 3rd class was vehicles with crossover bodyworks and Sports Utility Vehicles (SUV). The 4th class was utility vehicles and family vans weighing up to approximately 2.5 t. The last class was vans, trucks, and vehicles with more than 2 axles. Motorcycles were excluded from the classification. These 5 classes were also determined based on the composition of the vehicles (see Table 1) and their count crossing the campus area with a test platform.

Table 1. Image dataset.

Vehicle Type	Class	Train	Test
Hatchback	1	650	428
Sedan/Combo	2	650	384
SUV	3	650	304
MPV/Minivan	4	650	227
Van/Truck	5	650	376

Each CNN was retrained 5 times for the 6 epochs achieved, in equal conditions, with an accuracy of over 90% in the tested dataset. One epoch represents the processing of all training samples. After that, training samples were shuffled for the next epoch. Those CNNs were supervised and retrained by using a Graphic Processor Unit (GPU) with only 2 GB GDDR5 memory in previous research. The accuracy of the created dataset was enough for our purpose of classifying data from FBG sensor arrays [30].

Thus, the retrained CNNs for image classification were prepared to classify vehicle records from FBG arrays using the visual part of the records. Each record from the arrays was synchronized with 1 image from the industrial camera taking into consideration the distance between the entry ramp and the measured sensory area. The synchronized image dataset was divided into 2 identical datasets with resolutions of 224 × 224 px for GoogLeNet and 227 × 227 px for AlexNet classification. In 77.26% of the images, both CNNs were consistent. The accuracy of the CNNs used for visual classification is shown in Table 2. Other images were manually verified and included in the correct classes.

Table 2. Outputs from CNNs for image classification.

	AlexNet	GoogLeNet	ResNet-50
Achieved train validation	99.79%	90.67%	91.30%
Achieved test validation	90.2%	90.8%	89.2%

2.2.5. Annotated Dataset Based on FBG Sensor Data

The prepared dataset consisted of 5965 vehicle records recognized with only one line using vertically oriented FBG sensors divided into 5 classes. This dataset did not contain vehicle speed, wheelbase, or wheel size information. For simple classification, a neural network was created in the Integrated Development Environment (IDE) MATLAB® 2020a for image input in the Tagged Image File Format (TIFF) with a resolution of 600 × 5 px (600 time samples × 5 vertically oriented FBG sensors). These data were normalized into a range from 0 to 1 with eight decimal precision and were saved in TIFF format per each partial record without data compression.

2.2.6. Proposed CNN for Vehicle Classification

The structure of the CNN created is in Table 3 below. The CNN was tested for various dataset interclass combinations. Due to wheelbases and the speed of overpassing vehicles, up to 600 samples per record (1.2 s, see Figure 6) were recorded for all vehicles, trucks included. Most of the small vehicles' last wheel was on average recorded up to a time sample of 200 records (0.4 s) for speeds under 50 km/h and the last wheel up to a time sample of 500 records (1 s) per all vertical sensors with speeds under 10 km/h.

Table 3. Design of CNN for vehicle classification.

Layers	Parameters	Number of CFs *
Input	600 × 5 × 1	
Conv2D + ReLU	300 × 4	128
Conv2D + ReLU	100 × 4	64
Conv2D + ReLU	100 × 4	32
MaxPool2D	2 × 1	
Conv2D + ReLU	50 × 2	24
FC + Softmax	2, 3 or 5	
ClassOutput	2, 3 or 5	

* Convolution Filter (CF).

For that reason, the 1st 2D convolution layer was set to filter sizes from 300 to 4, covering at least a wheel per record in the 1st layer. Enlarging the filter size on the 1st layer during training did not show any improvements. After a 2D max pooling layer, a last 2D convolution layer was added with a filter size of 50 to 2. This design showed the best-achieved results for binary classification on our prepared dataset. After the last convolution laser, there was a fully connected layer with the softmax function to assign the result to only one of all output classes based on an overall number of trained classes.

For training purposes, there were 800 vehicle samples separated from the first 4 classes and 400 samples from the last truck class. Those samples were divided by a ratio of 9:1 for the input training set and validation set during training.

3. Experimental Results

For all CNNs, the training had the same option setup as training for 200 epochs with the batch size set to twenty. On the main diagonal in the confusion matrix, correctly classified vehicles of all tested vehicles are shown in Table 4. For the first class (hatchback class), forty-nine-point-six percent of vehicles were correctly classified (valid column), as shown in Table 5. For the second class (combo/sedan class), twelve-point-point-eight percent of vehicles were correctly classified. For the third class (SUV class), fifty-six-point-three percent of vehicles were correctly classified. For the fourth class (MPV/minivan class), twenty-six-point-eight percent of vehicles were correctly classified. For the last class (van/truck class), sixty-two percent of vehicles were correctly classified. An overall accuracy of 28.9% for all tested vehicles using the proposed CNN was achieved. Due to the classification into five classes and their similarities, a validation accuracy of only 28.9% was achieved.

Table 4. Results from the test part of the dataset from the CNN for vehicle classification. Final results for each class on the main diagonal in confusion matrix (highlighted as bold) are shown.

Class	1	2	3	4	5	Valid
1	**9.8%**	1.7%	4.7%	3.3%	0.3%	49.6%
2	17.8%	**25.8%**	19.5%	9.8%	1.8%	12.8%
3	2.5%	0.8%	**8.5%**	3.0%	0.4%	56.3%
4	1.4%	0.4%	2.3%	**1.7%**	0.5%	26.8%
5	0.1%	0.1%	0.3%	0.6%	**1.8%**	62%
Overall						28.9%

The proposed CNN was modified to three classes for better spatial separation of classes. On the main diagonal in the confusion matrix, correctly classified vehicles of all tested vehicles are shown in Table 5. For the first class (hatchback class), seventy-four-point-three percent of vehicles were correctly classified (valid column), as shown in Table 5. For the second class (SUV class), thirty-seven-point-eight percent of vehicles were correctly classified. For the third class (van/truck class), seventy-eight-point-nine percent of vehicles were correctly classified. An overall accuracy of 60.0% of all tested vehicles using the proposed CNN was achieved.

The proposed CNN was modified for classification between two classes (hatchback class to van/truck class). An overall accuracy of 92.7% for both tested vehicle classes, as shown in Table 6, using the proposed CNN was achieved.

Table 5. Results from the test part of the dataset from the CNN for vehicle classification reduced to 3 classes. Final results for each class on the main diagonal in confusion matrix (highlighted as bold) are shown.

Class	1	2	3	Valid
1	**38.9%**	10.9%	2.5%	74.3%
2	18.8%	**15.2%**	6.2%	37.8%
3	1.1%	0.5%	**5.9%**	78.9%
Overall				60%

Table 6. Results from the test part of the dataset from the CNN for vehicle classification reduced to 2 classes. Final results for each class on the main diagonal in confusion matrix (highlighted as bold) are shown.

Class	1	3	Valid
1	**83.2%**	4.2%	95.1%
3	3.0%	**9.6%**	76.1%
Overall			92.7%

Proposed CNN for Binary Vehicle Classification

For the improvement of the achieved validation for three classes in the combination of binary classification, we used the designed CNN for classification of three variations of the prepared dataset. Three classes were compared as binary, with one to the rest. Continuing, data from the first class were classified in opposition to the combined data of the other two and the second class in opposition to the first and third class. Finally, data from the third class were classified into the combined data from the first and second classes, as shown in Figure 15.

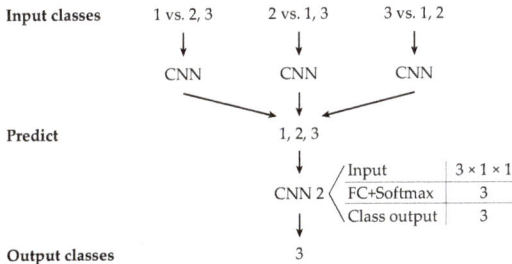

Figure 15. Process of the vehicle classification of overpassing vehicles.

In the first part, the proposed CNN for binary vehicle classification (small vehicles to the rest of the vehicles) using the training dataset was trained. This training dataset was modified to a ratio of 1:1 (800 records for each class). The results from the test dataset are shown in Table 7.

Table 7. Results from the test part of the dataset from the CNN for vehicle binary classification. Final results for each class on the main diagonal in confusion matrix (highlighted as bold) are shown.

Class	1	2,3	Valid
1	**42.7%**	9.7%	81.6%
2,3	16.9%	**30.8%**	64.6%
Overall			73.5%

In the second part, the proposed CNN for binary vehicle classification (SUV vehicles to the rest of the vehicles) using the training dataset was trained. This training dataset was modified to a ratio of 1:1 (800 records for each class). The results from the test dataset are shown in Table 8.

Table 8. Results from the test part of the dataset from the CNN for vehicle binary classification. Final results for each class on the main diagonal in confusion matrix (highlighted as bold) are shown.

Class	2	1,3	Valid
2	**21.7%**	18.3%	54.2%
1,3	10.9%	**49%**	81.8%
Overall			70.7%

In the third part, the proposed CNN for binary vehicle classification (truck vehicles to the rest of the vehicles) using the training dataset was trained. This training dataset was modified to a ratio of 1:2 (400 records for the van/truck class, 800 images for the rest of the vehicles). The results from the test dataset are shown in Table 9.

An improved process by binary predicting three classes as one to the rest achieved valid classification on the test dataset of 70.8%, as shown in Table 10. Each row in the confusion matrix represents a test group for the class, and the columns represent the category. Highlighted values in the diagonal show properly classified vehicles.

Table 9. Results from the test part of the dataset from the CNN for vehicle binary classification. Final results for each class on the main diagonal in confusion matrix (highlighted as bold) are shown.

Class	3	1,2	Valid
3	**4.5%**	3.1%	59.2%
1,2	2%	**90.5%**	97.8%
Overall			94.9%

Table 10. Confusion matrix of classified vehicles. Final results for each class on the main diagonal in confusion matrix (highlighted as bold) are shown.

Class	1	2	3	Valid
1	**40.9%**	10.9%	0.5%	78.1%
2	13.8%	**25.8%**	0.5%	64.3%
3	1.3%	2.1%	**4.1%**	54.9%
Overall				70.8%

Due to there being no information about vehicle speed, the wheelbase with axle configurations, and wheel sizes, the valid classification of 70.8% achieved is acceptable. These results are from one line of vertical FBG sensors with partially overpassing vehicles by one line of wheels. The speed and wheelbase similarities between the first and second classes created significant incorrect classifications, which can be reduced with a larger part of the dataset for the training of the designed CNN.

4. Discussion

The obtained results from the experimental platform, which consists of vertically oriented FBG sensor arrays, are presented. Due to the location of the testing platform on the two way access road into the university campus, because the sensor arrays are installed in the middle of this road, there was some limitation. More than 80% of recorded vehicles passed over the inbuilt sensors. Those vehicles were recorded only with one line of wheels without measurement by the horizontally oriented FBG sensors. A minimum of two lines of sensors is necessary for wheelbase distance determination and vehicle speed measuring. We focused on the classification of passing vehicles only from one line of vertical FBG sensors. The proposed neural network was capable of separating trucks from other vehicles with an accuracy of 94.9%. To classify three different classes, an accuracy of 70.8% was achieved. Based on experimental results, extending the sensor arrays is recommended.

The approach that solved the width of the vehicle's wheels is based on horizontal fiber optic sensors with a 45° orientation in the vehicle's direction over the test platform for the left and right side of the vehicle, separated as shown in Figure 16. This solution is limited by vehicle speed due to the double bridge construction over the road. Another technological approach, which is already widely used, is based on bending plates installed in a concrete block of the road. These sensors could be separated for the left and right side of the vehicle or be combined for weighing the whole vehicle's axle. With these realizations, there is no need to know the wheel width, because there is a whole contact area of the wheel with road sensors.

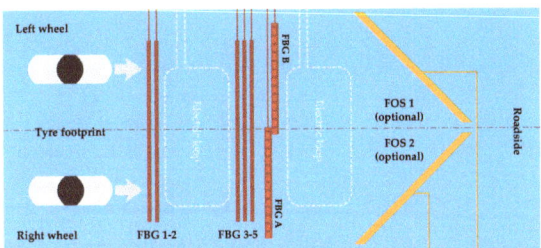

Figure 16. Proposed experimental platform with Fiber Optic Sensors (FOS) with 45° orientations (orange color).

In order to gain significant improvements of these results, it would be necessary to extend the sensor arrays to the full width of the road. An alternative solution will be a change from two way road management to one way.

Author Contributions: Conceptualization, M.F. and M.M.; methodology, P.K.; software, M.F. and M.B.; validation, M.F. and M.M.; visualization, J.D.; supervision, M.D. All authors read and agreed to the published version of the manuscript.

Funding: This work was funded by the Slovak Research and Development Agency under the project APVV-17-0631 and the Slovak Grant Agency VEGA 1/0840/18. This work was also supported by Project ITMS: 26210120021 and 26220220183, co-funded by EU sources and the European Regional Development Fund.

Conflicts of Interest: The authors declare no conflict of interest.

Abbreviations

The following abbreviations are used in this manuscript:

2D	Two-Dimensional
3D	Three-Dimensional
Avg	Average
CCN	Convolutional Neural Network
CCTV	Closed-Circuit Television
CF	Convolution Filter
Conv2D	Two-Dimensional Convolution layer
Conv2D R	Conv2D Reduce
COST	European Cooperation in Science and Technology
DAG	Directed Acyclic Graph
FBG	Fiber Bragg Grating
FC	Fully Connected
FOS	Fiber Optic Sensor
GPU	Graphic Processor Unit
HO	Horizontally Oriented
IDE	Integrated Development Environment
IoT	Internet of Things
MPV	Multi-Purpose Vehicle
NoR	Nothing-on-Road
px	pixel(s)
ReLU	Rectified Linear Unit
SUV	Sports Utility Vehicles
TIFF	Tagged Image File Format
VO	Vertically Oriented

References

1. Jeng, S.; Chu, L. Tracking Heavy Vehicles Based on Weigh-In-Motion and Inductive Loop Signature Technologies. *IEEE Trans. Intell. Transp. Syst.* **2015**, *16*, 632–641. [CrossRef]
2. Santoso, B.; Yang, B.; Ong, C.L.; Yuan, Z. Traffic Flow and Vehicle Speed Measurements using Anisotropic Magnetoresistive (AMR) Sensors. In Proceedings of the 2018 IEEE International Magnetics Conference (INTERMAG), Singapore, 23–27 April 2018; pp. 1–4.
3. Xu, C.; Wang, Y.; Bao, X.; Li, F. Vehicle Classification Using an Imbalanced Dataset Based on a Single Magnetic Sensor. *Sensors* **2018**, *18*, 1690. [CrossRef] [PubMed]
4. Lamas, J.; Castro-Castro, P.M.; Dapena, A.; Vazquez-Araujo, F. Vehicle Classification Using the Discrete Fourier Transform with Traffic Inductive Sensors. *Sensors* **2015**, *15*, 27201–27214. [CrossRef] [PubMed]
5. He, H.; Wang, Y. Simulation of piezoelectric sensor in weigh-in-motion systems. In Proceedings of the 2015 Symposium on Piezoelectricity, Acoustic Waves, and Device Applications (SPAWDA), Jinan, China, 30 October–2 November 2015; pp. 133–136.
6. Weigh-in-Motion Pocket Guide. 2018. Available online: https://www.fhwa.dot.gov/policyinformation/knowledgecenter/wim_guide/ (accessed on 10 August 2020).
7. Mardare, I.; Tita, I.; Pelin, R. Researches regarding a pressure pulse generator as a segment of model for a weighing in motion system. *IOP Conf. Ser. Mater. Sci. Eng.* **2016**, *147*, 012060. [CrossRef]
8. Al-Tarawneh, M.; Huang, Y.; Lu, P.; Bridgelall, R. Weigh-In-Motion System in Flexible Pavements Using Fiber Bragg Grating Sensors Part A: Concept. *IEEE Trans. Intell. Transp. Syst.* **2019**, 1–12. [CrossRef]
9. Bajwa, R.; Coleri, E.; Rajagopal, R.; Varaiya, P.; Flores, C. Development of a Cost Effective Wireless Vibration Weigh-In-Motion System to Estimate Axle Weights of Trucks. *Comput.-Aided Civ. Infrastruct. Eng.* **2017**. [CrossRef]
10. Wenbin, Z.; Qi, W.; Suo, C. A Novel Vehicle Classification Using Embedded Strain Gauge Sensors. *Sensors* **2008**, *8*, 6952–6971. [CrossRef]
11. Birgin, H.; Laflamme, S.; D'Alessandro, A.; García-Macías, E.; Ubertini, F. A Weigh-in-Motion Characterization Algorithm for Smart Pavements Based on Conductive Cementitious Materials. *Sensors* **2020**, *20*, 659. [CrossRef] [PubMed]
12. Gautam, A.; Singh, R.R.; Kumar, A.; Thangaraj, J. FBG based sensing architecture for traffic surveillance in railways. In Proceedings of the 2018 3rd International Conference on Microwave and Photonics (ICMAP), Dhanbad, India, 9–11 February 2018; pp. 1–2.
13. Caprez, M.; Doupal, E.; Jacob, B.; O'Connor, A.; OBrien, E. Test of WIM sensors and systems on an urban road. *Int. J. Heavy Veh. Syst.* **2000**, *7*. [CrossRef]
14. Thöns, S.; Limongelli, M.P.; Ivankovic, A.M.; Val, D.; Chryssanthopoulos, M.; Lombaert, G.; Döhler, M.; Straub, D.; Chatzi, E.; Köhler, J.; et al. *Progress of the COST Action TU1402 on the Quantification of the Value of Structural Health Monitoring. Structural Health Monitoring 2017*; DEStech Publications, Inc.: Stanford, CA, USA, 2017. [CrossRef]
15. Matos, J.; Casas, J.; Strauss, A.; Fernandes, S. COST ACTION TU1406: Quality Specifications for Roadway Bridges, Standardization at a European level (BridgeSpec)—Performance Indicators. In *Performance-Based Approaches for Concrete Structures—14th fib Symposium Proceedings*; fib: Cape Town, South Africa, 2016.
16. Casas, J.R.; Matos, J.A.C. Quality Specifications for Highway Bridges: Standardization and Homogenization at the European Level (COST TU-1406). *Iabse Symp. Rep.* **2016**, *106*, 976–983. [CrossRef]
17. Haus, J. *Optical Sensors: Basics and Applications*; Wiley-VCH: Weinheim, Germany, 2010.
18. Yin, S.; Ruffin, P.B.; Yu, F.T.S. (Eds.) *Fiber Optic Sensors*, 2nd ed.; Number 132 in Optical Science and Engineering; CRC Press: Boca Raton, FL, USA, 2008.
19. Venghaus, H. (Ed.) *Wavelength Filters in Fibre Optics*; Number 123 in Springer Series in Optical Sciences; Springer: Berlin, Germany; New York, NY, USA, 2006.
20. Krizhevsky, A.; Sutskever, I.; Hinton, G. ImageNet Classification with Deep Convolutional Neural Networks. *Neural Inf. Process. Syst.* **2012**, *25*. [CrossRef]
21. Phung, V.H.; Rhee, E.J. A High-Accuracy Model Average Ensemble of Convolutional Neural Networks for Classification of Cloud Image Patches on Small Datasets. *Appl. Sci.* **2019**, *9*, 4500. [CrossRef]
22. Kamencay, P.; Benco, M.; Mizdos, T.; Radil, R. A New Method for Face Recognition Using Convolutional Neural Network. *Adv. Electr. Electron. Eng.* **2017**, *15*. [CrossRef]

23. Szegedy, C.; Liu, W.; Jia, Y.; Sermanet, P.; Reed, S.; Anguelov, D.; Erhan, D.; Vanhoucke, V.; Rabinovich, A. Going deeper with convolutions. In Proceedings of the 2015 IEEE Conference on Computer Vision and Pattern Recognition (CVPR), Boston, MA, USA, 7–12 June 2015; pp. 1–9.
24. Han, X.; Zhong, Y.; Cao, L.; Zhang, L. Pre-Trained AlexNet Architecture with Pyramid Pooling and Supervision for High Spatial Resolution Remote Sensing Image Scene Classification. *Remote Sens.* **2017**, *9*, 848. [CrossRef]
25. Samir, S.; Emary, E.; El-Sayed, K.; Onsi, H. Optimization of a Pre-Trained AlexNet Model for Detecting and Localizing Image Forgeries. *Information* **2020**, *11*, 275. [CrossRef]
26. Wang, J.; Hua, X.; Zeng, X. Spectral-Based SPD Matrix Representation for Signal Detection Using a Deep Neutral Network. *Entropy* **2020**, *22*, 585. [CrossRef]
27. Kim, J.Y.; Lee, H.E.; Choi, Y.H.; Lee, S.J.; Jeon, J.S. CNN-based diagnosis models for canine ulcerative keratitis. *Sci. Rep.* **2019**, *9*. [CrossRef] [PubMed]
28. Lin, C.; Chen, S.; Santoso, P.S.; Lin, H.; Lai, S. Real-Time Single-Stage Vehicle Detector Optimized by Multi-Stage Image-Based Online Hard Example Mining. *IEEE Trans. Veh. Technol.* **2020**, *69*, 1505–1518. [CrossRef]
29. Liu, W.; Liao, S.; Hu, W. Perceiving Motion From Dynamic Memory for Vehicle Detection in Surveillance Videos. *IEEE Trans. Circuits Syst. Video Technol.* **2019**, *29*, 3558–3567. [CrossRef]
30. Frniak, M.; Kamencay, P.; Markovic, M.; Dubovan, J.; Dado, M. *Comparison of Vehicle Categorisation by Convolutional Neural Networks Using MATLAB*; ELEKTRO 2020 PROC; IEEE: Taormina, Italy, 2020; p. 4.

© 2020 by the authors. Licensee MDPI, Basel, Switzerland. This article is an open access article distributed under the terms and conditions of the Creative Commons Attribution (CC BY) license (http://creativecommons.org/licenses/by/4.0/).

Article

TADILOF: Time Aware Density-Based Incremental Local Outlier Detection in Data Streams

Jen-Wei Huang *, Meng-Xun Zhong and Bijay Prasad Jaysawal

Department of Electrical Engineering, National Cheng Kung University, Tainan City 701, Taiwan; s1993126@gmail.com (M.-X.Z.); bijay@jaysawal.com.np (B.P.J.)
* Correspondence: jwhuang@mail.ncku.edu.tw

Received: 16 August 2020; Accepted: 12 October 2020; Published: 15 October 2020

Abstract: Outlier detection in data streams is crucial to successful data mining. However, this task is made increasingly difficult by the enormous growth in the quantity of data generated by the expansion of Internet of Things (IoT). Recent advances in outlier detection based on the density-based local outlier factor (LOF) algorithms do not consider variations in data that change over time. For example, there may appear a new cluster of data points over time in the data stream. Therefore, we present a novel algorithm for streaming data, referred to as time-aware density-based incremental local outlier detection (TADILOF) to overcome this issue. In addition, we have developed a means for estimating the LOF score, termed "approximate LOF," based on historical information following the removal of outdated data. The results of experiments demonstrate that TADILOF outperforms current state-of-the-art methods in terms of AUC while achieving similar performance in terms of execution time. Moreover, we present an application of the proposed scheme to the development of an air-quality monitoring system.

Keywords: outlier detection; local outlier factor; data streams; air quality monitoring

1. Introduction

The expansion of Internet of Things is increasing the importance of outlier detection in streaming data. A wide range of tasks ranging from factory control charts to network traffic monitoring depend on the identification of anomalous events associated with intrusion attacks, system faults, and sensor errors [1,2]. Some outlier detection methods are designed to find global outliers, while some methods try to find local outliers [1,2].

The local outlier factor, LOF, proposed in [3], is a well-known density-based algorithm for the detection of local outliers in static data. LOF measures the local deviation of data points with respect to their K nearest neighbors, where K is a user-defined parameter. This kind of method can be useful in several applications, such as detecting fraudulent transactions, intrusion detection, direct marketing, and medical diagnostics. Later, the concept of LOF was extended for incremental databases [4], and for streaming environments [5,6]. However, recent advances in LOF-based outlier detection algorithms for data streams, MILOF [5] and DILOF [6], do not consider variations in data that may change over time. For example, there may appear a new cluster of data points over time in the data streams. In addition, algorithms for data streams need to avoid using outdated data. To handle the data streams, the algorithms utilize a fixed window size to limit the number of data points held in memory by summarizing previous data points. These recent studies base their summaries only on the distribution of previous data; i.e., they do not take the sequence of data into account. The fact that these methods lack a mechanism for the removal of outdated data can greatly hinder their performances. Imagine a situation where sensors installed near a factory are used to detect the emission of PM2.5 pollutants. If pollutants were emitted on more than one occasion (with an intermittent period of normal concentrations), then the fact that the initial pollution event is held in memory might prevent the detection of subsequent violations. In other words, if the previous pollution event is held in memory for longer time, the next pollution event will be treated as an inlier and the method could not detect next pollution event.

Moreover, limited memory and computing power impose limitations on window size and thus on model performance because limitations on memory capacity and computational power necessitate the elimination of some previous data points. However, setting an excessively small window size can degrade performance because we can only hold a few data points in memory and hence there may be lack of neighboring data points with similar features, which affects the outlier scores.

A data stream potentially contains an infinite number of data points: $S = \{s_1, s_2, ..., s_t, ...\}$. Each data point $S_t \in R^D$ is collected at time t. We need to consider the following constrains for applications in data stream environments.

- Continuous data points (usually infinite).
- Limited memory and limited computing power.
- Real time responses for processed data.

Our goal is to detect outliers by calculating the LOF score for each data point. In addition, we are focusing on detecting outliers in data stream. Therefore, the following constraints must be considered in the detection of outliers in a data stream.

- Memory limitations constrain the amount of data that can be held in memory. We need to consider this for handling unbounded data stream environment.
- The state of the current data point as an outlier/inlier must be established before dealing with subsequent data points. Note that we do not have any information related to subsequent data points appearing in the data stream.
- Adding new data may induce new clusters.
- Limited computing power needs to be utilized before new data arrives in the data stream. Therefore, the algorithms need to be efficient in terms of execution time.

In this study, we sought to resolve these issues by developing a (1) time-aware and density-summarizing incremental LOF (TADILOF) and (2) a method to approximate the value of LOF. For time-aware summarization, we include a time component, also termed time indicator, with each data point. The inclusion of a time component in the summary phase makes it possible to consider the sequential order of the data, and thereby deal with concept drift and enable the removal of outdated data points. Basically, every data point is assigned a time indicator referring to the point at which it was added to the streaming data. When a new data point arrives, the time indicators of K-nearest neighbor data points are updated if the newly added data point is not judged as an outlier. Using this strategy, the data points near to new data points are updated with the current time indicator and therefore these data points are less likely to be removed in the summarization phase. Thus, our proposed method is more likely to follow the variations in data that may change over time.

Furthermore, we propose a method to calculate approximate LOF score based on the summary information of previous data points. Note that this involves estimating the distances between newly-added data points and potential deleted neighbors (i.e., data points deleted in a previous summary phase). In the proposed method, LOF score is used to decide whether a newly added data point is an outlier or not in accordance with a LOF threshold. LOF score represents the outlierness of the data points based on the local densities defined using K-nearest neighbor data points. In addition, LOF score is able to adjust for the variations in the different local densities [2]. If the newly added data point is detected as an outlier as per LOF threshold, we use a second check based on proposed approximate LOF score to finally decide whether it is an outlier or not.

To maintain the data in the window, we use the concept of a landmark window strategy as used in the recent studies, MILOF [5] and DILOF [6], for local outlier detection in data streams. When the window is filled with the data points, we summarize the window to make space available for new data points by removing the old and less important data identified by the proposed summarization method. In our proposed summarization method, we summarize the data points of complete window using three quarters of the window. Then, one quarter size of window becomes available for new data points. We discuss the details of our summarization method in Section 3.2.

To limit the data to fit into available memory, the sliding window technique used in several applications for data streams is also an option. In the sliding window technique, all the old data points are deleted that cannot fit into memory. However, this may degrade the performance of local outlier detection because new events cannot be differentiated from some past events, and the accuracy of the estimated local outlier factor of data points will be affected if the histories of earlier data points are deleted [5]. Therefore, we use the landmark window strategy. In addition, the proposed strategies of using a time indicator and approximate LOF are suitable in combination with a landmark window for local outlier detection accuracy.

In addition, to evaluate the performance of our proposed method, we executed extensive experiments against the state-of-the-art algorithms on various real datasets. The results of experiments illustrate that the proposed algorithm outperforms state-of-the-art competitors in terms of AUC while achieving similar performance in terms of execution time. The results of experiments validate the effectiveness of the proposed method to use the time component and approximate LOF, which help to achieve better AUC.

Moreover, we applied the proposed method to a real-world data streaming environment for the monitoring of the air quality. The Taiwanese monitoring system referred to as the location-aware sensing system (LASS) employs 2000 sensors, each of which can be viewed as an individual data stream. We used the proposed system to detect outliers in each of these data streams. We call this type of outlier a temporal outlier because such outliers are compared with historical data points from the same device. We then combine the position of every device to facilitate the detection of spatial outliers and pollution events based on outliers from the neighboring devices.

The main contributions of this work are as follows.

- We developed a novel algorithm to detect outliers in data streams. The proposed approach is capable of adapting the changes in variations of data over time.
- We developed an algorithm to calculate approximate LOF score in order to improve model performance.
- Extensive experiments using real-world datasets were performed to compare the performance of the proposed scheme with those of various state-of-the-art methods.
- The efficacy of the proposed scheme was demonstrated in a real-world pollution detection system using PM2.5 sensors.

The rest of this paper is organized as follows. In Section 2, we discuss related works. Then, we introduce the proposed method in Section 3. In Section 4, we describe our experiments and a performance evaluation of the proposed method. Section 5 demonstrates a case study based on our proposed method for monitoring air quality and detection of pollution events. Finally, conclusions are presented in Section 6.

2. Background and Related Work

Outlier and anomaly detection on large datasets and data streams is a very important research area that has been useful for several applications [1,2,7]. Some studies focus on detecting global outliers, whereas other studies focus on detecting local outliers [1,2]. Different approaches have been studied for outlier detection, such as distance-based methods, density-based methods, and neural network-based methods [8].

In addition, clustering techniques can also be used for outlier detection. Therefore, we discuss some works on clustering and outlier detection based on clustering. In [9], the authors discussed a method for incremental K-means clustering. In the incremental database, this approach is better than traditional K-means. Similarly, the study in [10] proposes IKSC, incremental kernel spectral clustering, for online clustering in dynamic data. Another study in [11] discusses various machine learning approaches for real-world SHM (structural health monitoring) applications. The authors discuss the temporal variations of operational and environmental factors and their influences on the damage detection process. In [12], the authors propose enhancement of density-based clustering and outlier detection based on clustering. In addition, the authors discuss the approach for parameter reduction for density-based clustering. In [13], the authors propose a density-based outlier detection method using DBSCAN. First, the authors compute the minimum radius of an accepted cluster; then a revised version process of DBSCAN is used to further fit for data clustering and the decision of whether each point is normal or abnormal can be made. In [14], the authors provide survey of unsupervised machine learning algorithms that are proposed for outlier detection. In [15], the authors propose a cervical cancer prediction model (CCPM) for early prediction of cervical cancer using risk factors as inputs. The authors utilize several machine learning approaches and outlier detection for different preprocessing tasks.

The local outlier factor (LOF) [3] is a well-known density-based algorithm for the detection of local outliers in static data. This method can be useful in several applications, such as detecting fraudulent transactions, intrusion detection, direct marketing, and medical diagnostics [16–18]. Based on LOF, the study in [19] proposed a method to mine top-n local outliers. Later, the concept of LOF was extended for dynamic data—for instance, incremental LOF (iLOF) [4] was made for incremental databases, and MiLOF [5] and DILOF [6] were made for streaming environments. The application of LOF to incremental databases requires updating every previous data point and the recalculation of the LOF score, both of which are computationally intensive. iLOF reduces the time complexity to $O(n \log n)$ by updating the LOF score of data points affected by newly-added data points. Unfortunately, this approach is inapplicable to data streams with limited memory resources. MiLOF leverages the concept of K-means [20] to facilitate outlier detection in data streams by overcoming the space complexity of iLOF (i.e., $O(n^2)$). MiLOF uses a fixed window size to limit the number of data points held in memory by summarizing previous data points through the formation of K-cluster centers. Note, however, that MiLOF is prone to the loss of density information and a large number of points are required to represent sparse clusters. DILOF was developed to improve the summarization process using the nonparametric Rényi divergence estimator [21] to select minimum divergence subset from previous data points. However, neither MiLOF nor DILOF consider the concept-drift [22,23] of data in data streams to avoid using outdated data [24]. Furthermore, MiLOF and DILOF base their summaries only on the distribution of previous data; i.e., they do not take the sequence of data into account.

Some other methods based on LOF have been proposed for top-n outlier detection. In [25], the authors proposed the TLOF algorithm for scalable top-n local outlier detection. The authors proposed a multi-granularity pruning strategy to quickly prune search space by eliminating candidates without computing their exact LOF scores. In addition, the authors designed a density-aware indexing mechanism that helps the proposed pruning strategy and the KNN search. In [26], the authors proposed local outlier semantics to detect local outliers by leveraging kernel density estimation (KDE). The authors proposed a KDE-based algorithm, KELOS, for top-n local outliers over data streams. In [27], the authors proposed the UKOF algorithm for top-n local outlier detection based on KDE over large-scale high-volume data streams. The authors defined a KDE-based outlier factor (KOF) to measure the local outlierness score, and also proposed the upper bounds of the KOF and an upper-bound-based pruning strategy to reduce the search space. In addition, the authors proposed LUKOF by applying the lazy update method for bulk updates in high-speed large-scale data streams.

Since this study proposes a method to find local outliers in data streams, we discuss LOF, iLOF, MiLOF, and DILOF in the following subsections.

2.1. LOF and iLOF

LOF scores are computed for all data points according to parameter K (i.e., the number of nearest neighbors). The LOF score is calculated as follows:

Definition 1. *$d(p,o)$ is the Euclidean distance between two data points p and o.*

Definition 2. *K-distance(p), $d_K(p)$, is defined as the distance between data point p and its Kth nearest neighbor.*

Definition 3. *Given two data points p and o, reachability distance reach-dist$_K(p,o)$ is defined as:*

$$\text{reach-dist}_K(p,o) = \max\{d(p,o), K\text{-distance}(o)\} \quad (1)$$

Definition 4. *Local reachability density of data point p, $LRD(p)$, is derived as follows:*

$$LRD(p) = \left(\frac{1}{K} * \sum_{o \in N_K(p)} \text{reach-dist}(p,o) \right)^{-1} \quad (2)$$

where N_K is the set of K nearest neighboring data points of point p, and K is a user-defined parameter.

Definition 5. *Local outlier factor of data point p, $LOF(p)$, is obtained as follows:*

$$LOF(p) = \frac{1}{K} * \sum_{o \in N_K(p)} \frac{LRD_K(o)}{LRD_K(p)} \quad (3)$$

If the LOF score of a data point is greater than or equal to the threshold, then that data point is considered an outlier.

LOF is used to calculate the LOF scores only once. iLOF was developed to deal with the problem of data insertion, wherein we update only the previous data points that are affected by the new data point. Note that iLOF is not applicable to the detection of outliers in streaming data, due to the fact that there is no mechanism for the removal of outdated points. In addition, real-world applications lack the memory resources required to deal with the enormous (potentially infinite) number of data points generated by streaming applications.

Since LOF and iLOF are not suitable for data streams, MiLOF [5] was proposed for the detection of outliers in streaming data. We discuss MiLOF in the next subsection.

2.2. MiLOF

MiLOF [5] was developed for the detection of outliers in streaming data using limited memory resources. Essentially, MiLOF overcomes the memory issue by summarizing previous data points. MiLOF is implemented in three phases: insertion, summarization, and merging. Note that the insertion step of MiLOF is similar to that of iLOF. When the number of points held in memory reaches the limit imposed by window size b, the summarization step is invoked, wherein the K-means algorithm is used to find c cluster centers to represent the first $\frac{b}{2}$ data points, after which the insertion step is repeated iteratively. In the merging phase, weights are assigned to each cluster center based on the number of associated data points. The weighted K-means algorithm is then used to merge the new cluster center with the old cluster center. When using MiLOF, the total amount of data held in memory does not exceed $m = b + c$. MiLOF can be used to reduce memory and computation requirements; however, it does not preserve the density of the original dataset within the summary, which is crucial to detection accuracy.

2.3. DILOF

Being similar to MILOF, DILOF is a density-based local outlier detection algorithm for data streams that utilizes LOF score to detect outliers. DILOF is implemented in two phases: detection and summarization. The detection phase, which is called last outlier-aware detection (LOD), uses the iLOF technique to calculate LOF values when new data points are added to the dataset. DILOF then classifies the data points within the normal class or as an outlier. The summarization phase, which is called nonparametric density summarization (NDS), is activated when the number of data points reaches the limit defined by window size W. DILOF uses the nonparametric Rényi divergence estimator [21] to characterize the divergence between the original data and summary candidate. The gradient descent method is then used to determine the best summary combination. Summarization compiles half of the data $X = \{x_1, x_2, ..., x_{W/2}\}$ within a space one quarter the size of the window size $Z = \{z_1, z_2, ..., z_{W/4}\}$ by minimizing the loss function. There are four terms in the loss function. In the following, we introduce them one by one.

The first term is the Rényi diversity between the summary candidate and the original data. Renyi diversity is calculated using Equation (4), as follows:

$$\sum_{n=1}^{W/2} y_n \frac{p_K(x_n))}{v_K(x_n))} \tag{4}$$

In Equation (4), y_n is the binary decision variable of each data point x_n. Data point x_n is selected when y_n equals 1 and discarded when y_n equals 0. However, assessing every subset combination to determine the minimum loss values is impractical. NDS resolves this issue by relaxing the decision variable to produce an unconstrained optimization problem, where y_n becomes a continuous variable. Using the gradient descent method, NDS selects the best combination of x_n—i.e., the half of parameter set y_n with the highest values. $p_k(x_n)$ is the Euclidean distance between data point x_n and its Kth-nearest neighbor in X. $v_k(z_n)$ is the Euclidean distance between data point z_n and its Kth-nearest neighbor in Z. This term is given by the Rényi divergence estimator.

The second term is the shape term, which preserves the shape of the data distribution by selecting data points at the boundary of clusters, such that the data point within the boundary always has a higher LOF value. This term is shown as Equation (5).

$$-\sum_{n=1}^{W/2} y_n e^{LOF_K(x_n))} \tag{5}$$

The third and fourth terms are regularization terms. The third term is used to control y_n close to 0–1. It is important to avoid excessively high x_n values, which would render other data points ineffective. The fourth term is used to select half of all data points. These terms are shown in Equation (6).

$$\sum_{n=1}^{W/2} \psi_{0,1}(y_n) + \frac{\lambda}{2} (\sum_{n=1}^{W/2} y_n - \frac{W}{4})^2 \tag{6}$$

Combining all of the components, we obtain the loss function of DILOF as follows:

$$\min_{y} \sum_{n=1}^{W/2} y_n \frac{p_k(x_n)}{v_k(x_n)} - \sum_{n=1}^{W/2} y_n e^{LOF_k(x_n)} \\ + \sum_{n=1}^{W/2} \psi_{0,1}(y_n) + \frac{\lambda}{2} \left(\sum_{n=1}^{W/2} y_n - \frac{W}{4} \right)^2 \tag{7}$$

The gradient descent method is then used to obtain the optimal result as shown in Equation (8).

$$y_n^{(i+1)} = y_n^{(i)} - \eta \left\{ \sum_{x \in C_{K,n}} \frac{p_K(x)}{v_K(x)} + \frac{p_K(x_n)}{v_K(x_n)} - e^{LOF_K(x_n)} \right.$$
$$\left. + \psi_{0,1}'(y_n^i) + \lambda \left(\sum_{n=1}^{W/2} y_n^{(i)} - \frac{W}{4} \right) \right\} \quad (8)$$

In Equation (8), ψ is the learning rate, i is the number of iteration, and $C_{(K,n)}$ is a set of data points that have x_n as their Kth-nearest neighbor in Z. Interested readers are referred to the DILOF paper [6] for details on the calculation of $C_{(k,n)}$. After the decision variable has been updated, the larger half is selected as the summary point. Following this summarization phase, half of all data points are summarized into a quarter of all data points. This leaves a space equal to one quarter of the window size into which new data points can be inserted.

The DILOF method lacks a mechanism by which to remove outdated data or compensate for concept drift. NDS calculates only the difference in density in selection of a summary point. We therefore added the concept of time to differentiate outdated data points.

3. Proposed Method: TADILOF

In this section, we outline the proposed TADILOF algorithm and approximate LOF score. Algorithm 1 presents the pseudocode of the TADILOF algorithm. Our scheme also uses density to select the summary; therefore, we have two phases: detection and summarization. In the detection phase, we include a step in which previous information is used to obtain the approximate LOF, which is then used to determine whether the newly-added point is an outlier. This detection phase is referred to as ODA, outlier detection using approximate LOF. We add a time component to the summarization phase, and therefore refer to it as time-aware density summarization (TADS). We provide the details of procedures TADS and ODA in the following subsections. The approximate LOF score is calculated only when there is information from previous data points. Therefore, we introduce the time component before obtaining the approximate LOF score.

Algorithm 1 TADILOF algorithm

Input: *DS*: A data stream $D = \{d_1, d_2, ..., d_t, ...\}$,
　　　Window size: *W*,
　　　Number of neighbor: *K*,
　　　Threshold: θ,
　　　Step size: η,
　　　Regularization constant: λ,
　　　Maximum number of iteration: *I*
Output: The set of outliers in streams
 1: $dataInMemory = \{\}$;
 2: $outlierSet = \{\}$;
 3: **while** a new data point d_t is in stream **do**
 4: 　　$dataInMemory$.add(d_t)
 5: 　　$LOF_k(d_t)$ = ODA(d_t,$outlierSet$,θ)
 6: 　　**if** $LOF_k(d_t) > \theta$ **then**
 7: 　　　　$outlierSet$.add(d_t)
 8: 　　**if** $dataInMemory$.length $> W$ **then**
 9: 　　　　$dataInMemory$=TADS($dataInMemory$,η,λ,I)
10: **end while**

3.1. Time Component

Addition of a time component to this type of task allows the model to distinguish old data from new, thereby making it possible to recognize concept drift over time. For example, daytime readings might not be explicitly differentiated from nighttime readings in the PM2.5 data, despite the fact that time of day plays an important role in PM2.5 concentrations. Another example is the degree to which purchasing behavior varies over time as a function of the strength of the economy. The addition of a time component also provides a mechanism by which to remove outdated data, which might otherwise compromise model performance.

In this study, we include a time component in the summarization phase. Basically, every data point is assigned a time indicator t_i referring to the point at which it was added to the streaming data. In other words, the time indicators describe the age of every data point. The difference between t_i and the current time point corresponds to the length of time that data point d_i has existed in the dataset. The objective is to discard outdated data and preserve newer data points, which are presumed to more closely approximate the current situation. TADILOF refreshes data points close to the current data point and updates the time indicator of points neighboring the new data point, as shown in the following equation. Fortunately, this does not incur additional calculations due to the fact that we have already identified the neighbors of the new data in the LOF process.

$$t_i = t_{new}, \; if \; d_i \in N_K(d_{new}) \qquad (9)$$

Refreshing the time indicator of each data point enables our loss function to select data points that fit the current concept. Thus, a new model can be used to select data points in accordance with the density as well as the concept(s) represented by the current data streams. When TADS is triggered to summarize previous data points, it calculates the time difference t_diff between summarized time stamp t_s and the time stamp of data point d_i as follows:

$$t\_diff_i = max\{t_s - t_i - \alpha * W, 0\} \qquad (10)$$

In Equation (10), α is a hyperparameter indicating the amount of time that must elapse before TADILOF designates data as outdated and removes them. For example, $\alpha = \frac{W}{4}$ means that any data point with a time difference of less than one quarter of the window size is less likely to be selected for removal by the objective function. We present TADS in the next subsection.

3.2. Time-Aware Density Summarization (TADS)

Figure 1 presents the proposed TADS (in the TADILOF algorithm), which differs from NDS (in the DILOF algorithm). Note that NDS always retains the most recent half window of data points and summarizes the older half within a quarter size window. By contrast, TADS summarizes data points from three quarters of the window, and does not necessarily retain only the latest data. Rather, the TADS mechanism considers the density and the age of the data points. The time term is added to the TADS loss function as follows:

$$\min_y \sum_{n=1}^{W} y_n * t\_diff_n + \sum_{n=1}^{W} y_n \frac{p_K(x_n)}{v_K(x_n)} - \sum_{n=1}^{W} y_n e^{LOF_K(x_n)} \\ + \sum_{n=1}^{W} \psi_{0,1}(y_n) + \frac{\lambda}{2}(\sum_{n=1}^{W} y_n - \frac{3W}{4})^2 \qquad (11)$$

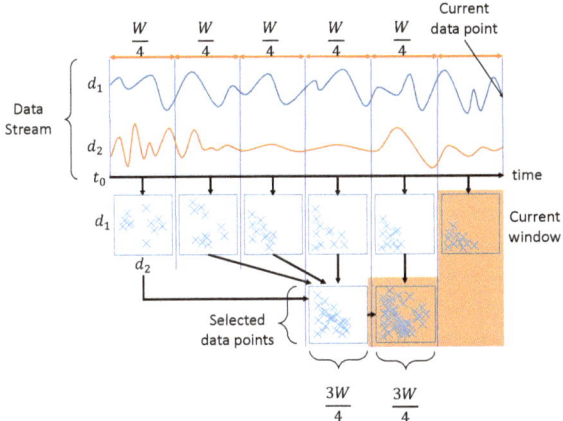

Figure 1. The summarization phase of TADILOF.

The details of the TADS procedure are shown in Algorithm 2.

Algorithm 2 Procedure TADS

Input: set of data point in memory $X = \{x_1, x_2, ... x_W\}$,
  Window size: W,
  Step size: η,
  Regularization constant: λ,
  Maximum number of iteration: I
Output: summary set
1: **for each** *data point* $x \in X$ **do**
2:     **if** $LOF_k(x) < historicalLOF(x)$ **then**
3:        update LOF, LRD and meanDistance
4: **end for**
5: $Y = \{y_1, y_2, ... y_W\}$
6: **for each** *decision variables* $y \in Y$ **do**
7:     $y = 0.75$
8: **end for**
9: **for** $i = 1{:}I$ **do**
10:     $\eta = \eta * 0.95$
11:     **for** $n = 1{:}W$ **do**
       ▷ Using objective function, calculate the score of each data point for selection in the summary set.
12:        $y_n^{(i+1)} = y_n^{(i)} - \eta \left\{ t\_diff_n + \sum_{x \in C_{K,n}} \frac{p_k(x)}{v_k(x)} - e^{LOF_k(x_n)} \psi_{0,1}'\left(y_n^i\right) + \lambda \left(\sum_{n=1}^{W} y_n^{(i)} - \frac{3W}{4} \right) \right\}$
13:     **end for**
14: **end for**
15: Project Y into binary domain
16: **for** $n=1{:}\frac{3W}{4}$ **do**
17:     $Z \leftarrow Z \cup \{x_n\}$
18: **end for**
19: Return Z

3.3. LOF Score and ODA (Outlier Detection Using Approximate LOF)

Limitations on memory capacity and computational power necessitate the elimination of some previous data points; however, setting an excessively small window size can degrade performance. Let us take an example shown in Figure 2 with two local clusters from the data stream. The symbols in different shape do not represent different kind of data points in a data stream. We have just make different symbols to represent two different local clusters of data points from data stream in Figure 2. In the example in Figure 2, new point A sits very close to cluster 1, but some of the points in that cluster were deleted in the previous summarization phase, with the result that the new point is unable to find a sufficient number of neighbors in cluster 1. This means that LOF must be calculated using points from cluster 2, which could present the new point as an outlier. We sought to overcome this issue by calculating approximate LOF scores, which are then saved with the LRD and the mean distance between each point to neighbors in every summarization phase. This saved information can then be used to calculate the reachability of potential neighbors.

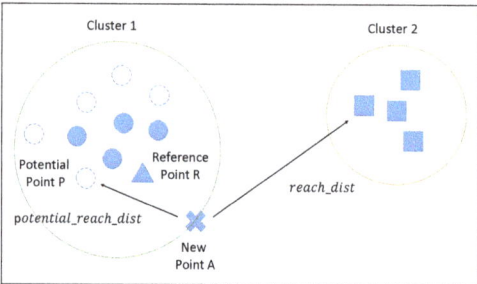

Figure 2. The case of new point to calculate LOF score with point from other cluster.

Assume that new point A is added to the dataset. If the calculated LOF exceeds the threshold, then the algorithm classifies it as an outlier. At the same time, historical information related to reference point R (a KNN neighbor of A) is used to find potential neighbor point P as a function of historical distance between R and its neighbors. Following the identification of the reference point R and its potential neighbor point P, the approximate LOF value is calculated to reassess whether the data point in question should be classified as an outlier or an inlier.

Calculation of the approximate LOF score requires preservation of some of the information in the previous window. In the summarization phase, the LOF score of any data point selected for inclusion in the first summary is retained as its historical LOF score. Note that its historical LRD and the mean distance to its neighbors are also preserved. For any data point selected for the initial and subsequent summarization, we compare the current LOF score with its historical LOF score. In cases where the current LOF score is lower, the associated information is updated. Note that a lower LOF score is indicative of the density typical of inliers.

Point A has K-nearest neighbors. Our aim is to identify the neighbor with the lowest product of historical LOF score and Euclidean distance between A and itself. That neighbor is then used as a reference point R by which to calculate the approximate LOF score of A.

We can use the historical LRD of R to obtain the mean reachability distance between R and P using the following equation:

$$mean\text{-}reach\text{-}dist(R,P) = \frac{1}{historical\,LRD(R)} \quad (12)$$

Our objective is to identify potential neighbors of new point A. Even though the current state indicates that A is an outlier, it may in fact be an inlier if some of its neighbors avoided deletion in the previous few windows.

There are three scenarios in which new point A, reference point R, and potential neighbor P, which represents a deleted data point, could be distributed in ODA. In Definition 1 $d(R,P)$ is used to represent the mean Euclidean distance between R and P. Using Definition 3, $reach\text{-}dist(R,P)$ indicates the mean reachability distance between R and P. Before we discuss these three scenarios, it is necessary to discuss the distribution of potential neighbors. Potential neighbor P can be in any position, including the space between the reference point and the new point. It is infeasible to record all potential neighbor positions; therefore, we use the case where the potential neighbor is located at the greatest distance between the new data point and itself. We then use the mean distance between R and its historical neighbors and the mean reachability distance to calculate the approximate reachability distance between A and P.

In the first scenario (Figure 3 left), reachability distance $reach\text{-}dist(R,P)$ is equal to Euclidean distance $d(R,P)$, which is larger than $K\text{-}distance(P)$. In this scenario, ODA can use $d(R,P) + d(R,A)$ to cast the mean approximate reachability distance between A and P. In the second scenario (Figure 3 middle), $reach\text{-}dist(R,P)$ is larger than $d(R,P)$ but less than $d(R,P) + d(R,A)$. In this case, ODA can also use $d(R,P) + d(R,A)$ to cast the mean approximate reachability distance between A and P. In the third scenario (Figure 3 right), $reach\text{-}dist(R,P)$ is larger than $d(R,P) + d(R,A)$. In this case, ODA can use $reach\text{-}dist(R,P)$ to represent the mean approximate reachability distance $reach\text{-}dist(A,P)$.

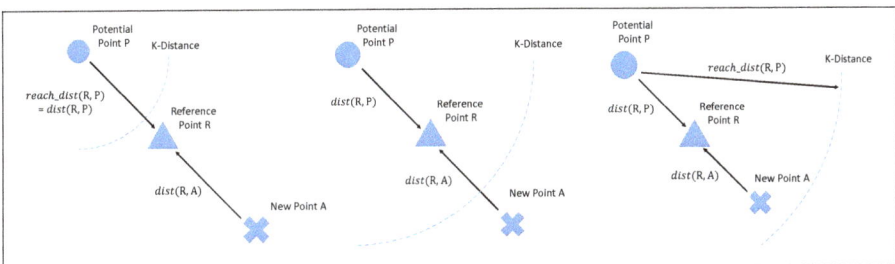

Figure 3. Three scenarios of a potential neighbor, a reference point, and a new point.

By assembling these, we can obtain the approximate mean reachability distance between point P and A using the following equation:

$$mean\text{-}reach\text{-}dist(A,P) =$$
$$max\left\{d(R,P) + d(R,A), \frac{1}{historical LRD(R)}\right\} \quad (13)$$

After obtaining the approximate mean reachability distance of point A, we can calculate the approximate LRD of A using Equation (2) (Definition 4), based on the fact that LRD is the reciprocal of the mean reachability distance.

$$Approximate LRD(A) = mean\text{-}reach\text{-}dist(A,P)^{-1} \quad (14)$$

ODA then calculates the sum of LRD of P using Definition 5, as follows:

$$mean\text{-}LRD(P) = historical LOF(R) * historical LRD(R) \quad (15)$$

The approximate reachability distance and average LRD of the potential neighbor are then used to compute the approximate LOF using Definition 5, as follows:

$$Approximate LOF(A) = \frac{mean\text{-}LRD(P)}{Approximate LRD(A)} \quad (16)$$

ODA can use this approximate LOF to determine whether A is an outlier or an inlier. The pseudocode of the ODA procedure is shown in Algorithm 3.

Algorithm 3 Procedure of ODA

Input: data point x_t
  set of data point in memory $X = \{x_1, x_2, ...x_t\}$,
  threshold: θ,
  set of detected outlier: $outlierSet$
Output: LOF score of x_t
 1: Using incremental LOF technique updates all reverse KNNs of x_t
 2: $N_K(x)$ = All KNNs of x_t
 3: **for each** $neighbor\ n \in N_K(x)$ **do**
 4:   updating time stamp of n
 5: **end for**
 6: Compute $LOF_k(x_t)$
 7: **if** $LOF_k(x_t) > \theta$ **then**
 8:   Reference Point R= $\arg\min_{r \in N_K A} historicalLOF(r) * d(r, A)$
 9:   Find the approximate reachability distance using Equation (12)
10:   Find the approximate LRD of (x_t) using Equation (13)
11:   Use historical LRD of R and historical LOF of R to find mean of LRD of potential neighbors by Equation (14)
12:   Find the approximate LOF of (x_t) using Equation (15)
13:   **if** approximate LOF of $(x_t) >$ Threshold **then**
14:     $outlierSet$.add(x_t)

3.4. Time and Space Complexity

In DILOF [6], the authors analyzed time complexity from the perspectives of summarization and detection separately. Note that time complexity of DILOF in the detection phase is $O(W)$, whereas time complexity of DILOF in the summarization phase is $O(\frac{W^2}{2})$. The space complexity of DILOF algorithms is $O(W*D)$, where D is the dimensionality of the data points.

In the following, we discuss the detection phase of the proposed algorithm, TADILOF, in which we calculate the approximate value of the points classified as outliers by the LOF score. Let us assume that z is the number of points that are classified as outliers. In our proposed detection phase, $O(K)$ is incurred in calculating the approximate LOF score for each point. Thus, $O(W + z*K)$ indicates the time complexity in the detection phase. However, the number of neighbors K is far less than window size W. Therefore, the cost incurred in the detection phase is $O(W)$.

The time complexity of TADILOF in the summarization phase is $O(W^2)$. TADILOF tends to require more time than DILOF. However, the execution times in the experiments were still very close.

The additional space complexity associated with the proposed method includes the time indicator, historical LOF, historical LRD, and mean neighbor distance. Note that the size of the data in the summary is $\frac{3W}{4}$. Therefore, the total cost is $O(3W)$. From this, we can see that the space complexity of TADILOF with approximate LOF is $O(W*(D+3))$.

4. Performance Evaluation

In this section, we compare the performance of TADILOF with the state-of-the-art, DILOF [6] and MiLOF [5] algorithms. In addition, we have included results of experiments from iLOF [4] algorithm on some datasets. We downloaded the implementation of DILOF and iLOF from URL provided in [6]. In [6], two versions of DILOF were implemented. One without "skipping scheme" and another with "skipping scheme". We discuss the *skipping scheme* and the related experiments in Section 4.4.

First, we describe the datasets and experiment settings, i.e., the parameters used in the experiments. We then examine the performance of each algorithm.

4.1. Datasets

The performance of the proposed method was evaluated by applying it to various datasets, which are shown in Table 1. We downloaded these preprocessed datasets from ODDS, Outlier Detection Datasets, Library [28]. These datasets were originally from UCI Machine Learning Repository (https://archive.ics.uci.edu/ml/index.php). ODDS Library provides preprocessed versions of these datasets. For the details about these datasets and information on preprocessing, we refer the readers to the ODDS Library website (http://odds.cs.stonybrook.edu/).

Table 1. Datasets.

Dataset	# Data Points	# Dimensions	# Outlier Data Points	Need to Shuffle
Annthyroid	7200	6	534	false
Cardio	1831	21	176	true
HTTP (KDD Cup 99)	567,498	3	2211	false
Letter Recognition	1600	32	100	true
Mnist	7603	100	700	true
Musk	3062	166	97	true
Pendigits	6870	16	156	false
Satellite	6435	36	2036	false
SMTP (KDD Cup 99)	95,156	3	30	false
Vowels	1456	12	50	true

4.2. Experiment Settings

The same set of hyperparameters were used for TADILOF and DILOF. The learning rate and maximum number of gradient descent iterations were set at 0.3 and 0.001, respectively. The K-nearest neighbors were 8 for all of the datasets. These parameters were suggested in DILOF [6] and we have used the same parameters in our experiments for comparisons to other algorithms. In addition, we ran another experiment for different K values. Some of the preprocessed datasets contained all the outliers grouped together (as a class) at the beginning or end. Some datasets had outliers scattered among inliers. We therefore shuffled datasets of the former kind before running the algorithms. The last column in Table 1 shows whether we shuffled the dataset or not, where "true" means we shuffled the dataset. We also assessed model performance using windows of various sizes, due to the importance of this parameter in terms of memory usage and computation time. For small datasets, we selected a small window size $W = \{100, 120, 140, 160, 180, 200\}$. Similarly, for larger datasets, we selected larger window size $W = \{100, 200, 300, 400, 500, 600, 700\}$. For LOF score thresholds, we use $LOF\_Thresholds = \{0.1, 1.0, 1.1, 1.15, 1.2, 1.3, 1.4, 1.6, 2.0, 3.0\}$ which were used in DILOF implementation. The same thresholds were used in the experiments, and false positive rate (FPR) and true positive rate (TPR) were calculated for each threshold. Then AUC in ROC space was calculated for all the algorithms. All experiments were performed on a PC with Intel Core i7-3770 3.4 GHz, 32 GB RAM, and Windows 10 64-bit operating system. The algorithms were implemented in C++ programming language.

4.3. Experimental Results

4.3.1. AUC, Execution Time, and Memory Usage

We evaluated MiLOF, DILOF, and TADILOF in terms of AUC and execution time on various datasets. As reported in [6], "DILOF without skipping scheme" had better performance than "DILOF with skipping scheme" in the datasets except for "HTTP KDD Cup 99" dataset. Therefore, we compare "DILOF without skipping scheme" with the proposed TADILOF in this section. We discuss the skipping scheme and related experiments on "HTTP KDD Cup 99" dataset in Section 4.4.

First we ran experiments on Pendigits, SMTP, and Vowels datasets to assess the results for different *K* values. The window size was set at 140 for Pendigits and Vowels dataset while the window size was set at 400 for SMTP dataset. Figures 4 and 5 show the results of this experiments, i.e., AUCs and execution times of MILOF, DILOF, and TADILOF algorithms. For the remaining experiments, we set *K* at 8, which was also used in DILOF [6].

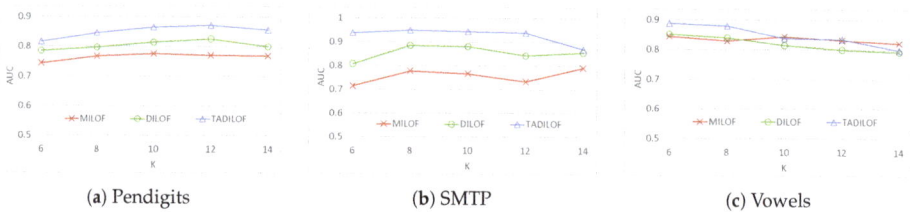

(a) Pendigits   (b) SMTP   (c) Vowels

Figure 4. AUC on various datasets for different *K* values.

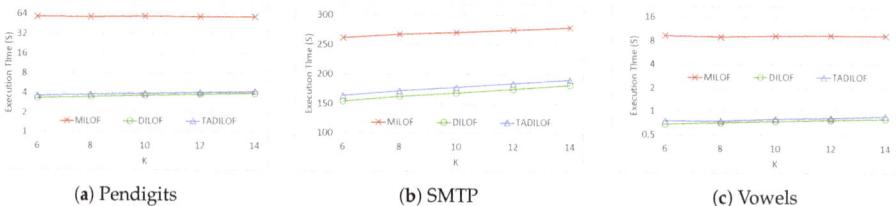

(a) Pendigits   (b) SMTP   (c) Vowels

Figure 5. Execution time on various datasets for different *K* values.

Next, we ran the experiments on various datasets to assess the performances of the algorithms for different window sizes. Figures 6 and 7 show the AUCs and execution timse of all the algorithms respectively. We can see that TADILOF outperformed MiLOF and DILOF in terms of AUC in most of the cases on various datasets. Next we discuss each experiment one by one.

Figure 6 illustrates that the AUC increases with the increase of window size on the Annthyroid, Letter Recognition, Mnist, Satellite, SMTP, and Vowels datasets. Similarly, the AUC decreases with the increase of window size on Cardio, Musk, and Pendigits datasets. In both the cases, TADILOF outperforms the competitors in terms of AUC for most of the window sizes on all these datasets. In terms of AUC, TADILOF is a clear winner on Cardio, Musk, Pendigits, Satellite, and Vowels datasets.

On the Annthyroid dataset, both MiLOF and TADILOF have similar AUCs for window sizes 100 and 120. However, in the case of window sizes larger than or equal to 140, TADILOF outperforms all the competitors.

On the Letter Recognition dataset, TADILOF outperforms DILOF in terms of AUC. Similarly, MiLOF outperforms DILOF. In addition, MiLOF outperforms TADILOF in the case of window sizes smaller than 140. However, in the case of window sizes larger than 140, TADILOF outperforms MiLOF.

On the Mnist dataset, TADILOF has higher AUCs for some window sizes, whereas for other window sizes MiLOF has higher AUCs. Both MiLOF and TADILOF outperform DILOF in terms of AUC on Mnist dataset.

On the SMTP dataset with a relatively small window size (100, 200, and 300), the performances of TADILOF and DILOF were similar. However, for the window sizes larger than 300, TADILOF clearly outperformed DILOF in terms of AUC. When the window size exceeds 400, the performance of DILOF dropped dramatically due to its inability to remove outdated data. Increasing the window size beyond 500 led to a slight drop in AUC of TADILOF. However, TADILOF maintained AUC at above 0.9 for larger windows that exceeded window size 300.

The reasons behind the better performance of TADILOF are as follows. The method removes outdated data which might otherwise have influence on new data points, thereby preventing the identification of outliers. The ability to follow the concept drift of the data using time indicator was also shown to enhance performance. In addition, approximate LOF score calculated with the historical information provides the second chance to judge the data point as outlier or inlier. Using the time component for time-aware summarization helps one to eliminate too-old data from the summary. Thus, it prevents the influence of data which are too old. However, due to window size limitation, some not-so-old data may also be deleted. Thus storing some statistics for K-neighbors from previous window helps to judge the new data by applying second check based on approximate LOF if the new data point is detected as outlier based on current LOF score.

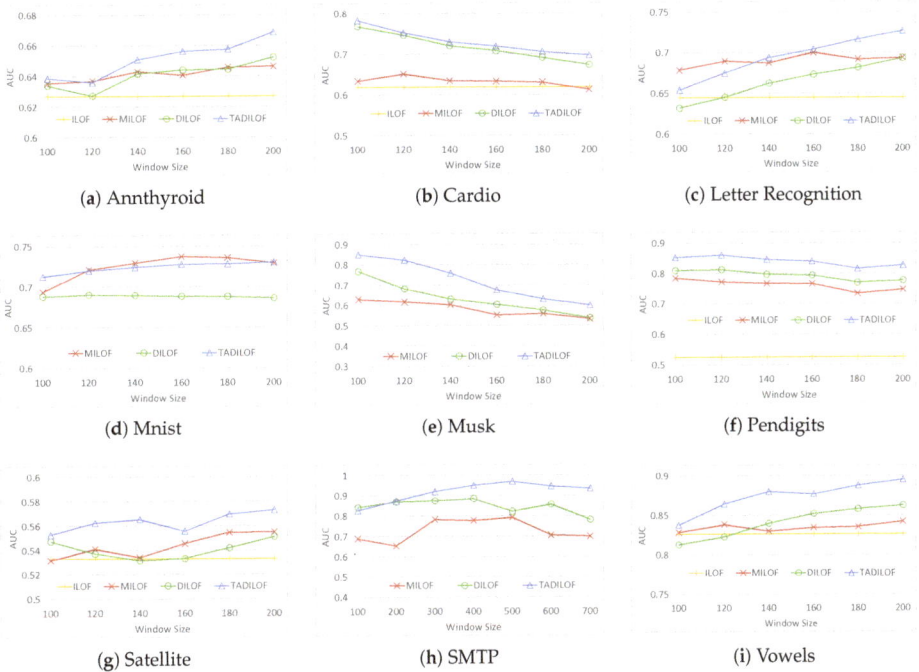

Figure 6. AUC on various datasets with different window sizes and $K = 8$.

Figure 7 shows the performances of the algorithms in terms of execution time. Note that the y-axis is in log scale of base 2 in the figures for Annthyroid, Mnist, Musk, Pendigits, and Satellite datasets. Both DILOF and TADILOF significantly outperform iLOF and MiLOF in terms of execution time. Overall, the time complexity of TADILOF matched the values estimated in Section 3.4. The time consumption of TADS was similar to that of the original NDS. The only difference was the fact that TADS calculated the Rényi divergence between all data points in memory and three quarters of the data points. In contrast, NDS computed half of all data points and a quarter of all data points. The approximation of LOF values increased execution time only slightly. Nevertheless, TADILOF had a similar performance to DILOF in terms of execution time. Overall, the proposed algorithm outperformed state-of-the-art competitors in terms of AUC while achieving similar execution times.

Similarly, Figure 8 shows the performances of DILOF and TADILOF on various datasets in terms of memory usage. We used Win32 API for reporting the memory usage of DILOF and TADILOF. Figure 8 demonstrates that in most of the cases, TADILOF used only a little more memory than DILOF. The results of experiments in terms of memory usage conformed with the theoretical analysis.

Nevertheless, we can see from the results of experiments that both DILOF and TADILOF do not take much memory and are suitable for data stream environment.

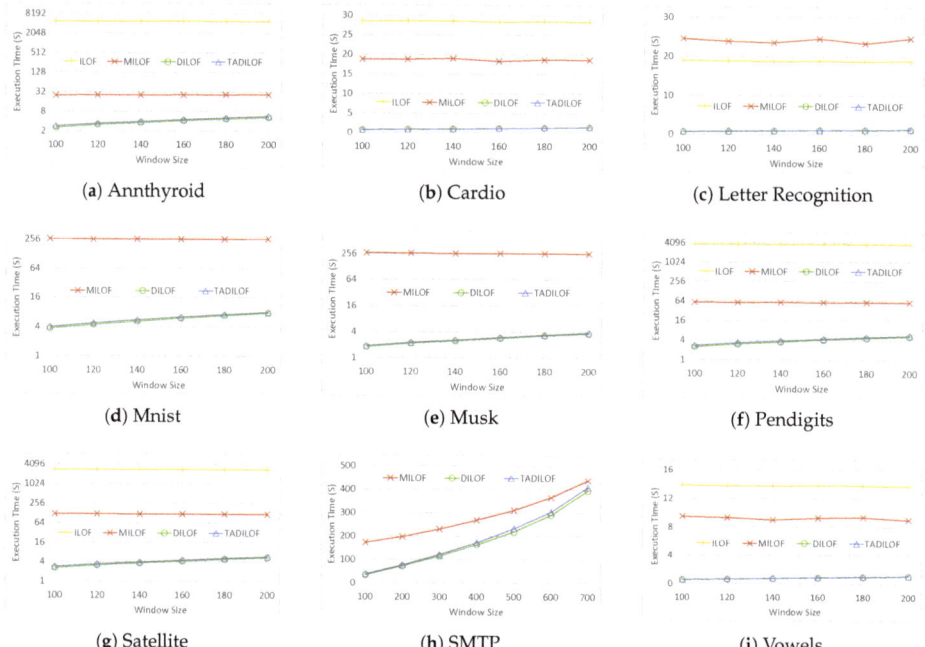

Figure 7. Execution time on various datasets with different window sizes and $K = 8$.

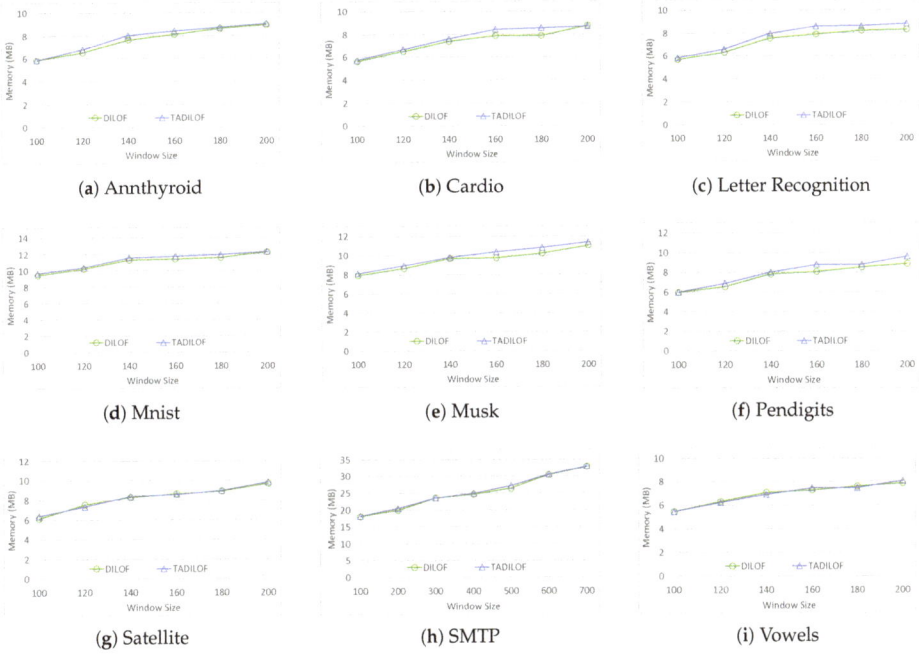

Figure 8. Memory usage on various datasets with different window sizes and $K = 8$.

4.3.2. Precision, Recall, and F1 Score

On the same datasets, we investigated the precision, recall and F1 score for different window sizes and $K = 8$. The Tables 2–10 show the precision, recall, and F1 score on various datasets for DILOF, TADILOF, and MILOF. In most cases, TADILOF had better precision and recall. Particularly, the recall values are much better than those of the other algorithms. Thus the F1 scores of TADILOF are the best. As for the precision, TADILOF performed better when the window size was larger.

Table 2. Precision, recall, and F1 score on Annthyroid dataset.

Window Size	Precision			Recall			F1 Score		
	DILOF	TADILOF	MILOF	DILOF	TADILOF	MILOF	DILOF	TADILOF	MILOF
100	0.259074	0.224178	0.2289622	0.350187	0.383895	0.3506741	0.188322	0.198476	0.1945009
120	0.264844	0.222732	0.2331385	0.355993	0.392697	0.3582022	0.191396	0.200518	0.1975793
140	0.259869	0.213771	0.2369482	0.367790	0.404307	0.3630711	0.195014	0.201042	0.2018931
160	0.257486	0.218562	0.2381993	0.378652	0.415730	0.3679961	0.197863	0.206054	0.2023676
180	0.258819	0.217542	0.2464307	0.375094	0.418352	0.3726779	0.196032	0.207163	0.2043856
200	0.264608	0.218433	0.2433799	0.380899	0.426779	0.3750937	0.199770	0.210723	0.2032031

Table 3. Precision, recall, and F1 score on Cardio dataset.

Window Size	Precision			Recall			F1 Score		
	DILOF	TADILOF	MILOF	DILOF	TADILOF	MILOF	DILOF	TADILOF	MILOF
100	0.3467338	0.3693657	0.3151908	0.3914205	0.4547727	0.3050568	0.2156009	0.2700938	0.1910918
120	0.3381342	0.3508179	0.3284806	0.3751136	0.4397159	0.3127273	0.2011454	0.2549688	0.1956810
140	0.3218308	0.3431242	0.3028065	0.3655682	0.4323864	0.2982955	0.1903846	0.2475356	0.1816338
160	0.3151459	0.3354626	0.3063062	0.3569886	0.4157387	0.3037500	0.1832554	0.2353132	0.1818394
180	0.3209461	0.3262367	0.3021858	0.3512500	0.4147726	0.3022158	0.1781800	0.2318135	0.1784863
200	0.3120879	0.3206106	0.2919695	0.3422159	0.4043182	0.2994319	0.1706907	0.2229609	0.1725868

Table 4. Precision, recall, and F1 score on Letter Recognition dataset.

Window Size	Precision			Recall			F1 Score		
	DILOF	TADILOF	MILOF	DILOF	TADILOF	MILOF	DILOF	TADILOF	MILOF
100	0.12697568	0.11241224	0.2220374	0.2059	0.2308	0.2593	0.06782202	0.08080138	0.1311881
120	0.14821722	0.16318930	0.2436584	0.2139	0.2443	0.2616	0.07340590	0.09222663	0.1351141
140	0.15472405	0.15568830	0.2298457	0.2193	0.2517	0.2618	0.07528581	0.09592009	0.1339924
160	0.17106840	0.17378370	0.2574958	0.2271	0.2625	0.2663	0.08395074	0.10338267	0.1392814
180	0.19773730	0.20144530	0.2839139	0.2335	0.2706	0.2718	0.08891001	0.11030273	0.1452346
200	0.20078190	0.19852930	0.2843155	0.2375	0.2732	0.2696	0.09236163	0.11257103	0.1431587

Table 5. Precision, recall, and F1 score on Mnist dataset.

Window Size	Precision			Recall			F1 Score		
	DILOF	TADILOF	MILOF	DILOF	TADILOF	MILOF	DILOF	TADILOF	MILOF
100	0.221380000	0.191549667	0.240385667	0.209047667	0.243714000	0.241381000	0.074075133	0.107531667	0.135068000
120	0.200322333	0.272608333	0.254099000	0.212190333	0.248285667	0.242285667	0.076509233	0.112503333	0.138804667
140	0.198687667	0.260456000	0.292730000	0.216428333	0.257047667	0.249047667	0.080405100	0.121062667	0.140650667
160	0.228525333	0.293933667	0.301560333	0.217190333	0.262143000	0.252381000	0.080920167	0.125967000	0.144668000
180	0.177270000	0.281288667	0.307995000	0.219428667	0.265905000	0.257190333	0.083189067	0.127457667	0.145774667
200	0.186638333	0.297026333	0.298204333	0.221857000	0.270571333	0.257428333	0.084071200	0.133201333	0.147606667

Table 6. Precision, recall, and F1 score on Musk dataset.

Window Size	Precision			Recall			F1 Score		
	DILOF	TADILOF	MILOF	DILOF	TADILOF	MILOF	DILOF	TADILOF	MILOF
100	0.4421690	0.4141334	0.4397151	0.3313403	0.5407216	0.2925772	0.2013681	0.3326063	0.1927423
120	0.4083079	0.4027774	0.4104153	0.2854639	0.4829896	0.2662887	0.1614765	0.2968274	0.1652040
140	0.3923583	0.3881677	0.4092076	0.2637113	0.4541238	0.2397939	0.1438728	0.2802120	0.1452295
160	0.3857042	0.3906538	0.3659905	0.2461858	0.4086599	0.2360825	0.1276172	0.2505055	0.1363073
180	0.4097445	0.3819757	0.3576605	0.2322681	0.3759795	0.2064948	0.1189272	0.2324761	0.1081695
200	0.3928690	0.3710396	0.3543731	0.2198969	0.3363918	0.1968042	0.1107008	0.2034848	0.1058940

Table 7. Precision, recall, and F1 score on Pendigits dataset.

Window Size	Precision			Recall			F1 Score		
	DILOF	TADILOF	MILOF	DILOF	TADILOF	MILOF	DILOF	TADILOF	MILOF
100	0.0540172	0.0955094	0.10309758	0.312179	0.445513	0.3918589	0.0699342	0.103071	0.11157027
120	0.0517353	0.1142970	0.08978204	0.322436	0.483333	0.3849999	0.0718471	0.111540	0.10553305
140	0.0582843	0.0868875	0.08765809	0.331410	0.481410	0.3944872	0.0731726	0.108052	0.10599671
160	0.0553464	0.0676196	0.07356239	0.330128	0.485897	0.3857051	0.0716655	0.104400	0.09873375
180	0.0429529	0.0734877	0.07456543	0.314103	0.478205	0.3844233	0.0598091	0.105238	0.09594913
200	0.0500194	0.0743026	0.08276458	0.328205	0.484615	0.3871795	0.0667915	0.102561	0.09896692

Table 8. Precision, recall, and F1 score on Satellite dataset.

Window Size	Precision			Recall			F1 Score		
	DILOF	TADILOF	MILOF	DILOF	TADILOF	MILOF	DILOF	TADILOF	MILOF
100	0.486230	0.488270	0.4720359	0.256925	0.333792	0.2466356	0.229341	0.303750	0.2279936
120	0.498198	0.496403	0.4636664	0.257122	0.341945	0.2488359	0.228337	0.307481	0.2281764
140	0.481029	0.494004	0.4694753	0.260806	0.332760	0.2601866	0.230814	0.295530	0.2381250
160	0.498065	0.492069	0.4886004	0.266994	0.325688	0.2682712	0.233976	0.286045	0.2438489
180	0.498793	0.507865	0.4879055	0.278340	0.339096	0.2791945	0.242351	0.298429	0.2519019
200	0.491820	0.505140	0.4673425	0.289293	0.341454	0.2788359	0.252402	0.297941	0.2501888

Table 9. Precision, recall, and F1 score on SMTP dataset.

Window Size	Precision			Recall			F1 Score		
	DILOF	TADILOF	MILOF	DILOF	TADILOF	MILOF	DILOF	TADILOF	MILOF
100	0.00265890	0.00164838	0.002386766	0.7633	0.7400	0.5029	0.00525291	0.00327270	0.004681796
200	0.00256851	0.00247500	0.002520710	0.7733	0.7900	0.5147	0.00507621	0.00491061	0.004933168
300	0.00332737	0.00344192	0.002814311	0.7933	0.8133	0.6179	0.00655182	0.00679199	0.005521916
400	0.00265894	0.00238982	0.002669190	0.7867	0.9133	0.6603	0.00525692	0.00475252	0.005260103
500	0.00191050	0.00313257	0.002218161	0.7467	0.9467	0.6417	0.00379399	0.00620751	0.004383276
600	0.00203202	0.00191777	0.001829777	0.7700	0.9767	0.5839	0.00403406	0.00382331	0.003621248
700	0.00168554	0.00185686	0.001824417	0.6767	0.9067	0.5933	0.00334839	0.00369957	0.003614118

Table 10. Precision, recall, and F1 score on Vowels dataset.

Window Size	Precision			Recall			F1 Score		
	DILOF	TADILOF	MILOF	DILOF	TADILOF	MILOF	DILOF	TADILOF	MILOF
100	0.14336408	0.1570996	0.1922093	0.3256	0.3898	0.4302	0.1121199	0.130352	0.179416
120	0.16889650	0.1551854	0.1959132	0.3476	0.4350	0.4202	0.1239128	0.148712	0.171690
140	0.16830130	0.1644227	0.2006647	0.3660	0.4604	0.4350	0.1329999	0.158371	0.175122
160	0.17210987	0.1958837	0.2394359	0.3756	0.4758	0.4384	0.1360716	0.166075	0.179563
180	0.16043631	0.1741156	0.2275521	0.3862	0.5022	0.4494	0.1367308	0.174471	0.182075
200	0.16436960	0.1809000	0.2074316	0.3914	0.5000	0.4348	0.1390244	0.173421	0.173099

4.4. Skipping Scheme for a Sequence of Outliers

In some cases, there may appear long sequence of outliers which can form a dense cluster of outliers. As reported in [6], in "HTTP KDD Cup 99" dataset there is a long sequence of outliers causing the algorithms to not perform well. In DILOF [6], the authors propose a skipping scheme to solve the sequence of outliers problem. Any point previously classified as an outlier point is set as the "last outlier," before calculation of the Euclidean distance between the new point and the last outlier. If the Euclidean distance exceeds the average of all points to its first nearest neighbor, then that point is classified as an outlier and excluded from the database. Note however that the last outlier is identified using a particular threshold. Under these conditions, the fact that a different threshold could give a different last outlier means that it would be unreasonable to calculate AUC, considering that the likelihood of registering a true positive (TP) or false positive (FP) does not necessarily vary with the threshold. In this situation, the area under the curve is recalculated (i.e., the ROC is not continuous), such that AUC is unable to accurately indicate the performance of the model. Nonetheless, we propose to fix the threshold at a particular value to deal with this issue.

Note that the skipping scheme proposed with DILOF does not necessarily perform well on dense datasets, due to the fact that many points belonging to dense clusters might be skipped. For example, when there are a small number of sparse clusters in the memory, a new denser cluster appears. The distance between the points associated with this cluster will be larger than the average distance of previous data points, with the result that all of the points from this cluster are immediately discarded by the skipping scheme.

Thus, we modified the skipping scheme to calculate the average distance between new data points and their K neighbors. We then conducted a comparison of the distance between the last outlier and the new data point. In the event that the former is larger than the latter, then we immediately designate the new data point as an outlier and discard it. We implemented this modified skipping scheme with TADILOF.

We set the threshold of last outlier to $T = \{2.5, 3.0\}$ with the number of neighbors set at 8, and a window size of $W = \{100, 200, 300, 400, 500, 600, 700\}$. The experimental results obtained using the HTTP KDD Cup 99 dataset are presented in Figure 9. Figure 9 illustrates that the modified skipping scheme achieved an AUC of more than 0.9 on the HTTP KDD Cup 99 dataset, regardless of the window size.

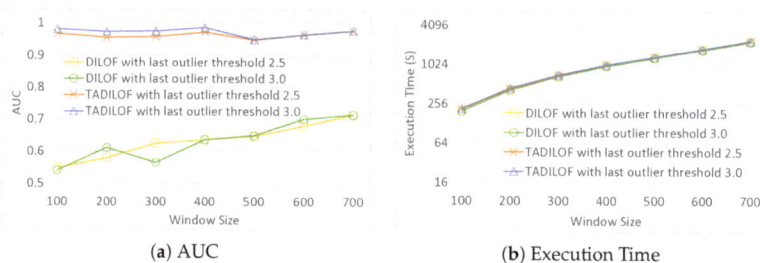

Figure 9. AUC and execution time on KDD 99 HTTP dataset using skipping scheme.

5. PM2.5 Sensors Case Study

In this section, we introduce the application of our proposed method that we used for monitoring air quality in Taiwan. There are several recent studies which have focused on air quality and PM2.5 forcasting [29–32], and anomaly detection in air quality [33].

In an effort to control air pollution in Taiwan, low-cost devices have been developed for monitoring air quality. These devices are referred to as LASSs. The Taiwanese government has initiated a project in cooperation with Edimax for the wide-scale deployment of LASS in elementary schools, high schools, and universities. The LASS used in this project are referred to as AirBox devices. Our objective in this study was to enable the real-time monitoring of all 2000 AirBox devices simultaneously.

We deployed a system in Taiwan for the detection of outliers in a large-scale dataset from PM2.5 sensors. This system provided 2000 data streams from 2000 sensors transmitting reading data at intervals of 5 min. The proposed method was used to detect outliers in each of the streams, with a focus on temporal outliers to compensate for inter-device variation in terms of quality and sensitivity. Following the identification of temporal outliers, we combined the positions of the devices with meteorological data to facilitate the detection of pollution events.

In addition, we used precision PM2.5 stations which are provided by the Environmental Protection Administration (EPA), Taiwan, to predict air quality. We integrated the data from precision PM2.5 sensors provided by EPA, Taiwan, because the quality of the data from precision PM2.5 sensors is better. However, there are only 77 PM2.5 stations in Taiwan and they provide an average PM2.5 value every hour. In this situation, we cannot find small pollution events. Therefore, we used low cost but large-scale PM2.5 devices for detecting pollution events. There are some advantages to using those PM2.5 sensors. The first benefit is that we can monitor air quality of Taiwan by a fine resolution on space because the number of active devices is more than 2000 regarding those that are deployed in Taiwan. The second benefit is that their sampling rate is 5 min. Therefore, we can also have a fine resolution on time domain to monitor air quality of Taiwan.

After getting fine resolution data based on both time and space, the challenge is how to use those data to detect pollution events. There are three challenges of using those data to detect pollution events. The first one is those devices are low cost and there is lack of maintenance. In general case, those kind of sensors need device correction every few month, so that the reading number is more accurate. The second challenge is there are numerous devices, and each device has a very high sampling rate which is every 5 min. We can see one of these devices as a data stream, and hence there are 2000 data streams. Therefore, we need to handle this large amount of data streams. Our proposed method has the capability to not only find outliers on different devices but also to deal with large number of data streams which have the high sampling rate. Next, we introduce how our method finds the pollution events in the following subsection.

Monitoring a PM2.5 Pollution Event

In this section, we introduce how to use the proposed method to monitor PM2.5 pollution event. First, we define spatial neighbors of devices using average wind speed of Taiwan. According to Central Weather Bureau (CWB) of Taiwan, the average wind speed of Taiwan is 3.36 km/h. Therefore, we define neighbor distance to be 1.5 km, which means any two devices are neighbors if the distance between these two devices is less than 1.5 km.

Each device produces a data stream because every device samples the concentration of PM2.5 at interval of every five minutes and the number of data values is unbounded. We implemented our proposed method on each data stream. Thus, we can detect outliers on different devices separately. We call this type of outlier a temporal outlier because such outliers are compared with historical data points from the same device. If proposed method detects any temporal outlier on devices, we add the device to a set called outlier-event-pool and set an expire time as 30 min. In next 30 min, if we can find two neighbor devices in the outlier-event-pool for any device in the outlier-event-pool, we call this event a pollution event. Otherwise, it represents a spatial outlier of the device.

Figure 10 shows an example of spatial outlier. A spatial outlier means that there is only one device which has a sudden rise/fall in the measurement value and other nearby devices do not have any such change in the measurement. In Figure 10, the data stream in blue represents a target device which shows outlier data points marked in red. Outliers from other data streams are not shown, i.e., not marked in red in this figure. Similarly, Figure 11 shows an example of pollution event. At the left side of the figure, the measurement value from a device has a sudden rise. Then the neighbor devices in the right side of the figure also has a rise in the measurement value in next few minutes. Since this event may have been started by nearby device shown in the left side of the figure. Thus, we can get the potential pollution event region.

Now, we discuss a use case related to a fire event, where we applied the proposed approach discussed above. In this case study, we targeted to track the pollution events where there is sudden increase in PM2.5 values. Our analysis targeted a fire event, which was reported at 17:51 2019/11/12 in Tainan city following reports of burning rubber. The Tainan EPB sent emergency notifications to Tainan citizens at 21:00. However, our system detected (and reports) the event at approximately 17:00. Figure 12 presents PM2.5 data for all devices in the vicinity of the fire throughout the day. We can see some flat lines in the readings. These are due to device malfunctions or reading errors (we have mentioned above about the issues related to the low-cost airbox devices). Similarly, we can see some bottom curves in Figures 11 and 12. These are there because of the placements of the airbox devices. Some of the devices were placed indoors whereas other devices were placed outdoors. The indoor devices had a different environment (such as air conditioned room) than the outdoor devices, which affected the readings among different devices. Therefore, bottom curves are different from the others.

Figure 13 shows the result that our implemented system detects the pollution event (fire event). In Figure 13, we can see that the proposed system sends the alert to subscribers at approximately 5 p.m.

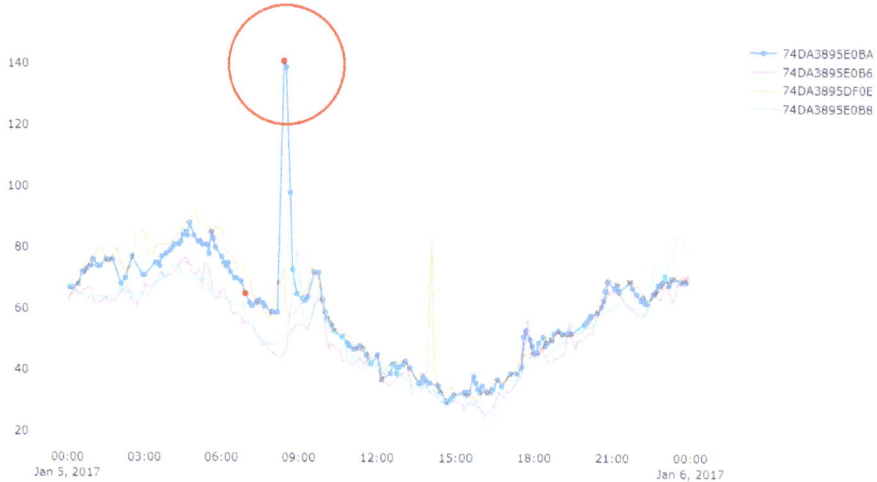

Figure 10. An example of a spatial (and temporal) outlier.

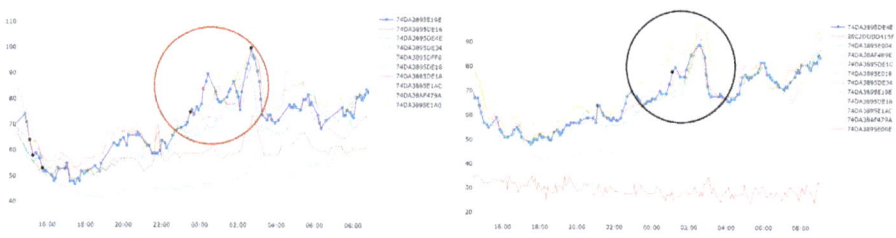

Figure 11. An example of a pollution event.

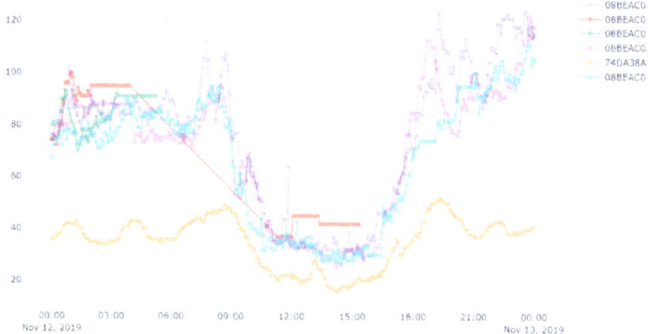

Figure 12. A case study of a fire event with PM2.5 sensors' data.

Figure 13. A case study on a fire event with PM2.5 sensors' data—detected event.

6. Conclusions

This paper presents a novel algorithm to detect local outliers in data streams using LOF score. In addition, we used a time indicator with data points to resolve the issue of concept drift in data streams with the aim of improving accuracy in the detection of outliers. Moreover, we developed a novel method by which historical information is used to calculate approximate LOF values to improve accuracy with only a negligible increase in memory cost. The results of experiments illustrate that the proposed method, TADILOF, outperforms the state-of-the-art competitors in terms of AUC in most of the cases on various datasets. In addition, a practical application of the proposed scheme to PM2.5 sensor data clearly demonstrated its efficacy.

Author Contributions: Conceptualization, J.-W.H., M.-X.Z., and B.P.J.; methodology, J.-W.H., M.-X.Z., and B.P.J.; software, M.-X.Z.; validation, J.-W.H., M.-X.Z., and B.P.J.; formal analysis, J.-W.H., M.-X.Z., and B.P.J.; writing—original draft preparation, M.-X.Z.; writing—review and editing, J.-W.H. and B.P.J.; visualization, M.-X.Z.; supervision, J.-W.H.; project administration, J.-W.H.; funding acquisition, J.-W.H. All authors have read and agreed to the published version of the manuscript.

Funding: The work is funded by Ministry of Science and Technology, Taiwan (MOST 105-EPA-F-007-004).

Conflicts of Interest: The authors declare no conflict of interest.

References

1. Chandola, V.; Banerjee, A.; Kumar, V. Anomaly Detection: A Survey. *ACM Comput. Surv. (CSUR)* **2009**, *41*. [CrossRef]
2. Aggarwal, C.C. *Outlier Analysis*; Springer: Cham, Switzerland, 2017.
3. Breunig, M.M.; Kriegel, H.P.; Ng, R.T.; Sander, J. LOF: Identifying Density-Based Local Outliers. In Proceedings of the 2000 ACM SIGMOD International Conference on Management of Data (SIGMOD '00), Dallas, TX, USA, 16–18 May 2000; Association for Computing Machinery: New York, NY, USA, 2000; pp. 93–104. [CrossRef]
4. Pokrajac, D.; Lazarevic, A.; Latecki, L.J. Incremental Local Outlier Detection for Data Streams. In Proceedings of the 2007 IEEE Symposium on Computational Intelligence and Data Mining, Honolulu, HI, USA, 1 March–5 April 2007; pp. 504–515. [CrossRef]
5. Salehi, M.; Leckie, C.; Bezdek, J.C.; Vaithianathan, T.; Zhang, X. Fast Memory Efficient Local Outlier Detection in Data Streams. *IEEE Trans. Knowl. Data Eng.* **2016**, *28*, 3246–3260. [CrossRef]
6. Na, G.S.; Kim, D.; Yu, H. DILOF: Effective and Memory Efficient Local Outlier Detection in Data Streams. In Proceedings of the 24th ACM SIGKDD International Conference on Knowledge Discovery & Data Mining (KDD '18), London, UK, 19–23 August 2018; Association for Computing Machinery: New York, NY, USA, 2018; pp. 1993–2002. [CrossRef]

7. Ramaswamy, S.; Rastogi, R.; Shim, K. Efficient Algorithms for Mining Outliers from Large Data Sets. In Proceedings of the 2000 ACM SIGMOD International Conference on Management of Data (SIGMOD '00), Dallas, TX, USA, 16–18 May 2000; Association for Computing Machinery: New York, NY, USA, 2000; pp. 427–438. [CrossRef]
8. Kieu, T.; Yang, B.; Jensen, C.S. Outlier Detection for Multidimensional Time Series Using Deep Neural Networks. In Proceedings of the 2018 19th IEEE International Conference on Mobile Data Management (MDM), Aalborg, Denmark, 25–28 June 2018; pp. 125–134.
9. Chakraborty, S.; Nagwani, N.K. Analysis and Study of Incremental K-Means Clustering Algorithm. In *International Conference on High Performance Architecture and Grid Computing*; Springer: Berlin/Heidelberg, Germany, 2011; pp. 338–341.
10. Langone, R.; Agudelo, O.M.; Moor, B.D.; Suykens, J.A. Incremental kernel spectral clustering for online learning of non-stationary data. *Neurocomputing* **2014**, *139*, 246–260. [CrossRef]
11. Figueiredo, E.; Park, G.; Farrar, C.R.; Worden, K.; Figueiras, J. Machine learning algorithms for damage detection under operational and environmental variability. *Struct. Health Monit.* **2011**, *10*, 559–572. [CrossRef]
12. Cassisi, C.; Ferro, A.; Giugno, R.; Pigola, G.; Pulvirenti, A. Enhancing density-based clustering: Parameter reduction and outlier detection. *Inf. Syst.* **2013**, *38*, 317–330. doi:10.1016/j.is.2012.09.001. [CrossRef]
13. Abid, A.; Kachouri, A.; Mahfoudhi, A. Outlier detection for wireless sensor networks using density-based clustering approach. *IET Wirel. Sens. Syst.* **2017**, *7*, 83–90. [CrossRef]
14. Domingues, R.; Filippone, M.; Michiardi, P.; Zouaoui, J. A comparative evaluation of outlier detection algorithms: Experiments and analyses. *Pattern Recognit.* **2018**, *74*, 406–421. [CrossRef]
15. Ijaz, M.F.; Attique, M.; Son, Y. Data-Driven Cervical Cancer Prediction Model with Outlier Detection and Over-Sampling Methods. *Sensors* **2020**, *20*, 2809. [CrossRef] [PubMed]
16. Lazarevic, A.; Kumar, V. Feature Bagging for Outlier Detection. In Proceedings of the Eleventh ACM SIGKDD International Conference on Knowledge Discovery in Data Mining (KDD '05), Chicago, IL, USA, 21–24 August 2013; Association for Computing Machinery: New York, NY, USA, 2005; pp. 157–166. [CrossRef]
17. Kriegel, H.P.; Kröger, P.; Schubert, E.; Zimek, A. LoOP: Local Outlier Probabilities. In Proceedings of the 18th ACM Conference on Information and Knowledge Management (CIKM '09), Hong Kong, 2–6 November 2018; Association for Computing Machinery: New York, NY, USA, 2009; pp. 1649–1652. [CrossRef]
18. Kriegel, H.P.; Kroger, P.; Schubert, E.; Zimek, A. Interpreting and Unifying Outlier Scores. In Proceedings of the 2011 SIAM International Conference on Data Mining, Mesa, AZ, USA, 28–30 April 2011; pp. 13–24. [CrossRef]
19. Jin, W.; Tung, A.K.H.; Han, J. Mining Top-n Local Outliers in Large Databases. In Proceedings of the Seventh ACM SIGKDD International Conference on Knowledge Discovery and Data Mining (KDD '01), San Francisco, CA, USA, 26–29 August 2001; Association for Computing Machinery: New York, NY, USA, 2001; pp. 293–298. [CrossRef]
20. Jain, A.K. Data clustering: 50 years beyond K-means. *Pattern Recognit. Lett.* **2010**, *31*, 651–666. [CrossRef]
21. Póczos, B.; Xiong, L.; Schneider, J. Nonparametric Divergence Estimation with Applications to Machine Learning on Distributions. In Proceedings of the Twenty-Seventh Conference on Uncertainty in Artificial Intelligence (UAI'11), Barcelona, Spain, 14–17 July 2011; AUAI Press: Arlington, VA, USA, 2011; pp. 599–608.
22. Hulten, G.; Spencer, L.; Domingos, P. Mining Time-Changing Data Streams. In Proceedings of the Seventh ACM SIGKDD International Conference on Knowledge Discovery and Data Mining (KDD '01), San Francisco, CA, USA, 26–29 August 2001; Association for Computing Machinery: New York, NY, USA, 2001; pp. 97–106. [CrossRef]
23. Tsymbal, A. The problem of concept drift: Definitions and related work. Technical report. *Comput. Sci. Dep. Trinity Coll. Univ. Dublin* **2004**, *106*, 58.
24. Fan, W. Systematic Data Selection to Mine Concept-Drifting Data Streams. In Proceedings of the Tenth ACM SIGKDD International Conference on Knowledge Discovery and Data Mining (KDD '04), Seattle, WA, USA, 22–25 August 2004; Association for Computing Machinery: New York, NY, USA, 2004; pp. 128–137. [CrossRef]
25. Yan, Y.; Cao, L.; Rundensteiner, E.A. Scalable top-n local outlier detection. In Proceedings of the 23rd ACM SIGKDD International Conference on Knowledge Discovery and Data Mining, Halifax, NS, Canada, 13–17 August 2017; pp. 1235–1244.

26. Qin, X.; Cao, L.; Rundensteiner, E.A.; Madden, S. Scalable Kernel Density Estimation-based Local Outlier Detection over Large Data Streams. In Proceedings of the 22nd International Conference on Extending Database Technology (EDBT), Lisbon, Portugal, 26–29 March 2019; pp. 421–432.
27. Liu, F.; Yu, Y.; Song, P.; Fan, Y.; Tong, X. Scalable KDE-based top-n local outlier detection over large-scale data streams. *Knowl.-Based Syst.* **2020**, *204*, 106186. [CrossRef]
28. Rayana, S. ODDS Library. 2016. Available online: http://odds.cs.stonybrook.edu/ (accessed on 18 June 2020).
29. Zheng, Y.; Liu, F.; Hsieh, H.P. U-Air: When Urban Air Quality Inference Meets Big Data. In Proceedings of the 19th ACM SIGKDD International Conference on Knowledge Discovery and Data Mining (KDD '13), Chicago, IL, USA, 11–14 August 2013; Association for Computing Machinery: New York, NY, USA, 2013; pp. 1436–1444. [CrossRef]
30. Hsieh, H.P.; Lin, S.D.; Zheng, Y. Inferring Air Quality for Station Location Recommendation Based on Urban Big Data. In Proceedings of the 21th ACM SIGKDD International Conference on Knowledge Discovery and Data Mining (KDD '15), Sydney, Australia, 10–13 August 2015; Association for Computing Machinery: New York, NY, USA, 2015; pp. 437–446. [CrossRef]
31. Zheng, Y.; Yi, X.; Li, M.; Li, R.; Shan, Z.; Chang, E.; Li, T. Forecasting Fine-Grained Air Quality Based on Big Data. In Proceedings of the 21th ACM SIGKDD International Conference on Knowledge Discovery and Data Mining (KDD '15), Sydney, Australia, 10–13 August 2015; Association for Computing Machinery: New York, NY, USA, 2015; pp. 2267–2276. [CrossRef]
32. Soh, P.W.; Chang, J.W.; Huang, J.W. Adaptive Deep Learning-Based Air Quality Prediction Model Using the Most Relevant Spatial-Temporal Relations. *IEEE Access* **2018**, *6*, 38186–38199. [CrossRef]
33. Chen, L.; Ho, Y.; Hsieh, H.; Huang, S.; Lee, H.; Mahajan, S. ADF: An Anomaly Detection Framework for Large-Scale PM2.5 Sensing Systems. *IEEE Internet Things J.* **2018**, *5*, 559–570. [CrossRef]

Publisher's Note: MDPI stays neutral with regard to jurisdictional claims in published maps and institutional affiliations.

© 2020 by the authors. Licensee MDPI, Basel, Switzerland. This article is an open access article distributed under the terms and conditions of the Creative Commons Attribution (CC BY) license (http://creativecommons.org/licenses/by/4.0/).

MDPI
St. Alban-Anlage 66
4052 Basel
Switzerland
Tel. +41 61 683 77 34
Fax +41 61 302 89 18
www.mdpi.com

Sensors Editorial Office
E-mail: sensors@mdpi.com
www.mdpi.com/journal/sensors

www.ingramcontent.com/pod-product-compliance
Lightning Source LLC
LaVergne TN
LVHW070219100526
838202LV00015B/2061